W0051286

Advances in Environmental Science

D.C. Adriano and W. Salomons, Editors

Editorial Board:

Series Editors:

D.C. Adriano
University of Georgia's Savannah
 River Ecology Laboratory
P.O. Box E
Aiken, South Carolina 29801
USA

W. Salomons
Delft Hydraulics Laboratory
Institute for Soil Fertility
P.O. Box 30003
NL-9750 RA Haren (GR)
The Netherlands

D.C. Adriano, Coordinating Editor

Associate Editors:

B.L. Bayne, Institute for Marine Environmental Research, Plymouth
 PL1 3DH, UK
M. Chino, University of Tokyo, Tokyo 113, Japan
A.A. Elseewi, Southern California Edison, Rosemead, CA 91770, USA
M. Firestone, University of California, Berkeley, CA 94720, USA
U. Förstner, Technical University of Hamburg-Harburg, 2100 Hamburg
 90, FRG
B.T. Hart, Chisholm Institute of Technology, Victoria 3145, Australia
T.C. Hutchinson, University of Toronto, Toronto M5S 1A4, Canada
S.E. Lindberg, Oak Ridge National Laboratory, Oak Ridge, TN 37831,
 USA
M.R. Overcash, North Carolina State University, Raleigh, NC 27650, USA
A.L. Page, University of California, Riverside, CA 92521, USA

Acidic Precipitation

Volume 5

International Overview and Assessment

Edited by A.H.M. Bresser and W. Salomons

With 110 Illustrations

Springer-Verlag
New York Berlin Heidelberg
London Paris Tokyo Hong Kong

Volume Editors:

A.H.M. Bresser
National Institute of Public
   Health and Environmental
   Protection
NL-3720 BA Bilthoven
The Netherlands

W. Salomons
Delft Hydraulics Laboratory
Institute for Soil Fertility
NL-9750 RA Haren (GR)
The Netherlands

Library of Congress Cataloging-in-Publication Data
International overview and assessment.
   (Advances in environmental science; vol. 5)
   1. Acid rain—Environmental aspects. 2. Acid
precipitation (Meteorology)—Environmental aspects.
I. Bresser, A. H. M. II. Salomons, W. (Willem),
1945– . III. Series.
TD195.42.I58 1989   363.73'86   89-11513

Printed on acid-free paper.

© 1990 Springer-Verlag New York, Inc.
Softcover reprint of the hardcover 1st edition 1990
All rights reserved. This work may not be translated or copied in whole or in part without the written permission of the publisher (Springer-Verlag New York, Inc., 175 Fifth Avenue, New York, NY 10010, USA), except for brief excerpts in connection with reviews or scholarly analysis. Use in connection with any form of information storage and retrieval, electronic adaptation, computer software, or by similar or dissimilar methodology now known or hereafter developed is forbidden.
The use of general descriptive names, trade names, trademarks, etc., in this publication, even if the former are not especially identified, is not to be taken as a sign that such names, as understood by the Trade Marks and Merchandise Marks Act, may accordingly be used freely by anyone.

Typeset by Compositors Corporation.

9 8 7 6 5 4 3 2 1

ISBN-13: 978-1-4613-8943-9   e-ISBN-13: 978-1-4613-8941-5
DOI: 10.1007/978-1-4613-8941-5

Preface to the Series

In 1986, my colleague Prof. Dr. W. Salomons of the Institute for Soil Fertility of the Netherlands and I launched the new *Advances in Environmental Science* with Springer-Verlag New York, Inc. Immediately, we were faced with a task of what topics to cover. Our strategy was to adopt a thematic approach to address hotly debated contemporary environmental issues. After consulting with numerous colleagues from Western Europe and North America, we decided to address *Acidic Precipitation*, which we view as one of the most controversial issues today.

This is the subject of the first five volumes of the new series, which cover relationships among emissions, deposition, and biological and ecological effects of acidic constituents. International experts from Canada, the United States, Western Europe, as well as from several industrialized countries in other regions, have generously contributed to this subseries, which is grouped into the following five volumes:

Volume 1 *Case Studies*
(D.C. Adriano and M. Havas, editors)

Volume 2 *Biological and Ecological Effects*
(D.C. Adriano and A.H. Johnson, editors)

Volume 3 *Sources, Depositions, and Canopy Interactions*
(S.E. Lindberg, A.L. Page, and S.A. Norton, editors)

Volume 4 *Soils, Aquatic Processes, and Lake Acidification*
(S.A. Norton, S.E. Lindberg, and A.L. Page, editors)

Volume 5 *International Overview and Assessment*
(A.H.M. Bresser and W. Salomons, editors)

From the vast amount of consequential information discussed in this series, it will become apparent that acidic deposition should be seriously addressed by many countries of the world, in as much as severe damages have already been inflicted on numerous ecosystems. Furthermore, acidic constituents have also been shown to affect the integrity of structures of great historical values in various places of the world. Thus, it is hoped that this up-to-date subseries would increase the "awareness" of the world's citizens and encourage governments to devote more attention and resources to address this issue.

The series editors thank the international panel of contributors for bringing this timely series into completion. We also wish to acknowledge the very insightful input of the following colleagues: Prof. A.L. Page of the University of California, Prof. T.C. Hutchinson of the University of Toronto, and Dr. Steve Lindberg of the Oak Ridge National Laboratory.

We also wish to thank the superb effort and cooperation of the volume editors in handling their respective volumes. The constructive criticisms of chapter reviewers also deserve much appreciation. Finally, we wish to convey our appreciation to my secretary, Ms. Brenda Rosier, and my technician, Ms. Claire Carlson, for their very able assistance in various aspects of this series.

Aiken, South Carolina   *Domy C. Adriano*
                                                                        Coordinating Editor

Preface to *Acidic Precipitation*, Volume 5 *(Advances in Environmental Science)*

Acidification research has been ongoing for several decades. It was not until the 1980s, however, that scientists began to recognize the complexity of the factors causing the decline in forest growth and deterioration of fish populations in acidified lakes. The general feeling, based on correlative research, was that long-range transported air pollution was the main cause. Proof, however, was difficult to obtain because of complex interactions of various stress factors including natural ones. It was then realized that integrated research efforts are necessary among countries of the world mostly affected by acidic precipitation. In several countries extensive research programs have been set up since.

In this volume an overview and assessment are presented of the research of a number of countries. Due to differing initiation dates, variation in available funds, and differences in perception of the acidification problem, the programs show a great diversity in scope and depth. It is not the intent of the editors to judge the contents of the various programs presented here. Nor has it been the purpose to present a complete overview of the acidification research in the world. Although chapters from some major industrial countries are not included here for several reasons, indepth coverage of certain aspects of acidification research from these places is discussed in other volumes of the series.

In this volume the authors present an overview of acidification research in their respective countries, with emphasis on major research findings. The various chapters show the intensive research effort under way to study and unravel the causes and effects of acidification.

Bilthoven, The Netherlands   *A.H.M. Bresser*

Haren, The Netherlands   *W. Salomons*

Contents

Acidic Precipitation Research in France.. 307
M. Bonneau and C. Elichegaray

Contributors

P. Anttila, Finnish Research Project on Acidification (HAPRO), Ministry of the Environment, E. Esplanadi 18 A, P.O. Box 399, SF-00121 Helsinki, Finland

M. Bonneau, Institut National de la Recherche Agronomique, INRA Centre de Nancy, Laboratoire Sols et Nutrition des Arbes Forestiers, Champenoux 54280, Seichamps, France

A.H.M. Bresser, National Institute of Public Health and Environmental Protection (RIVM), P.O. Box 1, 3720 BA Bilthoven, The Netherlands

D. Camuffo, National Research Council (CNR-ICTR), Corso Stati Uniti 4, I-35020 Padova, Italy

P.J. Dillon, Ontario Ministry of the Environment, Dorset Research Centre, P.O. Box 39, Dorset, Ontario P0A 1E0, Canada

C. Elichegaray, Agence pour la Qualité de l'Air, Tour GAN, Cédex 13, 92082 Paris La Défense 2, France

H.D. Gregor, Umweltbundesamt, Bismarckplatz 1, D-1000 Berlin 33, FRG

L. Horváth, Institute for Atmospheric Physics, H-1675 Budapest, P.O. Box 39, Hungary

T. Katoh, Hokkaido Research Institute for Environmental Pollution, Nisi-12cyoume, 19zyo, Kita-ku, Sapporo, Japan (T060)

P. Kauppi, Finnish Research Project on Acidification (HAPRO), Ministry of the Environment, E. Esplanadi 18 A, P.O. Box 399, SF-00121 Helsinki, Finland

J.C. Keith, Ontario Ministry of the Environment, Dorset Research Centre, P.O. Box 39, Dorset, Ontario P0A 1E0, Canada

K. Kenttämies, Finnish Research Project on Acidification (HAPRO), Ministry of the Environment, E. Esplanadi 18 A, P.O. Box 399, SF-00121 Helsinki, Finland

K. Kienzl, Federal Environmental Agency (Umweltbundesamt), Radetzkystrasse 2, A-1031 Vienna, Austria

T. Konno, National Agriculture Research Centre, Kannondai, Tukuba-shi, Ibaraki, Japan (T̄305)

I. Koyama, Tokyo Metropolitan Research Institute for Environmental Protection, Shinsuna, 1–7–5, Kota-ku Tokyo 136, Japan

H. Makino, Kanagawa Prefectural Environmental Center 2–14 Toyoharmati, Hiratsuka-si, Kanagawa, Japan (T̄245)

R. Orthofer, Environmental Planning Department, Austrian Research Center, A-2444 Seibersdorf, Austria

I. Pais, University of Horticulture and Food Industry, H-1114 Budapest, Villányi u. 29/31, Hungary

L.G. Solovyev, Vernadsky Institute of the Academy of Sciences of the USSR, Kosigin 19, Moscow, USSR

Z. Strzyszcz, Polish Academy of Sciences, Institute of Environmental Engineering, 41–800 Zabrze, Skłodowskiej Curie 34, Poland

H. Tsuruta, Yokohama Environmental Research Institute, Takigashira 1–2–15, Isogo-ku, Yokohama-shi, Kanagawa, Japan (T̄233)

H. Turner, Swiss Federal Research Institute for Forest, Snow, and Landscape, Birmensdorf CH 8903, Switzerland

Acidic Precipitation Research in Canada

J.C. Keith* and P.J. Dillon*

Abstract

The deposition rate of strong acid in eastern Canada is similar to or greater than that documented in Norway and Sweden. Since the mid-1970s, Canadian provincial and federal agencies have been studying the effects of acidic deposition on atmospheric, terrestrial, and aquatic systems. Research results have consistently supported the Canadian position that reductions in sulphur dioxide emissions are needed across eastern North America to protect sensitive environments.

I. Introduction

A. History of the Acidic Precipitation Problem in Canada

For several decades, Canadian scientists have been aware of the environmental damage to terrestrial and aquatic systems that can result from SO_2 emissions from point sources. In fact, the first reported case (Trail Smelter Arbitral Tribunal, 1941) was a transboundary pollution problem. Sulphur dioxide emissions from a smelter located in Trail, British Columbia, caused environmental damage to areas across the border in the state of Washington (see Figure 1–1 for a map of Canada and its areas that are sensitive to acidic deposition). This case set a legal precedent that will be discussed later in this section.

The acidification of lakes in Canada was first described in the province of Ontario in the Sudbury area in the early 1950s. Nickel and copper smelting operations began in the Sudbury geological basin in 1888. Smelter production increased throughout the 1930s, then again in the 1950s and the 1960s, resulting in continuous increases in the emissions of

---

*Ontario Ministry of the Environment, Dorset Research Centre, P.O. Box 39, Dorset, Ontario, Canada, P0A 1E0.

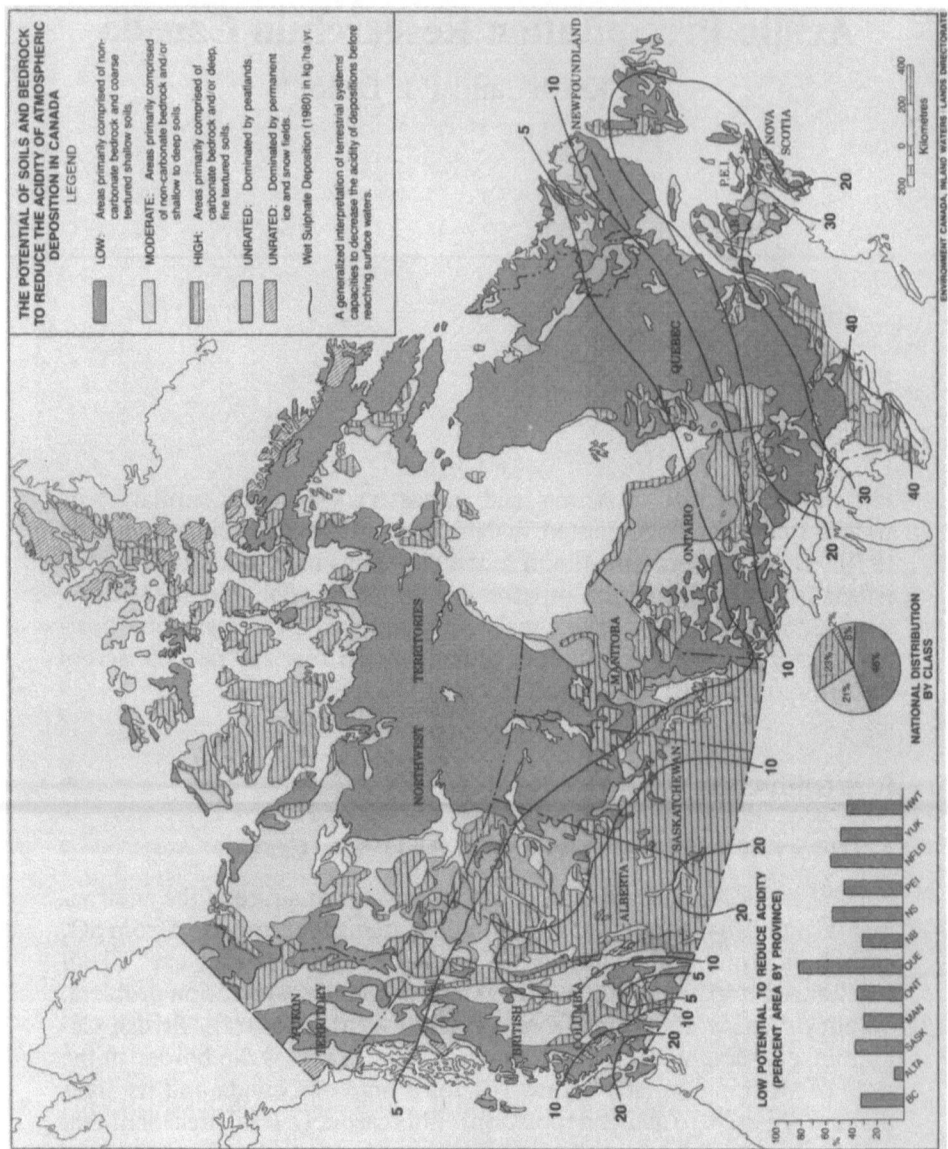

Figure 1–1. Acidic rain: A national sensitivity assessment—a national evaluation of surface water resources at risk based on the potential of soils and bedrock to reduce acidity. (Source: Environmental Fact Sheet 88-1, Inland Waters and Lands Directorate.)

sulphur compounds and metals, especially copper and nickel. The INCO Limited smelter was, for many years, the world's single largest point source of SO_2 emissions and remains the largest in North America. Nearly a century of smelting operations resulted in the destruction of the local forests and other vegetation as a consequence of the high SO_2 concentrations, and severe damage to many of the local lakes. After the lakes had exhausted their acid neutralizing capacities, acidification to pHs of approximately 4.0 and elevation of metal (Cu, Ni, Zn) concentrations to toxic levels resulted.

In the 1960s, studies began to document the extent of the damage. Beamish and Harvey (1972) showed that 70 of 150 lakes in the La Cloche Mountain area (approximately 50 to 100 km southwest of Sudbury) had pH levels less than 5.5. These lakes had lost populations of lake trout and white sucker (Beamish, 1974a, 1974b; Beamish, Lockhart, Van Loon, and Harvey, 1975). In addition, many lakes to the northeast of Sudbury were found to be equally acidic.

Pollution from a local source was also responsible for the acidification of the surrounding environment at Wawa, Ontario. In 1939, the Algoma Steel Corporation began open-pit iron ore mining and smelting operations in the area. With the discovery of high-quality underground deposits, mining activities expanded in 1949, 1954, and 1958, resulting in the abandonment of the open-pit mines in favor of the underground iron reserves. The SO_2 fumigations and associated acidic precipitation resulted in acute toxic effects to the birch, spruce, and pine forests within approximately 900 km², as well as degradation of lake chemistry and losses of fish populations within 65 m². Within a zone of 200 km², deposition of acid has maintained lake pH levels between 3.0 to 4.0 for nearly 20 years (Somers and Harvey, 1984).

These cases of environmental damage from acidic emissions have all resulted from local point sources. It was not until the mid-1970s that acidification was also considered to be a regional problem resulting from the long-range transport of air pollutants.

In 1975, the Ontario government began a detailed study of lakes in Muskoka-Haliburton (a popular recreational area approximately 250 km northeast of Toronto) with the objective of determining the effects of cottage development on water quality. Researchers soon discovered that atmospheric loadings of acid were elevated in this area and that there was evidence of water quality effects including loss of alkalinity and lowering of pH in lakes located in the region (Dillon et al., 1978).

Continued research in Ontario and the other eastern Canadian provinces soon demonstrated that the Muskoka-Haliburton area in Ontario was not the only region affected by long-range transport of sulphur oxides. A large portion of eastern Canada is underlain by the Precambrian Shield, which is characterized by granitic and siliceous bedrock of limited acid neutralizing capacity. Areas in Canada considered sensitive to acidic deposition based on underlying bedrock are shown in Figure 1–1.

Another important factor in determining the effect of acidic deposition on a specific area is the area's location with respect to emission sources. Over most of northeastern North America, the prevailing winds travel in a northeasterly direction and carry with them the pollutants generated in the midwestern American industrialized heartland. The wind pattern does vary to some extent with the season, and there are often air flows to the south in the winter. As a result, pollutants generated in Canada can also be transported across the border to the United States. Still, the transboundary flux of pollutants from the United States to Canada exceeds the flux from Canada to the United States by a factor of approximately 3 or 4 (RMCC, 1986), principally because of the fact that total U.S. emissions of both sulphur and nitrogen are much greater (RMCC, 1986).

The extreme sensitivity of the Canadian aquatic environment has resulted in great concern within Canada about the environmental consequences of continued high loadings of strong acids and their precursors. The transboundary nature of the problem increases the complexity of solving the problem but does not make a resolution impossible to achieve.

A legal precedent regarding transboundary pollution was established by the Trail Smelter case in 1941. Sulphur dioxide emissions from a smelter constructed in Trail, British Columbia, in the late 1890s resulted in extensive damage locally, as well as in areas across the border in the state of Washington. The initial dispute involved a private nuisance claim against a corporation in Canada by residents of the state of Washington. Although the dispute did not directly concern the two governments, they agreed to submit it to an Arbitral Tribunal established pursuant to an international convention. It was accepted that Canada would be responsible for any damage caused by the corporation and the United States was the proper claimant to represent claims by its citizens. In April 1935, a preliminary agreement on compensation to the residents was reached, and a final resolution on pollution control at the smelter was established in 1941. The Trail Smelter Arbitration (1941) established the following important international legal principle:

> ... no State has the right to use or to permit the use of its territory in such a manner as to cause injury by fumes in or to the territory of another or the properties or persons therein when the case is of serious consequence and the injury is established by clear and convincing evidence.[1]

Whatever the international interpretation of this law, it has a special claim to recognition by judicial and administrative tribunals in the United States and Canada.

---

[1]Trail Smelter Arbitral Tribunal, 1941.

B. History of Negotiations on Transboundary Air Pollution with the United States

In 1976, the Canadian Network for Sampling Precipitation (CANSAP) began monitoring the acidity of rainfall. The results showed that the deposition rate of strong acids in Canada was similar to that documented in Norway and Sweden (Summers and Whelpdale, 1976). Simultaneously, research in southern Ontario (Dillon et al., 1978) showed that deposition of acid and sulphate greatly exceeded that provided by any local sources.

In 1978, in recognition of the transboundary nature of the problem, the governments of the United States and Canada established a United States/Canada Bilateral Research Consultation Group to study the long-range transport of air pollutants (LRTAP) phenomenon and to coordinate research between the two countries. Furthermore, in November of that same year, the United States sent Canada a diplomatic note requesting an informal discussion of a congressional resolution calling for a cooperative agreement with Canada on transboundary air pollution. The United States was concerned about the potential environmental impacts of a new coal-fired power plant at Atikokan in northwestern Ontario on the Boundary Waters Canoe Area in Minnesota.

In 1979, Canada and the United States issued a joint statement on transboundary air quality announcing the intention of both governments to develop a cooperative agreement on air quality. After continued discussions between the two countries, Canada and the United States signed a Memorandum of Intent (MOI) in August of 1980 concerning transboundary air pollution. In this document, both countries declared their intention

> to develop a bilateral agreement which will reflect and further the development of effective domestic control programs and other measures to combat transboundary air pollution[2]

and

> to promote vigorous enforcement of existing laws and regulations as they require limitation of emissions from . . . existing facilities in a way which is responsive to the problems of transboundary air pollution.[3]

In addition, five working groups were established to provide the scientific basis of this agreement. While the working groups continued to meet and discuss the relevant scientific findings, several negotiating sessions were held between the two countries. At the February 25, 1982, negotiating meeting, the Canadian federal environment minister announced that Canada was prepared to reduce its SO_2 emissions by 50% east of the Manitoba/Saskatchewan border, contingent on a similar commitment by

---

[2]Canada/United States Memorandum of Intent, August 5, 1980.
[3]Ibid.

the United States. At the June 15, 1982, session, U.S. negotiators rejected Canada's proposal. As a result, the Canadian federal environment minister decided that future negotiations with the United States would be fruitless and broke off official negotiations.

However, the final work group reports were released in February of 1983 and subsequently reviewed. The U.S. peer review, which was conducted by the National Academy of Sciences, called for reductions in acidic emissions and rejected the argument that more research was needed before control action could be taken. The Canadian peer review report was released by the Royal Society of Canada and concluded that the scientific evidence was sufficient to warrant prompt introduction of abatement measures.

With the failure of the MOI negotiations and the continued concern expressed by the Canadian government, President Reagan agreed to the appointment of special envoys by the two governments to review and assess the international environmental problems associated with the long-range transport issue and to recommend actions to resolve them. The Joint Report of the Special Envoys on Acid Rain was released in January 1986 prior to the second summit meeting between Prime Minister Mulroney and President Ronald Reagan. It called for the implementation of a five-year, $5 billion control technology demonstration program by the U.S. government with joint funding between the government and private industry (Lewis and Davis, 1986). In addition, it recommended that both countries review existing air pollution legislation to identify opportunities for addressing transboundary environmental concerns and that they establish a bilateral advisory and consultative group on transboundary air pollution. Both countries accepted the Envoys' Report at the 1986 summit meeting.

Since that time, several meetings have been held by the Bilateral Advisory Group, and the commitment to the Envoys' Report was reaffirmed at the 1987 summit. However, the U.S. administration continued to express its reluctance to take abatement action, insisting that more research was needed before emissions reductions could be mandated. In 1985, Canada announced its intention to reduce emissions by at least 50% east of the Manitoba/Saskatchewan border by 1994 with the provinces of Quebec, Ontario, and Manitoba subsequently issuing regulations to major SO_2 emitters within their jurisdiction in order to fulfill their commitments. By 1988, comparable action had not been taken by the United States.

II. Atmospheric Studies

A. Emissions

1. SO_2 and NO_x Emissions Inventories

The compilation of up-to-date emissions inventories for acid-producing pollutants is necessary for several reasons. Historic and current databases allow us to determine if any emission trends are developing. Detailed and accurate emission inventories are needed for input to atmospheric transport models and for the development and implementation of successful control programs.

The 1980 emissions inventory of SO_2 and NO_x from anthropogenic sources is currently the most complete for both Canada and the United States. Total SO_2 emissions in Canada in 1980 were estimated at 4.6 million tonnes (t) (RMCC, 1986). In eastern Canada, six large copper and nickel smelters (two in Manitoba, two in Ontario, and two in Quebec) and one iron ore processing plant (in Ontario) together produced about 45% of the total (Figure 1–2). Electric utilities accounted for another 16.5% with substantial emissions in three provinces: Ontario, New Brunswick, and Nova Scotia. Regionally, the provinces east of the Manitoba/ Saskatchewan border accounted for more than 80% of the Canadian total (RMCC, 1986).

Total NO_x emissions in Canada in 1980 were 1.7 million t (RMCC, 1986). Mobile sources (cars, light-duty trucks, etc.) are the most significant source of NO_x in Canada, accounting for about two-thirds of the emissions in eastern Canada, with the remainder caused by power plants and other sources. NO_x emissions are more evenly distributed spatially than SO_2 emissions, but the eastern part of the country is still the major source area at approximately 60% (RMCC, 1986).

The national inventory also includes estimates of other species such as VOCs (volatile organic compounds), primary sulphates, and particulates.

Work is nearing completion on the development of a complete emissions inventory for 1985 for both Canada and the United States.

Due to the shortage of actual measurements, there is a degree of uncertainty in the estimates of natural emissions of SO_2 and NO_x. In 1980, it was estimated that sulphur emissions from natural sources were approximately 0.5 million t, principally from the biogenic activity of soils (RMCC, 1986).

Estimates for NO_x emissions from natural sources in Canada are currently not available. The U.S. National Oceanic and Atmospheric Administration estimated the total annual natural NO_x emissions in North America at 0.9 million t (N), 0.3 million t (N) as a result of lightning, and 0.6 million t (N) from soils (RMCC, 1986).

2. Trends in SO_2 and NO_x Emissions

Canadian SO_2 emissions in 1955 were approximately 4.6 million t. They rose to a peak of 6.7 million t between the mid-1960s and 1970, then decreased back to 4.6 million t in 1980 (RMCC, 1986). Eastern Canada continued to be the major contributor to these SO_2 emissions throughout this period. Subsequent to 1980, emissions dropped in eastern Canada. SO_2 emissions in eastern Canada were approximately 3.2 million t in

CANADIAN 1980 EMISSIONS OF SULPHUR DIOXIDE
10³ TONNES

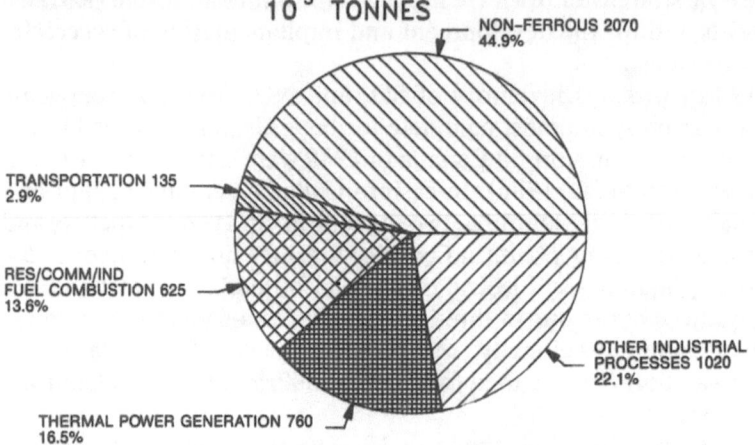

CANADIAN 1980 EMISSIONS OF NITROGEN OXIDE
10³ TONNES

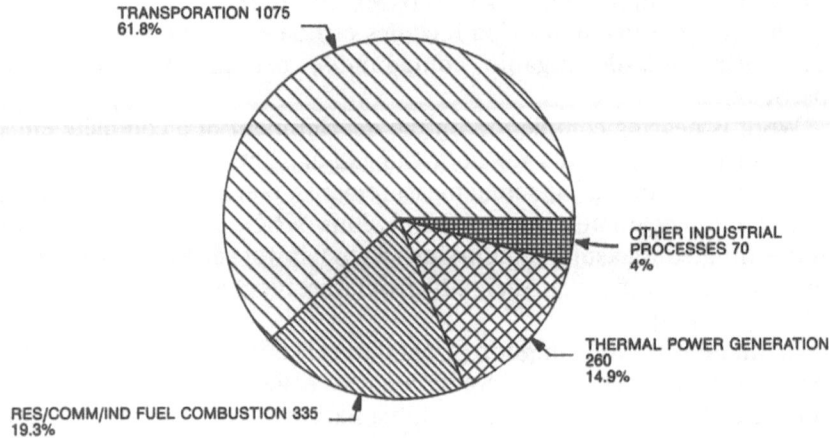

Figure 1-2. Anthropogenic emissions of SO_2 and NO_x in Canada in 1980. (Source: RMCC: Atmospheric Sciences, August 1986.)

1984, down from 3.8 million t in 1980 (Dillon, Lusis, Reid, and Yap, 1988; RMCC, 1986; Figure 1-3).

During the same time period, Canadian NO_x emissions increased from 0.6 million t in 1955 to 1.7 million t in 1980. Unlike SO_2 emissions, Canadian NO_x emissions have remained relatively constant since 1980 (Dillon et al., 1988; RMCC, 1986; Figure 1-3).

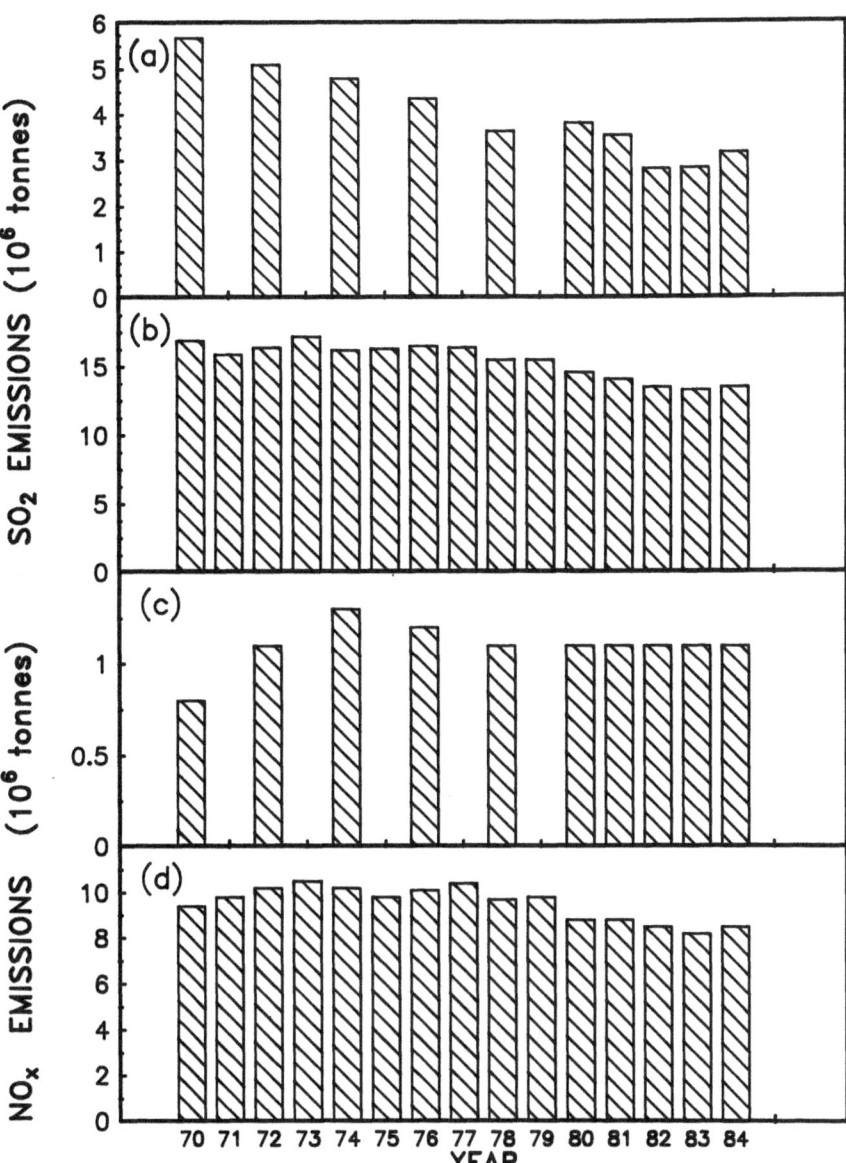

Figure 1–3. SO_2 emissions on (a) eastern Canada and (b) the eastern United States, and NO_x emissions in (c) eastern Canada and (d) the eastern United States. (Modified from Dillon et al., 1987.)

With the recent announcements of regulations mandating SO_2 emissions reductions by several of the eastern Canadian provinces, it is estimated that SO_2 emissions in eastern Canada (east of the Manitoba/Saskatchewan border) will be reduced by at least 50% by 1994 to 1.85

million t. Futhermore, with the new motor vehicle standards for NO_x that were effective in the fall of 1987 on 1988 model-year vehicles, NO_x emissions are expected to level off in Canada.

B. Deposition Monitoring

1. Wet and Dry Deposition

Several federal and provincial networks have been established to monitor the concentration and deposition rates of chemical species in precipitation on both local and regional scales (Table 1–1). Short-term as well as long-term spatial and temporal variations in wet deposition are under study. A National Atmospheric Data Base (Natchem) is being implemented as a repository for the Canadian data.

Figure 1–4 shows the wet deposition patterns of sulphate across eastern North America from the combined results of Canadian and American monitoring networks. The southern portions of the provinces of Ontario and Quebec currently receive the highest deposition rates of strong acid in Canada with the maritime provinces receiving higher deposition rates than the western provinces (where wet sulphate deposition is on average below 10 kg ha^{-1} yr^{-1}).

Several of the monitoring networks described in Table 1–1 originally attempted to collect dryfall as well, using both "wet" and "dry" buckets. However, an analysis of the results suggested that for sulphur and nitrogen compounds, the dryfall thus determined was not representative of dry deposition. At present, dry deposition is inferred from air concentration data that are measured at numerous sites across Canada by both federal and provincial agencies. The dry deposition rate is then calculated by multiplying the air concentration by a deposition velocity, which is determined for each substance taking into account information on local meteorology and surface characteristics.

In Canada, the APN, (Air and Precipitation Monitoring Network) was set up specifically to obtain ambient air concentrations of selected S and N compounds and precipitation chemistry at rural sites. This network was expanded and became the Canadian Air and Precipitation Monitoring Network (CAPMoN). Results from the APN sites showed that ambient SO_2 concentrations decreased with increasing distance from the major sources. An analysis of 4 years of air and precipitation chemistry measurements (Summers, Bowersox, and Stensland, 1986) at six locations in eastern Canada indicated that dry deposition of sulphur accounted for 22% of the total sulphur deposition, and dry deposition of nitrogen accounted for 21% of the total nitrogen deposition. On average, dry deposition accounted for about 40% of the total deposition at the southernmost stations, but less than 20% in remote areas.

Table 1-1. Federal and provincial deposition monitoring networks in Canada.

Network	Period of operation, number, and location of sites	Number of sites (1986)	Sampling period	Parameters measured
Canadian Network for Sampling Precipitation (CANSAP)	Began in 1977 with 45 sites across Canada. Operation ended in January 1986. Network replaced by CAPMon.	0	Monthly samples collected on last day of month	pH, conductivity, SO_4, NO_3, NH_4, Cl, K, Na, Ca, Mg
Canadian Air and Precipitation Monitoring Network (APN)	Began in November 1978 with 5 sites. 4 sites became part of CAPMoN.	1	Daily samples	pH, SO_4, NO_3, NH_4, Cl, K, Na, Ca, Mg, total phosphorus
Canadian Air and Precipitation Monitoring Network (CAPMoN)	Mid-1983 to present. Sites across Canada.	21	Daily precipitation samples (8 locations also monitor air quality)	pH, conductivity, SO_4, NO_3, NH_4, Cl, K, Na, Ca, Mg, alkalinity, acidity
British Columbia Provincial Precipitation Monitoring Network	Began operation in 1983 with 7 stations.	12	Weekly samples	pH, total acidity, total alkalinity, SO_4, NO_3, Cl, NH_4, Na, K, Ca, and Mg
Alberta Precipitation Quality Monitoring Network	April 1978–present.	10	Monthly samples	pH, acidity/alkalinity, conductivity, SO_4, NO_3, Cl, PO_4, NH_4, Na, K, Ca, and Mg

Table 1-1. (*Continued*)

Network	Period of operation, number, and location of sites	Number of sites (1986)	Sampling period	Parameters measured
Summer Rain Sampling Network in Northern Saskatchewan	1983–present.	7	Daily samples (May 15 to October 15)	pH, acidity, alkalinity, SO_4, NO_3, Cl, NH_4, K, Ca, Na, Mg, metal ions (2 samples/station/ season)
Manitoba Network for Precipitation Collection	1980 (2 stations)–present.	6	Daily samples	pH, conductivity, acidity SO_4, NO_3, Cl, Ca, NH_4, Mg, Na, and K
Acidic Precipitation in Ontario Study (APIOS) Cumulative Network	September 1980 (monthly samples)–present.	37	28-day samples as of January 1982	pH, SO_4, NO_3, NH_4, Cl, K, Na, Ca, Mg, total Kjeldahl nitrogen, total phosphorus, trace metals
Acidic Precipitation in Ontario Study (APIOS) Daily Network	September 1980–present.	16	Daily samples	pH, conductivity, SO_4, NO_3, NH_4, Cl, Na, Ca, Mg, acidity
Acidic Precipitation in Ontario Study (APIOS) Cumulative Air Network	November 1981–present	25	28-day cumulative samples	Particulate SO_4, NO_3, NH_4, K, Na, Ca, Mg, and trace metals. Gaseous HNO_3 and SO_2

Network	Period	Sites	Sampling	Parameters measured
Acidic Precipitation in Ontario Study (APIOS) Daily Air Network	July 1980–present.	4	Daily samples	Particulate SO_4, NO_3 and NO_4, K, Na, Ca, **Mg.** Gaseous SO_2 and HNO_3
Réseau d'echantillonage des précipitations du Québec	June 1981–present.	42	Weekly samples	pH, Ca, Mg, Na, K, NH_4, SO_4, NO_3, Cl, HCO_3, F
New Brunswick Precipitation Monitoring Network	1980–present.	7	Monthly samples to October 1986; weekly since then	pH, acidity, alkalinity, conductivity, various cations, anions, and metals
Nova Scotia Precipitation Monitoring Network	1978–present.	5	Weekly samples	pH, acidity, alkalinity, conductivity, SO_4, NO_3, NH_4, Cl, K, Na, Ca, Mg, and some heavy metals
Newfoundland Environment Acid Precipitation Network	1980–present	7	Weekly samples (3 sites co-located with Environment Canada)	pH, alkalinity, conductivity, SO_4, NO_3, Cl, NH_4, Na, Ca, Mg, and K

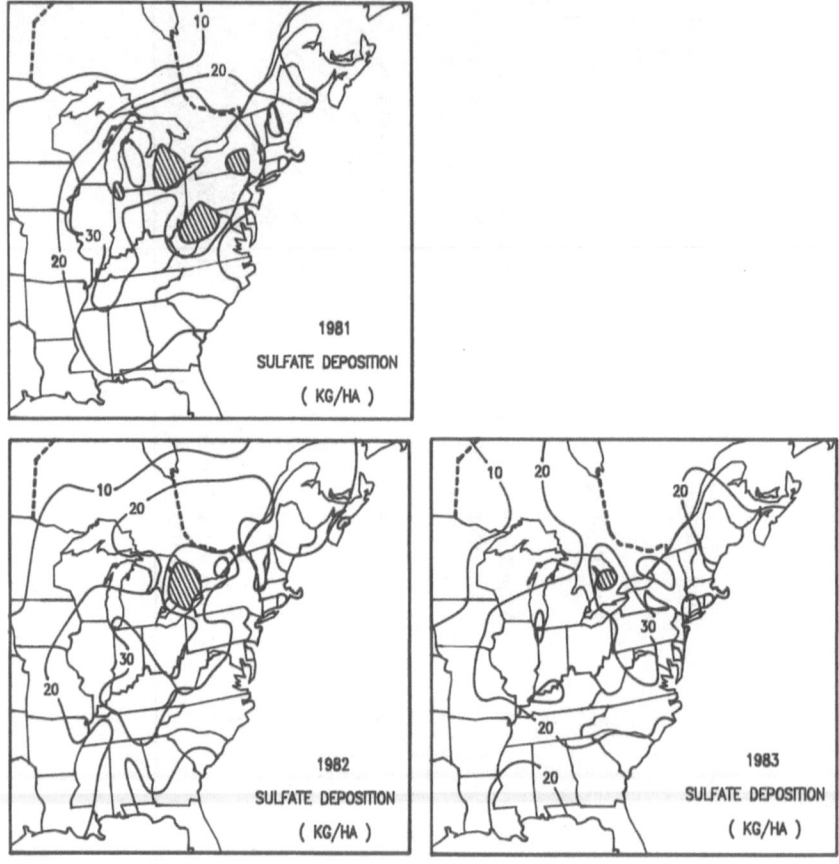

Figure 1–4. Annual sulphate deposition (kg ha^{-1} yr^{-1}) in eastern North America for the years 1981, 1982, and 1983. (Source: Summers et al., 1986.)

Ontario's airborne particulate and SO$_2$ sampling network consists of more than 20 sites with low-volume sampling carried out over four-week intervals to determine ambient concentrations of various particulate and gaseous compounds, as well as daily measurements at four sites. In southern Ontario, the dry deposition is about one-half of the wet deposition rate, whereas in central and northern Ontario it is about one-quarter to one-fifth of the wet deposition rate (Tang, Ahmed, and Lusis, 1986).

There are various additional special studies of atmospheric deposition in progress in Canada, including eddy correlation studies of dry deposition of SO$_2$ and O$_3$ to forest stands at Borden, Ontario (den Hartog, Neumann, and King, 1987), and of NO$_2$ and of SO$_2$ to agricultural crops and snow surfaces (Edwards and Ogram, 1986, Ogram, Northrup, and Edwards, 1988), instrument intercomparisons and evaluations at a number of sites, and the sampling of trace metals in air and precipitation at Muskoka in order to assess the impact of smelting activities in Sudbury.

2. Temporal Variability in Deposition

Deposition of sulphate and many other ions exhibits a strong seasonal cycle that is related to the seasonal nature of many meteorological and chemical processes. Studies by Tang et al. (1986) at Dorset, Ontario, and Summers, Bowersox, and Stensland (1986) in the central and maritime Canadian regions show that maximum deposition of sulphate occurs during the summer months, and that nitrate shows minimal variation between winter and summer periods.

In North America, major precipitation episodes with high rates of wet deposition can account for a large fraction of the annual total deposition. A four-day event in 1981 in Muskoka, Ontario, contributed 7 kg of wet sulphate (28%) to the annual total (Kurtz, Tang, Kirk, and Chan, 1984).

Over a long period of time (a year or more), it appears that, for many eastern Canadian locations, 20% of these storm events can contribute between 47% and 70% of the total wet and dry deposition of sulphate and nitrate (Barrie and Sirois, 1986). For sulphate, the contribution by single events is higher for wet deposition than for dry, particularly at remote locations, whereas for nitrate, the reverse is true (Barrie and Sirois, 1986).

The amount of atmospheric deposition varies from year to year due to several factors including changes in emission rates and meteorological variability, thus making trend analyses very difficult. However, Canadian agencies that have sufficient data available have conducted trend analyses.

Analysis of data from five rural sites in Nova Scotia between 1978 and 1984 (Underwood, 1985) showed some temporal trends with decreases in deposition of sulphate, nitrate, and hydrogen ion over recent years.

In an analysis conducted by Dillon et al. (1988), the results of 10 years of bulk deposition monitoring in the Muskoka-Haliburton area were reported. The recent decrease in SO_2 emissions in eastern North America was strongly correlated with a concomitant decrease in both bulk deposition rates and concentration of sulphate and hydrogen ions. Similar results were reported by Hedin, Likens, and Bormann (1987) for the northeastern United States. These long-term data records demonstrate the benefits of emissions controls.

Recent sampling of fogs and low clouds has also shown high concentrations of strong acidic anions and extremely low pHs. Measurements made by aircraft in central Ontario show a strong tendency for the lowest pH to occur at the cloud base with pH values typically ranging from 3.0 to 5.0 (Summers, Bowersox, and Stensland, 1986). The associated sulphate and nitrate concentrations range from 10 to 50 mg L^{-1} in polluted air masses. Extreme values of 60 mg L^{-1} have been recorded in summer convective clouds in Ontario (Leaitch, Strapp, Wiebe, and Isaac, 1986).

In late 1985, a network of mountain monitoring sites was established (CHEF—Chemistry of High Elevation Fogs) as a forest project with similar sites in the United States to monitor air, fog, and precipitation chemistry. Preliminary data from the Canadian sites indicate that fog (cloud on the mountain) water pH values (mean 3.7) near the summits are much lower than precipitation pH values (mean 4.3) at the same location (Schemenauer, 1986; Schemenauer and Winston, 1988).

C. Meteorological and Modeling Studies

1. Meteorological Studies

Numerous studies have been carried out related to acidic deposition and long-range transport. One example is an overview study on meteorological analyses of precipitation events in Ontario for the period 1976 to 1983 (Yap and Kurtz, 1986) that indicated that for southern and central Ontario, precipitation events most commonly occur with pre-warm front and cyclonic weather situations, and air parcel trajectories from the south and southwest octants. Trajectories from these octants, which include the heavily industrialized regions of the Ohio Valley and southwestern Ontario, also account for most of the wet deposition of sulphur and nitrogen.

This technique was used to assess the impact of a major reduction in SO_2 emissions on atmospheric deposition (Lusis et al., 1986). During the period from June 1982 to March 1983, the INCO and Falconbridge smelters at Sudbury, Ontario, were shut down. The results of the analysis indicated that for receptors in a 400 km radius of the smelters, the smelter contribution to wet deposition of sulphate is expected to be less than 10% to 15% of the total deposition. For dry deposition of sulphate, the estimated contribution was greater, being about 20% to 30% of the total or less.

Continued collection of pertinent meteorological data is needed to provide support for the analysis of emission reduction effects on atmospheric deposition and for other special studies.

2. Development of Simple Models

Both the provinces of Ontario and Quebec as well as the federal Atmospheric Environment Service have been involved in the development of simple linear atmospheric models where physical processes are represented by a small number of variables. These models can be used to evaluate various emission reduction scenarios by predicting the resultant changes in deposition patterns.

In Ontario, two simple long-range transport models have been developed: a statistical model and a Lagrangian or "trajectory puff" model. The statistical model is used to compute long-term average deposition and air concentration of a chemical specie based on statistical analysis of meteorological variables collected over many years. The Lagrangian model simulates deposition on a monthly and seasonal time scale (Ellenton, Ley, and Misra, 1985; Ellenton, Misra, and Ley, 1988).

Both the statistical and Lagrangian models have been extensively evaluated and used in the evaluation of control strategies. Results from modeling exercises have also been presented at various hearings and legal interventions in the United States.

These models have been used to evaluate the benefits of emission controls. As part of the Canadian commitment to reduce SO_2 emissions by 50% by 1994, Ontario announced its control program in December 1985. This program requires the province's four major producers of sulphur dioxide to reduce their emissions from the 1980 level of 1993 kilotonnes to a maximum of 665 kilotonnes of sulphur dioxide by 1994.

Ontario's statistical model was used to predict the environmental benefits of the province's control program. Deposition of wet sulphate to the province's sensitive Muskoka region would be reduced from the 1980 base case level of 32 kg ha^{-1} yr^{-1} to 27.6 kg, a 14% reduction. The model also predicted significant decreases in deposition in neighboring regions. For instance, deposition in the sensitive area of southern Quebec would decrease by 10%, in the Adirondacks by 8%, and in New Hampshire by 7%.

Simple models have been criticized for numerous reasons, most frequently for their neglect of the nonlinear aspects of atmospheric chemistry. Recently, however, these models have been subjected to extensive reviews that concluded that the simple trajectory models appear to simulate wet deposition of sulphur over at least a one-year time period with reasonable accuracy. On this time scale, the nonlinear effects of atmospheric sulphur chemistry are not critical (RMCC, 1986).

3. Eulerian Model Development

As a result of the numerous criticisms of the usefulness of simple models to predict source-receptor relationships for emission control strategy development, the Ontario Ministry of the Environment and the Canadian federal government entered into a cooperative agreement with the Federal Republic of Germany in 1983 for the development of an Eulerian model, the "Acid Deposition and Oxidants Model" (ADOM). This model will attempt to incorporate all relevant physical and chemical processes in as much detail as possible.

The Canadian Eulerian model is being developed under contract to ERT (Environmental Research and Technology) and includes three-dimensional transport and dispersion, cloud physics, gas and aqueous phase chemistry, and detailed treatments of wet and dry scavenging (Venkatram and Misra, 1988). As a result, this model will be applicable for episodic studies.

A working version of the model has been installed at the Canadian Meteorological Center (CMC) in Montreal. It has also been installed on the University of Toronto CRAY XMP computer where it is being used to investigate the nonlinearity of sulphur chemistry in the atmosphere and the impact of Canadian and U.S. emission reductions.

Both the Canadian model and its American counterpart, RADM (Regional Atmospheric Deposition Model), are to be evaluated with monitoring data to be collected during an ambitious field study in eastern North America during a two-year period starting in June 1988. The objective of this study is not only to determine if the models are predicting acidic deposition rates at the observed levels, but also to determine if they are making the right predictions for the right reasons (i.e., if the chemical and physical processes are correctly simulated).

Once evaluated, ADOM will be applied to investigate the importance of nonlinearity in sulphur deposition and to determine the most effective means of controlling various pollutants, including sulphur, nitrogen, and hydrocarbons.

III. Terrestrial Effects Studies

A. Soil Studies

1. Field Studies

The degradation of soils has been documented around major point sources of pollution such as the Sudbury and Wawa, Ontario, areas (Whitby, Stokes, Hutchinson, and Myslik, 1976). The deposition of toxic metals can also affect the soil to the extent that no tree growth occurs (Linzon and Temple, 1980). In addition, sulphur deposition can acidify soils resulting in the loss of nutrients and the mobilization of aluminum and other toxic elements, thereby limiting growth.

The possibility that continued acidic deposition could deplete these nutrients and result in less productive forest sites has serious socioeconomic consequences, since forestry is a $10 billion per year industry in Canada.

Soils common in Canada include luvisols, podzols, and brunisols. Although podzols are naturally acidic through the action of carbonic and organic acids, the addition of dilute sulphuric and nitric acids can still result in appreciable effects on water draining these naturally acidic soils, including increased aluminum mobilization (Reuss and Johnson, 1985). Acidic deposition is more likely to affect soils directly that are less naturally acidic such as luvisols and brunisols.

2. Laboratory Studies

Field evidence gathered to date suggests that acidic deposition can cause increased base cation leaching and Al solubilization (RMCC, 1986). Several Canadian studies have been conducted using column-lysimeter leaching experiments to determine the effects of acidic deposition on soils common to Canada.

Early experiments that used extremely high loadings of acid showed free movement of hydrogen ion through soils. However, more recent studies have shown that natural soils have considerable ability to buffer hydrogen ion additions. For example, Morrison (1981) observed that both podzol and brunisol soils consumed a considerable amount of hydrogen ion and that throughput of hydrogen ion took place only after soil bases had been substantially depleted. Other Canadian researchers (e.g., Hutchinson, 1980; Rutherford, van Loon, Mortensen, and Hern, 1985), using lysimeter studies, have demonstrated that bases are displaced by hydrogen ion, then transported by sulphate and other anions once the soils are saturated with sulphate.

Additional column-lysimeter work has addressed the question of increased solubilization of aluminum with increased acidity. Aluminum seems to be readily mobilized to high concentrations when the pH of the leaching solution falls sufficiently. In an experiment using a hardwood soil from Ontario and coniferous horizons from Quebec, leaching was limited principally to the upper soil horizons, and the solubilized Al was redeposited in the lower soil horizons (Rutherford, van Loon, Mortensen, and Hern, 1985).

Since it is difficult to determine the influence of acidic deposition on soil chemical properties and processes in the field, these experiments allow for the isolation of changes in soil chemistry or processes in the laboratory for comparison with the results of field studies. The combined field and laboratory results should provide a good indication of what is actually occurring in natural soils.

B. Vegetation Studies

1. Lichens and Mosses as Bioindicators

Both laboratory and field studies have indicated that precipitation pH is an important factor in lichen and moss survival. Deposition of airborne pollutants to mosses and lichens in rural areas of Canada has been measured and compared with levels in remote areas (Zakshek and Puckett, 1986). Sulphur and lead were the only elements with consistently higher levels in the lichen *Cladina rangiferina* in eastern Canada when compared with levels measured in the Northwest Territories. A comparison of measured sulphur levels in this lichen showed highest values in central Ontario and Quebec, with distinct east-west gradients.

Metal levels in lichens from approximately 50 locations across Canada were strongly influenced by their proximity to major sources of metal-contaminated particulates. The sulphur levels of lichens, however, exhibited decreases in concentrations from south to north (Case, 1985).

Futhermore, in an experiment subjecting a mature jack pine forest near Kirkland Lake, Ontario, to twice monthly applications of acidic rain ranging from 2.5 to 5.6 pH over a six-year period, the dominant ground cover, the feather moss *Pleurozium schreberi,* showed significant decrease in percentage cover, growth rate, and biomass accumulation within two years of all pH treatments below pH 4.0 (Hutchinson and Scott, 1986). Substantial decline also was evident in caribou lichens, *Cladina* spp., over the five years of spray applications.

The loss of lichen and moss ground cover will result in exposure of tree roots and soil microorganisms to drought and other stresses, directly influencing soil chemistry.

2. Laboratory/Greenhouse Studies

Numerous laboratory and greenhouse experiments have been conducted where different crop and tree species have been subjected to simulated acidic rain treatments to determine injury symptoms and to establish visible injury thresholds.

Visible foliar injury is usually manifested by the formation of necrotic spots, pits, or lesions on adaxial leaves. Eventually leaves can become deformed or reduced in area and their ability to photosynthesize can become inhibited. This type of injury has been documented in laboratory experiments only at pH of 3.0 or lower.

Greenhouse studies (Enyedi and Kuja, 1986) showed that crops varied in their sensitivity to acidic rain with visible injury occurring on the leaves, stems, and flowers of all plants at treatments of pH 3.0. Plant response was also dependent on the total dose of acidity, as well as the characteristics (intensity, frequency, and quantity) of the rain event.

Since foliar injury is not a reliable means of assessing actual effects on crop yield, Kuja, Jones, and Enyedi (1986) conducted a study to determine the effects of simulated acidic rain on seed germination, seedling establishment, and early stages of plant growth. Again, visible foliar injury was only observed at pHs between 3.0 and 2.6. Cultivar responses varied, indicating that, for yield or growth effects, plant response is not only species-dependent but also strongly cultivar-dependent.

Abouguendia, Baschak, and Goodwin (1986) conducted similar controlled growth experiments with six Saskatchewan forest and crop species and showed that simulated acidic rain with pH 3.6 or higher had minimal effects on the crop and forest species studied.

Laboratory studies to determine the effects of simulated acidic rain on tree growth have produced conflicting results. Visible injury is often considered the primary indication of dose response or impact. A review of the species studied to date shows that the upper pH limit for injury is about 3.0 for both conifers and deciduous species (RMCC, 1986). With the annual average pH of rain in the forested areas of eastern Canada currently above 4.0, there is limited likelihood for direct foliar injury at this time. However, more research is needed in this area using field-grown material to ensure that these assumptions are correct and that there are no extenuating factors.

Atmospheric pollutants may, however, directly affect reproductive processes. Cox (1983) demonstrated that pollen is significantly affected by pHs below 5.6. Furthermore, the pH where a 50% probability of death occurred for all pollen tested was within the range of rainfall acidity in eastern Canada. These results show that the natural regeneration of Canadian forests is at risk.

Cox (1985) tested 11 Canadian forest flora species for the combined effects of copper, lead or zinc, and pH on pollen function. The results indicated that, at current copper deposition levels, only the more sensitive species such as sugar maple and yellow birch experienced any effects on pollen germination or tube growth.

Another area of concern is the potential toxic effects of increased aluminum concentrations on tree roots. The possibility that aluminum toxicity is responsible for the dieback currently occurring in eastern North America cannot be ignored. Hutchinson, Bozic, and Munoz-Vega (1986) conducted a study to determine the relative sensitivity of five economically important eastern coniferous species to aluminum. The growth of red and white spruce was somewhat inhibited at 5 mg Al L^{-1} and severely inhibited at concentrations greater than 20 mg L^{-1}. The growth of the black spruce seedlings was also severely inhibited at concentrations between 20 mg and 150 mg Al L^{-1}. In contrast, growth of white pine was actually stimulated by concentrations between 5 and 20 mg Al L^{-1}. It was consistently more tolerant to aluminum toxicity than the other four species. Jack pine was intermediate in its response.

Laboratory studies have resulted in an increased understanding of many of the direct effects of acidic rain but the importance of these effects still needs to be established for individual species and regions.

3. Exclusion Canopy Studies

At the Brampton Laboratory of the Ontario Ministry of the Environment, an exclusion canopy system has been used to assess the impact of acidic deposition and associated air pollutants on commercially important crop and tree species (Kuja, Jones, and Enyedi, 1986).

The system was used to study the effects of simulated acidic rain treatments on soybean (cv. Hodgson) as measured by seed yield (kg ha^{-1}) or components of seed yield. The results did not show any significant yield reduction due to increased acidity of simulated acidic rain (Kuja, 1988). However, additional research is needed to assess the synergistic effects of acidic rain and other pollutants, specifically ozone.

In 1987, a multiyear study was initiated to investigate the potential effects of simulated acidic rain, ozone, and soil nutrient status on sugar maple (*Acer saccharum*) and spruce seedlings grown on the rain exclusion canopy treatment plots. The results from this experiment should provide insight into the possibility that acidic rain is having a major detrimental effect on sugar maples in eastern Canada.

C. Forest Decline Studies

The central hypothesis in forest decline is that stress of either biotic or abiotic origin alters tree health and renders forests susceptible to further loss of vigor (Manion, 1981). Disease organisms ultimately attack the weakened trees and result in their further demise (Houston, 1981).

For several years, maple syrup producers in Quebec have been concerned by the apparent decline in the condition of an increasing number of sugar bushes (Carrier, 1986). Maple decline is not a new problem for the province. Lachance (1985), in his summary of historical maple decline in Quebec, documented a seasonal and localized decline in the maple bushes of Beauce County in 1934. This is the same area where the current decline was documented for the first time in 1978.

During the summer of 1982, several cases of sugar maple decline were reported in the province of Quebec. As a result, during the summers of 1983, 1984, and 1985, surveys were conducted in order to determine the scope and severity of the maple decline problem in the province. By 1985, surveys that covered a total area of 25,100 km^2 showed that close to 60% of the trees could be classified as "healthy," or "slightly affected," 35.4% as "lightly affected," 4.2% as "moderately affected," and 0.9% as "heavily affected" (Carrier, 1986). The decline symptoms varied but included earlier fall coloring of small leaves, branch dieback, bark loss from main branches, longer healing time for trunk tap holes, reduced radial increment growth, and tree death. In addition, virtually all maple plots showed increased incidence of decline between 1983 and 1985. Average annual tree increment in maple forest association stands decreased between 1981 and 1985.

Results from a comparison of data from 129 sample plots established in the summers of 1983 and 1984 (Gagnon, Robitaille, Roy, and Gravel, 1985) showed that maple decline was most severe on the wettest (37%) or the driest sites (23%). Also, maple woods containing basswood had a lower severity of decline. Soils were analyzed from the various sites and their pH ranged from 4.0 to 4.7. The area of most severe sugar maple dieback coincided with the occurrence of high levels of nickel, chromium, and cobalt in the soil and with magnesium-rich till. Soil potassium deficiencies were also noted in some areas of decline.

Gagnon, Roy, Gravel, and Gagné (1986) detected a decrease in several nutrients over a period of 15 years in soil compositions from 53 maple stands, and Bernier, Brazeau, Camire, and Brousseau (1985) in a study of foliage of Appalachian maple stands showed that average concentrations of nitrogen and phosphorus were low and those of potassium and calcium were very low. Both the soil and foliar analyses indicate that the nutritional cycle of these maple stands has been disrupted and that these nutritional stress factors have been operating for several years. As a result, several researchers have been conducting fertilization experiments based on soil and foliar analyses for the identification of nutrient deficiencies.

Hypotheses postulated for the decline include insect infestations and climate changes but none of them could be identified as the universal causative factor. Recently, acidic precipitation and air pollution have been suggested as the most likely causes since the affected region receives a high annual wet sulphate loading of 40 kg ha^{-1} yr^{-1}.

In Ontario, several maple syrup producers in the Muskoka area expressed concern in 1984 about the apparent increase in dieback and mortality of sugar maple trees in the previous six years. This area has received 30 to 45 kg SO_4 ha^{-1} yr^{-1} over this time period (Dillon et al., 1988) and the lakes in the region are well known to be affected adversely.

In 1984, permanent observation plots were established in seven maple woodlots in the Muskoka area and in one woodlot near Thunder Bay in northwestern Ontario as a control plot. The 1984 results indicated that the current decline outbreak first began in 1978. Symptoms appeared throughout the Muskoka region with no consistent pattern apparent (McLaughlin, Linzon, Dimma, and McIlveen, 1985). When the data from all seven plots were combined, 58% of the trees were considered healthy, 20% were experiencing light to moderate decline symptoms, and 22% were exhibiting severe decline symptoms. Decline symptoms appeared to be most pronounced on older trees and on tapped or wounded trees. Site nutrient deficiencies did not appear to be implicated in the sugar maple decline.

The soil at the Muskoka sites had a mean pH of 4.8, and had high concentrations of exchangeable aluminum, ranging from 6 mg to 40 mg kg^{-1} in mineral soil (McLaughlin et al., 1985). Declining trees suffered extensive root death with significantly higher aluminum concentrations in the fine roots. Foliar tissue leaching was apparent with reduced elemental concentrations in the tops of the tree crowns. Annual growth increments were very small in all trees during the forest tent caterpillar defoliations in 1976 and 1977; however, growth recovered in the healthy trees after the collapse of the epidemic but not in the declining trees. Furthermore, incremental growth in the declining trees appeared to be decreasing 20 years prior to the epidemic.

Although the prime factors for this decline probably include the severe insect infestations in 1976 and 1977, and the spring droughts in 1976, 1977, and 1983, acidic precipitation is an additional environmental stress on the maple bushes in the Muskoka area (McLaughlin et al., 1985).

Forest declines are also occurring in the maritime provinces. In 1985, in the course of general surveillance, unexplained white spruce decline was observed in northern New Brunswick (Canadian Forestry Service, 1986). The condition involved areas of several hundred hectares in Restigouche County with the greatest decline apparent in mature trees but some effects also evident on other age-classes. In addition, red spruce deterioration was observed in the southern part of the province. Although spruce budworm infestations were responsible for defoliation in the past, they do not adequately explain the conditions observed. Both of these problems are being investigated further.

Since 1979, early leaf browning and premature leaf fall have been occurring annually in white birch in southern New Brunswick and occasionally in Nova Scotia in western Cumberland County along the Bay of Fundy (Canadian Forestry Service, 1986). Although mainly a condition of white birch, similar symptoms were observed in 1980 and 1981 on other deciduous species. The condition appears to be limited to a coastal strip of 1 to 15 km in width which extends inland about 30 km, mainly in low-lying areas. In 1985, detailed weekly observations were made in an attempt to determine the problem's cause. Although it was determined that several organisms were associated with this condition, none of them, alone or in combination, could satisfactorily explain the situation. This area is frequently subjected to acidic fog episodes, and intensive monitoring of air pollutants in coastal fog is now in progress.

Currently, the relative importance of acidic deposition and air pollution stress in relation to other stresses such as diseases, insects, weather extremes, and climate, still needs to be clarified for the major eastern Canadian declines. However, it is likely that the added stress of air pollution has increased the trees' susceptibility to other stresses.

D. Wildlife Effects

1. Amphibians

A number of field and laboratory studies have been undertaken to assess the actual and potential effects of acidic deposition on amphibian distribution and abundance.

Surveys of amphibian breeding habitats in Canada in areas unaffected by local sources have shown that many of these habitats are very acidic (Dale, Freedman, and Kerekes, 1985; Gascon and Planas, 1986; Pough, 1976).

Poor reproductive success has been documented as resulting in smaller populations in very acidic environments. Increased embryonic mortality at low pH (pH <5.0) has been measured in both natural populations and in situ bioassays (Clark, 1986; Clark and Hall, 1985; Freda and Dunson, 1986; Saber and Dunson, 1978). Elevated aluminum concentrations in acidic natural waters are also toxic and can further reduce embryonic survival (Clark and Hall, 1985; Clark and LaZerte, 1985; Glooschenko, Weller, and Eastwood, 1985).

Even though amphibian larvae are more acid-tolerant than embryos, they are more vulnerable to indirect effects of acidification due to potentially limited food resources. Because they are feeding very actively during the larval stage, food resources and competitive relationships are very important.

The food resource for the larval stage consists of benthic invertebrates and plankton whose community structure and species composition can be altered at pH <5.6 (Mierle, Clark, and France, 1986).

2. Waterfowl

Because of their dependence on the aquatic environment for nest sites, brood protection, and food, the loss or degradation of this environment by acidification could have serious implications for the future of the waterfowl resource in eastern Canada. The severity of such effects will depend on the sensitivity of the nesting habitats.

McNicol, Bendell, and Ross, (1985) conducted comparisons of waterfowl breeding success in central and northern Ontario and showed that productivity was significantly lower in the headwater lakes receiving deposition greater than 30 kg of SO_4 ha^{-1} yr^{-1}. Also, breeding success rates were affected by the abundance of fish stocks and the fish community structure.

The diets of some fish and ducks overlap in simple headwater lake systems. With the critical life stage for food availability being the early brood development period, survival can also be affected by competition with fish for available aquatic insect prey, especially since the community and population structure of benthic invertebrates can be altered at pH <5.6 (Mierle et al., 1986).

IV. Aquatic/Catchment Studies

The effects of acidic deposition on aquatic systems in Canada have been described and reviewed in detail in three major reports: Environment Canada (1983), Harvey, Pierce, Dillon, Kramer, and Whelpdale (1981), and RMCC (1986). In addition, Canadian information has been included in reviews prepared by American agencies including National Academy of Sciences (1986), Environmental Protection Agency (1985), and Cook (1988), and in technical review articles (e.g., Dillon, 1983; Dillon, Yan, and Harvey, 1984; Haines, 1981). The information presented in these reports is not repeated here; instead, a summary is provided highlighting more recent findings.

A. Chemical Studies

1. Surveys: Extent of Sensitivity/Damage

There are an estimated 700,000 lakes in Canada below 52°N east of the Ontario-Manitoba border, which corresponds approximately to the zone where SO_4 deposition is greater than approximately 1 g (wet) SO_4 m^{-2} yr^{-1} (Kelso, Minns, Gray, and Jones, 1986).

Based on chemical surveys done in each province, Minns and Kelso (1986) and Kelso et al. (1986) estimated that 50% of these lakes had alkalinity <50 µeq L^{-1}, a level that indicates sensitivity to acidic deposition.

Futhermore, an estimated 20% (~150,000 lakes had pH ≤6.0, indicating that biological effects may have occurred, while ~14,000 lakes were estimated to have pH <4.7, a level at which biological effects are expected to be very severe (Mierle et al., 1986; Mills and Schindler, 1986). Although the proportion of these lakes with pH <6.0 that are dystrophic (high concentrations of dissolved organic acids) is not known, it is highly unlikely that the low pHs can be explained by this factor (see later discussion), except possibly for the maritime provinces. Minns and Kelso also predicted, based on a simple model derived from the Henriksen (1979, 1982) and Wright (1983) relationships, that ~9% of these 700,000 lakes would ultimately have pH <5.0 if "current" (1980) SO_4 deposition remained constant. However, the model is not validated, nor is the "constant deposition" assumption valid; SO_4 deposition has dropped in at least part of eastern Canada since 1980 (see Dillon et al., 1988).

The sensitivity and extent of effects on lakes in eastern Canada has also been assessed on a regional basis by Jeffries, Wales, Kelso, and Linthurst, (1986) and Neary and Dillon (1988). Both of these studies showed that not only was the extent of affected lakes great in eastern Canada and Ontario, respectively, but that organic acids did not play a significant part in the distributions of lake pH or alkalinity. For example, the highest organic acidic levels in Ontario were found in the zone containing the lakes with the highest average pHs and alkalinities where S deposition was lowest (Figure 1–5). In fact, in Ontario at least, there are no known lakes in zones where S deposition is <0.5 g m^{-2} yr^{-1} that have high dissolved organic carbon (DOC) and pH <6.0.

Both the Ontario and the eastern Canadian surveys also demonstrated that the pattern of SO_4 concentration in lakes followed the S deposition pattern, a result consistent with the findings of the survey of lakes in the eastern United States (Landers, Overton, Linthurst, and Brakke, 1988). Neary and Dillon (1988) also found that, in Ontario, there was a strong relationship between both lake alkalinities and pHs and S deposition, provided that only the lakes with conductivity <50 µS were considered. These data provided a strong indication of both widespread reductions in alkalinity and pH and a cause-effect relationship with S deposition.

There are relatively few data available to assess the effects on groundwaters in Canada. Azzaria, Gelinas, Robitaille, and Wilhelmy,

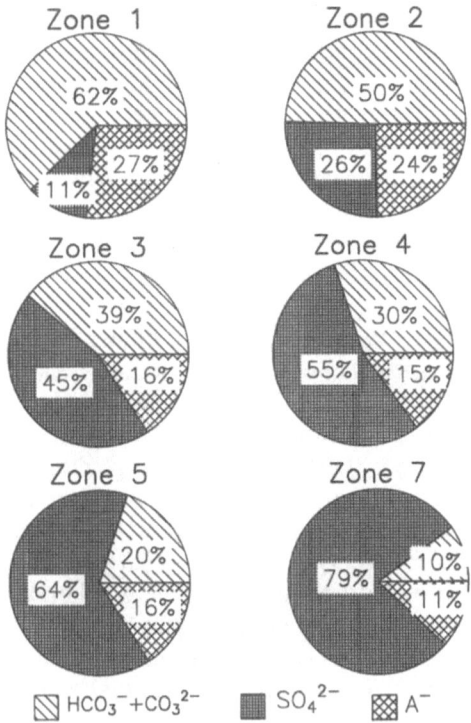

Figure 1-5. Relative importance of SO_4^{2-}, organic anions (A^-), and HCO_3^- ($+ CO_3^{2-}$) in lakes in Ontario with conductivity <50 µS in six S deposition zones:
Zone 1, <0.25 gS m^{-2} yr^{-1};
Zone 2, $0.25-0.50$ gS m^{-2} yr^{-1};
Zone 3, $0.50-0.75$ gS m^{-2} yr^{-1};
Zone 4, $0.75-1.0$ gS m^{-2} yr^{-1};
Zone 5, $1.0-1.25$ gS m^{-2} yr^{-1};
Zone 7, >1.25 gS m^{-2} yr^{-1}.

(1982) reported that in the Lac Laflamme (Quebec) catchment, shallow groundwater in zones of rapid infiltration had pH <4.5, whereas Craig, Johnston, and Bottomley (1986), Johnston, Bottomley, Craig, Inch, and Chew, (1985) and Bottomley, Craig, and Johnston (1984) found that shallow groundwater in the Turkey Lakes watershed (Ontario) demonstrated pH decreases (from ~5.2 to 4.6) in response to precipitation events.

2. Trends in Lakes and Streams

In Canada, as elsewhere, there is a paucity of historical water quality data useful for determining both the extent and rate of acidification of surface waters. As mentioned earlier, until the late 1970s acidic deposition was considered to be a local (e.g., Sudbury, Wawa, Trail) rather than a regional problem with effects apparent only in areas adjacent to major SO_2 sources. With very few exceptions, detailed monitoring programs designed to measure acidification rates directly and to monitor chemical trends were not established until the 1980s. As a result, assessment of the chemical trends in acidification has relied largely on indirect evidence, for example, paleolimnological relationships, geochemical relationships, or comparison of old and new data collected a number of years apart. The latter approach is usually compromised by changes in analytical methods.

The only reported examples of direct continuous measurement of lakes in high S deposition zones are Plastic and Harp Lakes in Ontario (Dillon, Reid, and de Grosbois, 1987). Plastic Lake's alkalinity declined by 2.1 ± 0.4 μeq L^{-1} yr^{-1} between 1979 and 1985, while its pH dropped 0.035 ± 0.007 units yr^{-1} (Figure 1–6); Harp Lake, on the other hand, demonstrated no significant change in either pH or alkalinity. The difference between these lakes was attributed to the fact that the Plastic Lake catchment is covered by very thin (<1 m) sandy basal till with many exposed bedrock (orthogneiss) ridges, whereas the Harp Lake catchment is covered by a generally thicker overburden of glacial till >1 m deep on $\sim$50% of the catchment and $\sim$10 m thick in the valley bottoms. Thus, Plastic Lake was typified as a "sensitive" lake; Harp Lake was considered to be relatively insensitive.

The loss of alkalinity in Plastic Lake was offset by a decrease in base cations, principally Ca, Na, and K, but not an increase in strong acidic anions (SO_4^{2-}, NO_3^{-}), indicating that depletion of available cations may have occurred in the catchment. The situation was complicated by the fact that over the period of study S deposition was decreasing. Unlike Plastic Lake, the Harp Lake SO_4^{2-} concentration decreased as a result of the declining S deposition.

Other sites with long-term records, especially Rawson Lake and other Experimental Lakes Area (ELA) lakes (Schindler, 1988) exist, but these are situated in a region where S deposition is low, and changes in lake chemistry are not expected.

Although a number of paleolimnological studies based on either diatom or chrysophyte -pH relationships have been conducted in areas where local sources have acidified lakes (e.g., Dickman and Thode, 1985; Dixit, 1986), there is little information yet available for any sensitive lakes in the rest of eastern Canada. The Paleoecological Investigation of Recent Lake Acidification PIRLA project (Charles et al, 1986) included three lakes in Algonquin Park, Ontario, but none have pH <6.0. Griffiths, Carney, Nakamoto, and Nicholls, (1990) have produced a calibration for Ontario lakes based on a large sample size ($\sim$60 lakes), and work is in progress to test this on a number of lakes including Plastic Lake.

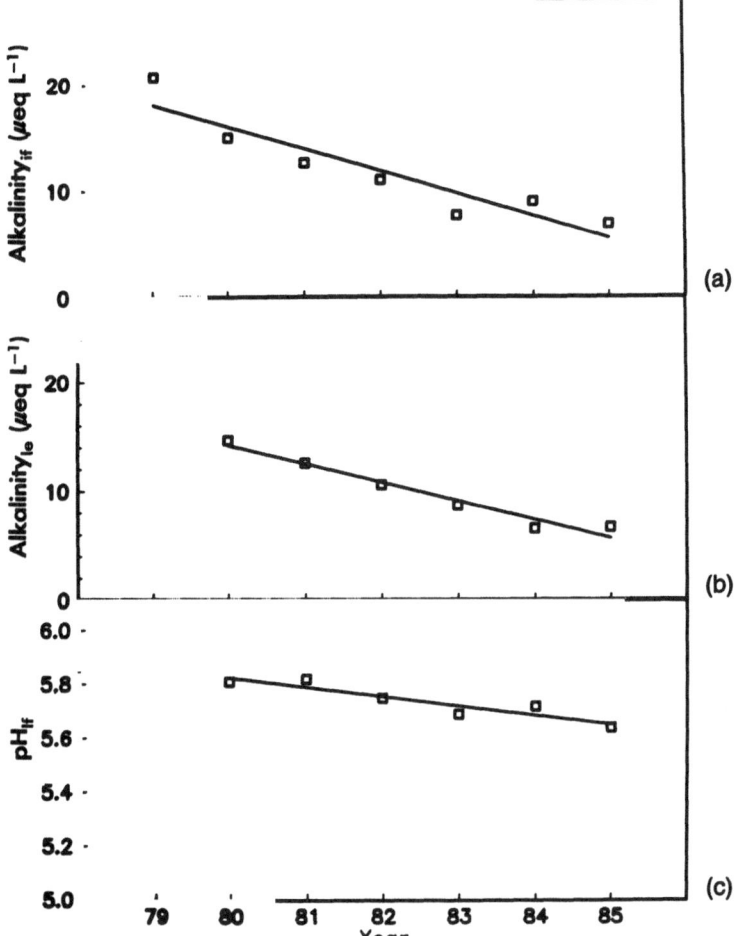

Figure 1-6. Annual alkalinity in the (a) ice-free and (b) ice-covered season, and (c) pH in Plastic Lake, 1979–1985.

B. Catchment Studies

Much of the acidic rain research in Canada has concentrated on the effects of acidic deposition on whole catchments, that is, on both their aquatic and terrestrial portions. A summary of the major catchment studies currently in progress in eastern Canada is given in Table 1–2; their locations are shown in Figure 1–7.

In each of these studies, hydrologic and elemental fluxes are measured in one or more catchments and lakes. These catchment studies also include soil and vegetation measurements and some tree condition and regeneration studies.

Table 1-2. Canadian catchment studies.

Location	S deposition (μeq m^{-2} yr^{-1})	Duration of studies	Monitoring sites	References
Experimental Lakes Area (Ontario)	14[a]	1969–present	5 lakes (control lakes for acidification experiments)	Schindler 1987, 1988
Turkey Lakes Watershed Study (Ontario)	54.8[b]	1981–present	4 lakes, 26 streams	Jeffries et al, 1986 Semkin and Jeffries, 1986
Muskoka-Haliburton (Ontario)	50–90[b,c]	1976–present	8 lakes, 24 streams	Dillon et al, 1982 LaZerte and Dillon, 1984 Seip et al., 1985 Dillon et al, 1987
Lac Laflamme Calibrated Catchment (Quebec)	46[a]	1980–present	Lac Laflamme and 3 nearby lakes	Jones and Deblois, 1987
Kejimkujik (Nova Scotia)	34[a]	1978–present	3 lakes	Kerekes and Freedman, 1986; Kerekes et al., 1982

[a] wet only
[b] total
[c] range from 1976 to 1986

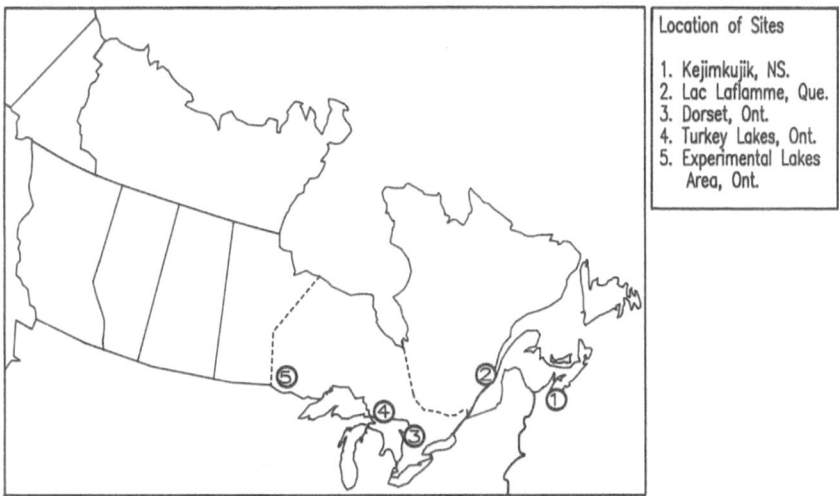

Figure 1–7. Location of Canadian catchment studies.

The sensitivity of the sites varies. The Experimental Lakes Area in northwestern Ontario has shallow podzolic basal tills with extensive bedrock exposure on hillcrests, but deep soils in low areas. Some of these sites are of moderate sensitivity to acidification, whereas others are very sensitive.

The Turkey Lakes basin is located on the Precambrian Shield but is only moderately sensitive, since the till is of variable thickness and contains a small amount of calcium carbonate (1% to 2%).

The catchments being studied by the Dorset Research Centre are located on the Precambrian Shield with overburden consisting mostly of thin till and rock ridges. The soils at Plastic Lake watershed are generally thin, acidic, weakly developed podzols. Most of these catchments are extremely sensitive as a result.

The Lac Laflamme catchment, within the Montmorency Experimental Forest, also lies on the Precambrian Shield and is extremely sensitive as well.

The poorly buffered oligotrophic watersheds combined with the low bicarbonate buffering capacity of the lakewaters also make the Kejimkujik catchments extremely sensitive to acidification.

These whole catchment studies provide valuable data on the interaction between terrestrial and aquatic systems.

C. Biological Effects

Most of the early concern relating to biological effects focused on fish, particularly sports fisheries. The decline of the Atlantic salmon fishery in a number of Nova Scotia rivers with low pH is well documented (Watt, Scott, and White, 1983) and supported by laboratory (Peterson, 1984) and field (Lacroix, 1985) studies. Adverse effects on several fish populations in Ontario were documented by Harvey (1982); Beggs and Gunn (1986) showed regional effects on lake trout and brook trout populations in Ontario.

More recently, evidence has accumulated from both field and laboratory studies that other organisms lower in the food web (including cyprinids, mollusks, crustaceans) are affected at higher pHs than many of the sports fish species (Holtze and Hutchinson, 1989; Mills and Schindler, 1986; Schindler et al., 1985). Because these lower trophic level organisms are usually the major food resource of the "sports" fish, their disappearance or reduction in abundance at higher pHs may have a major effect on the fish populations. In other words, the principal effects on fish are more likely to be indirect rather than acute toxicological response to pH (or aluminum).

Much of the information relating to the biological effects of acidic deposition in Canada is derived from the experimentally acidified lakes (L.223, L.302) in the Experimental Lakes Area (Schindler et al., 1985; see *Can. J. Fish. Aquat. Res.* 37, March 1980 and *Can. J. Fish. Aquat. Res.* 44, Supplement 1, 1987).

Acidification of L.223 from pH 6.8 to 5.0 over an eight-year period resulted in changes to many components of the food web (Table 1–3; Mills and Schindler, 1986). Diversity of most algae, zooplankton, benthos, and fish declined, while some species were lost and others appeared for the first time. Significant changes began at a pH just below 6.0, for example, decline in *Mysis relicta* (Nero and Schindler, 1983), and decline in fathead minnow populations. The appearance of filamentous algae (*Mougeotia* spp.) occurred at pH of 5.6, consistent with the observations

Table 1–3. pH thresholds of biotic changes for selected L. 223 organisms. (Modified from Mills and Schindler, 1986; Schindler and Turner, 1982.)

Biota	L. 223 pH
Mysis relicta abundance decline	5.93
Fathead minnow abundance decline	5.93
Mougeotia mats appear	5.64
Asterionella ralfsii appears	5.64
Orconectes virilis abundance declines	5.59
Lake trout recruitment ceases	5.59
White sucker recruitment ceases	5.02
Pearl dace abundance declines	5.09
Leeches become rare	5.09
Mayfly *Hexagenia* disappears	5.13

of Jackson (1985) who surveyed 40 lakes in Ontario. Lake trout recruitment failure at pH ~5.6 and white sucker recruitment failure at 5.0 were consistent with results of laboratory studies (Hutchinson, Holtze, Munro, and Pawson, 1989). Recruitment failure of the crayfish *Orconectes virilis* in L.223 at pH ~5.6 was similar to the results of laboratory and field toxicity tests (Berrill, Hollett, Margosian, and Hudson, 1985) with two other species of *Orconectes*.

As Plastic Lake has acidified (Dillon et al., 1987), a variety of biological effects have been observed and attributed to its declining pH. Harvey and Lee (1982) observed a dead and dying pumpkinseed sunfish population in the littoral zone of the lake each spring. Stephenson and Mackie (1986) found a depauperate population of *Hyalella azteca,* a common and normally very abundant amphipod, in Plastic Lake in 1983. In the following year no recruitment was observed, and the population subsequently disappeared. The population of a common and usually abundant snail, *Amnicola limnosa,* had reduced growth rate and abundance in 1980 (Rooke and Mackie 1984a, and 1984b), and is probably now extinct. Many other effects common to acidic lakes have been observed (blooms of filamentous algae, reduced zooplankton richness, loss of zooplankton species, loss of crayfish), but as yet they are unpublished.

There are very few data available allowing assessment of the effects of acidic deposition on stream biota. Hall and Ide (1987) showed that several mayfly and stonefly species disappeared from 1937 to 1942 and from 1984 to 1985 from several streams in Algonquin Park, Ontario, that demonstrated pH decreases of >1 unit during snowmelt. Neighboring streams that showed no pH decline had not lost the same benthic invertebrate species.

Although only a few case studies provide direct documentation of biological effects in Canada, the results of these few studies are consistent with both laboratory toxicology studies and with the whole-ecosystem experiments at the Experimental Lakes Area.

V. Summary

Acidic rain research conducted by Canadian federal and provincial agencies since the late 1970s has contributed immensely to our overall understanding of the acidic rain problem.

Atmospheric studies have shown that the southern portions of the provinces of Ontario and Quebec receive the highest deposition rates of strong acid, that there are seasonal and episodic trends to sulphate deposition, that fogs and low clouds have high concentrations of strong acidic anions and extremely low pHs, and that atmospheric models can be used to simulate deposition.

The terrestrial studies have shown that acidic deposition can affect tree growth and regeneration by a combination of factors including loss of nutrients, Al toxicity, and direct effects on reproductive processes. The added stress of air pollution can increase the trees' susceptibility to other stresses (insects, disease, weather).

The aquatic studies have documented the acidification of a sensitive lake (Plastic Lake in Ontario) and observed a variety of biological effects as a result of the declining pH. Surveys in Ontario and other parts of eastern Canada have demonstrated that the pattern of SO_4 concentration in lakes follows the S deposition pattern. There is widespread reduction in alkalinity and pH as a result of S deposition.

These data continue to support the Canadian federal/provincial decision that resulted in regulations to reduce SO_2 emissions by 50% by 1994.

References

Abouguendia, Z.M., L.A. Baschak, and R.C. Goodwin. 1986. *Effects of simulated acidic precipitation on Saskatchewan crop and forest species—results of the 1985/86 experiments.* SRC Technical Report No. 191.

Azzaria, L.M., P.J. Gelinas, R. Robitaille, and J.F. Wilhelmy. 1982. *Etude géologique et hydrogéologique du lac Laflamme, Parc des Laurentides, Québec.* Report for Environment Canada, Inland Waters Directorate, Dept. of Geology, U. of Laval, Ste. Foy, Québec, 95 pp.

Barrie, L.A., and A. Sirois. 1986. Wet and dry deposition of sulphates and nitrates in eastern Canada: 1979–1982. *Water Air Soil Pollut* 30: 303–310.

Beamish, R.J. 1974a. The loss of fish populations from unexploited remote lakes in Ontario, Canada, as a consequence of atmospheric fallout of acid. *Water Res* 8: 85–95.

Beamish, R.J. 1974b. Growth and survival of white suckers (*Catostomus commersoni*) in an acidified lake. *J Fish Res Board Can* 31: 49–54.

Beamish, R.J., and H.H. Harvey. 1972. Acidification of the La Cloche Mountain Lakes, Ontario, and resulting fish mortalities. *J Fish Res Board Can* 29: 1131–1143.

Beamish, R.J., W.L. Lockhart, J.C. Van Loon, and H.H. Harvey. 1975. Long-term acidification of a lake and resulting effects on fishes. *Ambio* 4: 98–102.

Beggs, G.L., and J.M. Gunn. 1986. Response of lake trout (*Salvelinus namaycush*) and brook trout (*S. fontinalis*) to surface water acidification in Ontario. *Water Air Soil Pollut* 30: 711–717.

Bernier, B., M. Brazeau, C. Camire, and A. Brousseau. 1985. *Recherche sur le dépérissement et le status nutritif de l'érablière des Appalaches au Québec.* Rapport d'étape sur les travaux réalisés en 1984–1985: 62 pp.

Berrill M., L. Hollett, A. Margosian, and J. Hudson. 1985. Variation in tolerance to low environmental pH by the crayfish *Orconectes rusticus, O. propinquis,* and *Cambarus robustus. Can J Zool* 63: 2586–2589.

Bottomley, D.J., D. Craig, and L.M. Johnston. 1984. Neutralization of acid runoff by groundwater discharge to streams in Canadian Precambrian Shield watersheds. *J Hydrol* 75: 1–26.

Canada/U.S. Memorandum of intent between the government of Canada and the government of the United States of America concerning transboundary air pollution. 1980.

Canadian Forestry Service. 1986. *Forest insect and disease conditions in Canada 1985.* ISBN 0-662-14890: 107 pp.

Carrier, L. 1986. *Decline in Québec's forests: Assessment of the situation.* ISBN 2-550-16816-X. Dépôt légal, Bibliothèque nationale du Québec.

Case, J.W. 1985. *Mapping trace element deposition in Ontario using lichens and mosses as biological monitors.* Poster Presentation at International Symposium on Acidic Precipitation, Muskoka '85 Conference, Muskoka, Canada.

Charles, D.F., et al. 1986. The PIRLA Project (Paleoecological Investigation of Recent Lake Acidification): Preliminary Results for the Adirondacks, New England, N. Great Lakes States, and N. Florida. *Water Air Soil Pollut* 30: 355–365.

Clark, K.L. 1986. Distributions of anuran populations in central Ontario relative to habitat acidity. *Water Air Soil Pollut* 30: 727–734.

Clark, K.L., and R.J. Hall. 1985. Effects of pH and aluminum on amphibian embryo-larval survival. *Can J Zool* 63: 116–123.

Clark, K.L., and B.D. LaZerte. 1985. A laboratory study of the effects of aluminum and pH on amphibian eggs and tadpoles. *Can J Fish Aquat Sci* 42: 1544–1551.

Cook, R.B. (ed.). 1988. *The effects of acidic deposition on aquatic resources in Canada: An analysis of past, present, and future effects.* ORNL/TM-10405.

Cox, R.M. 1983. Sensitivity of forest plant reproduction to long range transported air pollutants: In vitro sensitivity of pollen to simulated acid rain. *New Phytol* 95: 269–276.

Cox, R.M. 1985. *Determination of the sensitivity of forest flora reproductive processes to wet deposited trace elements and acid due to long range transported air pollutants.* Canadian Forestry Service Contract Report No. OS082-00289: 160 pp.

Craig, D., L.M. Johnston, and D.J. Bottomley. 1986. *Acidification of shallow groundwaters due to acid shock events.* Manuscript Report, National Hydrology Res. Inst., Env. Canada, Ottawa, Canada, 16 pp.

Dale, J., B. Freedman, and J.J. Kerekes. 1985. Acidity and associated water chemistry of amphibian habitats in Nova Scotia. *Can J Zool* 63: 97–105.

Den Hartog, G., H.H. Neumann, and K.M. King. 1987. *Measurements of ozone, sulphur dioxide and carbon dioxide fluxes to a deciduous forest.* Reprinted from the Proceedings of the 18th Conference on Agricultural and Forest Meteorology and 8th Conference on Biometeorology and Aerobiology, September 14–18, 1987.

Dickman, M.D., and H.G. Thode. 1985. The rate of lake acidification in four lakes north of Lake Superior and its relationship to downcore sulphur isotope ratios. *Water Air Soil Pollut* 26: 233–253.

Dillon, P.J. 1983. Chemical alterations of surface waters by acidic deposition in Canada. *Water Qual Bull* 8: 127–132.

Dillon, P.J., D.S. Jeffries, and W.A. Scheider. 1982. The use of calibrated lakes and watersheds for estimating point source. *Water Air Soil Pollut* 18: 241–258.

Dillon, P.J., D.S. Jeffries, W. Snyder, R. Reid, N.D. Yan, D. Evans, J. Moss, and W.A. Scheider. 1978. Acidic precipitation in south-central Ontario. *J Fish Res Board Can* 35: 809–815.

Dillon, P.J., M. Lusis, R. Reid, and D. Yap. 1988. Ten-year trends in sulphate and nitrate deposition in central Ontario. *Atmos Environ* 22: 901–905.

Dillon, P.J., R.A. Reid, and E. de Grosbois. 1987. The rate of acidification of aquatic ecosystems in Ontario, Canada. *Nature* 329: 45–48.

Dillon, P.J., N.D. Yan, and H.H. Harvey. 1984. Acidic precipitation: Effects on aquatic ecosystems. *CRC Critical Reviews in Environmental Control* 13: 167–194.

Dixit, Sushil S. 1986. Diatom-inferred pH calibration of lakes near Wawa, Ontario. *Can J Bot* 64: 1129–1133.

Edwards, G.C., and G.L. Ogram. 1986. Eddy correlation measurements of dry deposition fluxes using a tunable diode laser absorption spectrometer gas monitor. *Water Air Soil Pollut* 30: 187–194.

Ellenton, G., B. Ley, and P.K. Misra. 1985. A trajectory puff model of sulphur transport for eastern North America. *Atmos Environ* 19: 727–737.

Ellenton, G., P.K. Misra, and B. Ley. 1988. The relative roles of emissions changes and meteorological variability in variation of wet sulphur deposition: A trajectory model study. *Atmos Environ* 22: 547–556.

Environment Canada. 1983. Canada-United States, Memorandum of Intent on Transboundary Air Pollutants, Impact Assessment Final Report.

Environmental Protection Agency. 1985. *The acidic deposition phenomenon and its effects.* Critical Assessment Document. EPA/600/8-85/001.

Enyedi, A.J., and A.L. Kuja. 1986. Assessment of relative sensitivities during early growth stages of selected crop species subjected to simulated acidic rain. *Water Air Soil Pollut* 31: 325–335.

Freda J., and W.A. Dunson. 1986. Effects of low pH and other chemical variables on the total distribution of amphibians. *Copeia*: 454–466.

Gagnon, G., L. Robitaille, G. Roy, and C. Gravel. 1985. *Le dépérissement des érablières: Comportement de certaines variables écologiques.* La terre de Chez Nous (dossier). 3: 28 pp.

Gagnon, G., G. Roy, C. Gravel, and J. Gagné. 1986. *Etat des recherches sur le dépérissement au ministère de l'Energie et des Ressources.* Journée d'information sur l'acériculture: 47–85.

Gascon, C., and D. Planas. 1986. Spring pond water chemistry and the reproduction of the wood frog, *Rana sylvatica. Can J Zool* 64: 543–550.

Glooschenko, V., W. Weller, and C. Eastwood. 1985. *Amphibian abundance and distribution with respect to environmental contaminants and pH stress adjacent to Sudbury, Ontario.* Technical report of Wildlife Branch, Ontario Ministry of Natural Resources, Queen's Park, Toronto. 30 pp.

Griffiths, R.W., E. Carney, L. Nakamoto, and K.H. Nicholls. 1990. *Parameterization of calibration equations relating diatom microfossils to surface water pH and alkalinity in central Ontario lakes.* In prep.

Haines, T.A. 1981. Acidic precipitation and its consequences for aquatic ecosystems: A review. *Transactions of the American Fisheries Society* 110: 669–707.

Hall, R.J., and F.P. Ide. 1987. Evidence of acidification effects on stream insect communities in central Ontario between 1937 and 1985. *Can J Fish Aquat Sci* 44: 1652–1657.

Harvey, H.H. 1982. Population responses of fish in acidified waters. In R.E. Johnson (ed.), *Acid rain/fisheries: Proceedings of an International Symposium on Acidic Precipitation and Fishery Impacts in Northeastern North America,* American Fisheries Society, Bethesda, MD.

Harvey, H.H., and C. Lee. 1982. *Historical fisheries changes related to surface water pH changes in Canada.* Acid rain/fisheries: Proceedings of an International Symposium on Acidic Rain and Fishery Impacts in Northeastern North America, Cornell University, Ithaca, NY.

Harvey, H.H., R.C. Pierce, P.J. Dillon, J.P. Kramer, and D.M. Whelpdale. 1981. *Acidification in the Canadian aquatic environment.* Publ. NRCC No. 18475 of the Environmental Secretariat, National Research Council, Canada.

Hedin, L.O., G.E. Likens, and F.H. Bormann. 1987. Decrease in precipitation acidity resulting from decreased sulphate concentration. *Nature* 325: 244–246.

Henriksen, A. 1979. A simple approach for identifying and measuring acidification of freshwater. *Nature* 278: 542–545.

Henriksen, A. 1982. *Preacidification pH—values in Norwegian rivers and lakes.* NIVA Report 3/1982.

Holtze, K.E., and N.J. Hutchinson. 1989. Lethality of low pH and Al to early life stages of six fish species inhabiting PreCambrian Shield waters in Ontario. *Can J Fish Aquat Sci* 46: 1188–1202.

Houston, D.R. 1981. *Stress-triggered tree disease: The diebacks and declines.* NF-INF-41-81. U.S. Department of Agriculture, Forest Service, Broomall, PA.

Hutchinson, N.J., K.E. Holtze, J.R. Munro, and T.W. Pawson. 1989. Modifying effects of life stage, ionic strength and post-exposure mortality on lethality of H^+ and Al to lake trout and brook trout. *Aquat Toxic* 15: 1–26.

Hutchinson, T.C. 1980. Effects of acid leaching on cation loss from soils. In T.C. Hutchinson and M. Hava (eds), *Effects of acid precipitation on terrestrial ecosystems.* NATO Conf. Series, New York: Plenum Press.

Hutchinson, T.C., L. Bozic, and G. Munoz-Vega. 1986. Responses of five species of conifer seedlings to aluminum stress. *Water Air Soil Pollut* 31: 283–294.

Hutchinson, T.C., and M. Scott. 1986. *Effects of long-term simulated acid precipitation on forest ecosystems: The Burt Lake and maple decline in Ontario projects.* Presented at the Forest Decline Workshop, Oct. 20–22, 1986.

Jackson, M. 1985. *Filamentous algae in Ontario softwater lakes.* Poster presentation at Internat. Symp. on Acidic Precipitation, Sept. 15–20, 1985, Muskoka, Canada.

Jeffries, D.S., R.G. Semkin, R. Neureuther, and M. Seymour. 1986. Influence of atmospheric deposition on lake mass balances in the Turkey Lakes Watershed, Central Ontario. *Water Air Soil Pollut* 30: 1033–1044.

Jeffries, D.S., D. Wales, J.R.M. Kelso, and R.A. Linthurst. 1986. Regional chemical characteristics of lakes in North America: Part I—eastern Canada. *Water Air Soil Pollut* 31: 551–567.

Johnston, L.M., D.J. Bottomley, D. Craig, K. Inch, and H. Chew. 1985. *The role of groundwater in modifying the effects of acidic precipitation on surface waters.* Manuscript Report, National Hydrology Res. Inst., Env. Canada, Ottawa, Canada.

Jones, H.G., and C. Deblois. 1987. Chemical dynamics of N-containing ionic species in a boreal forest snowcover during the springmelt period. *Hydrol Proc* 1: 271–282.

Kelso, J.R.M., C.K. Minns, J.E. Gray, and M.L. Jones. 1986. Acidification of surface waters in eastern Canada and its relationship to aquatic biota. *Can Spec Publ Fish Aquat Sci* 87, 42 pp.

Kerekes, J., and B. Freedman. 1986. *Physical, chemical, and biological characteristics of three watersheds in Kejimkujik National Park, Nova Scotia.* Environ Can. Report.

Kerekes, J., G. Howell, S. Beauchamp, and T. Pollock. 1982. Characterization of three lake basins sensitive to acid precipitation in central Nova Scotia (June 1979 to May 1980). *Int. Revue. ges. Hydrobiol.* 67: 679–694.

Kuja, A. 1988. *A Canadian evaluation of NAPAP's interim assessment of the effects of acidic deposition on agricultural crops.* Presented at the APCA Conference in January 1988.

Kuja, A., R. Jones, and A. Enyedi. 1986. A mobile rain exclusion canopy system to determine dose-response relationships for crops and forest species. *Water Air Soil Pollut* 31: 307–315.

Kurtz, J., A.J.S. Tang, R.W. Kirk, and W.H. Chan. 1984. Analysis of an acid rain deposition episode at Dorset, Ontario. *Atmos Environ* 18: 387–394.

Lachance, D. 1985. Répartition géographique et intensité du dépérissement de l'érable à sucre dans les érablières au Québec. *Phyto-protection* 66: 83–90.

Lacroix, G.L. 1985. Survival of eggs and alevins of Atlantic salmon (*Salmo salar*) in relation to the chemistry of interstitial water in redds in some acidic streams of Atlantic Canada. *Can J Fish Aquat Sci* 42: 292–299.

Landers, D.H., W.S. Overton, R.A. Linthurst, and D.F. Brakke. 1988. Eastern lake survey. *Environ Sci Technol* 22: 128–135.

LaZerte, B.D., and P.J. Dillon. 1984. Relative importance of anthropogenic versus natural sources of acidity in lakes and streams of central Ontario. *Can J Fish Aquat Sci* 41: 1664–1677.

Leaitch, W.R., J.W. Strapp, H.A. Wiebe, and G.A. Isaac. 1986. Chemical and microphysical studies of non-precipitating and precipitating cloud. *J Geophysical Research* 91 (D11): 11821–11831.

Lewis, Drew, and William Davis. 1986. *Joint report of the special envoys on acid rain.* 35 pp.

Linzon, S.N., and P.J. Temple. 1980. Soil resampling and pH measurements after an 18-year period in Ontario. *Proceedings of International Conf. on the Ecological Impact of Acid Precipitation,* Sandefjord, Norway: 176–177.

Lusis, M.A., A.J.S. Tang, W.H. Chan, D. Yap, J. Kurtz, and P.K. Misra. 1986. Sudbury smelter impact on atmospheric deposition of acidic substances in Ontario. *Water Air Soil Pollut* 30: 897–908.

Manion, P.D. 1981. *Tree disease concepts.* Englewood Cliffs, NJ: Prentice-Hall.

McLaughlin, D.L., S.N. Linzon, D.E. Dimma, and W.D. McIlveen. 1985. *Sugar maple decline in Ontario.* Report No. ARB-144-85-Phyto, Ontario Ministry of the Environment, Toronto, Canada: 18 pp.

McNicol, D.K., B.E. Bendell, and R.K. Ross. 1985. *Waterfowl and aquatic ecosystem acidification in northern Ontario.* Manuscript report, Canadian Wildlife Service, Env. Canada, Ottawa, Canada.

Mierle, G., K. Clark, and R. France. 1986. The impact of acidification on aquatic biota in North America: A comparison of field and laboratory results. *Water Air Soil Pollut* 31: 593–604.

Mills, K.H., and D.W. Schindler. 1986. Biological indicators of lake acidification. *Water Air Soil Pollut* 30: 779–789.

Minns, Charles K., and John R.M. Kelso. 1986. Estimates of existing and potential impact of acidification on the freshwater resources of eastern Canada. *Water Air Soil Pollut* 31: 1079–1090.

Morrison, I.K. 1981. Effect of simulated acid precipitation on composition of percolate from reconstructed profiles of two northern Ontario forest soils. *Can For Serv Res Notes* 1: 6–8.

National Academy of Sciences. 1986. Committee on Monitoring and Assessment of Trends in Acid Deposition. 1986. *Acid deposition long-term trends*. National Academy Press, Washington, DC.

Neary, B.P., and P.J. Dillon. 1988. Effects of sulphur deposition on lake water chemistry in Ontario. *Nature* 333: 340–343.

Nero, R.W., and D.W. Schindler. 1983. Decline of Mysis relicta during acidification of lake 223. *Can J Fish Aquat Sci* 40: 1905–1911.

Ogram, G.L., F.J. Northrup, and G.C. Edwards. 1988. Fast-time response—tunable diode laser measurements of atmospheric trace gases for eddy correlation. *J Atmos Oceanogr Techn*. In press.

Peterson, R.H. 1984. Influence of varying pH and some inorganic cations on the perivitelline potential of eggs of Atlantic Salmon (*Salmo salar*). *Can J Fish Aquat Sci* 41: 1066–1069.

Pough, F.H. 1976. Acid precipitation and embryonic mortality of spotted salamanders, *Ambystoma maculatum*. *Science* 192: 68–70.

Reuss, J.O., and D.W. Johnson. 1985. Effect of soil processes on the acidification of water by acid deposition. *J Environ Qual* 14: 26–31.

RMCC. 1986. *Assessment of the state of knowledge on the long-range transport of air pollutants and acid deposition*. Part 2. Atmospheric sciences: 108 pp. Part 3. Aquatic effects: 57 pp. Part 4. Terrestrial effects: 80 pp. Federal/Provincial Research and Monitoring Coordinating Committee (RMCC).

Rooke, J.B., and J.B. Mackie. 1984a. Molluscs of six low-alkalinity lakes in Ontario. *Can J Fish Aquat Sci* 41: 777–782.

Rooke, J.B., and J.B. Mackie. 1984b. Growth and production of three species of molluscs in six low alkalinity lakes in Ontario, Canada. *Can J Zool* 62: 1474–1478.

Rutherford, G.K., G.W. van Loon, S.F. Mortensen, and J.A. Hern. 1985. Chemical and pedogenetic effects of simulated acid precipitation on two eastern Canadian forest soils. II. Metals. *Can J For Res* 15: 848–854.

Saber, P.S., and W.A. Dunson. 1978. Toxicity of bog water to embryonic and larval anuran amphibians. *J Exp Zool* 204: 33–42.

Schemenauer, R. 1986. Acidic deposition to forests. The 1983 chemistry of high elevation fog (CHEF) project. *Atmosphere-Ocean* 24: 303–328.

Schemenauer, R., and C. Winston. 1988. *The 1986 chemistry of high elevation fog project*. Presented at the 81st APCA Annual Meeting, June 1988.

Schindler, D.W. 1987. Detecting ecosystem responses to anthropogenic stress. *Can J Fish Aquat Sci* 44, Suppl. 1: 6–25.

Schindler, D.W. 1988. Effects of acid rain on freshwater ecosystems. *Science* 239: 149–157.

Schindler, D.W., K.H. Mills, D.F. Malley, D.L. Findlay, J.A. Shearer, I.J. Davies, M.A. Turner, G.A. Linsey, and D.R. Cruikshank. 1985. Long-term ecosystem stress: The effects of years of experimental acidification on a small lake. *Science* 228: 1395–1401.

Schindler, D.W., and M.A. Turner. 1982. Biological, chemical and physical responses of lakes to experimental acidification. *Water Air Soil Pollut* 18: 259–271.

Seip, H.M., R. Seip, P.J. Dillon, and E. de Grosbois. 1985. Model of sulphate concentration in a small stream in the Harp Lake catchment, Ontario. *Can J Fish Aquat Sci* 42: 927–937.

Semkin, P.G., and D.S. Jeffries. 1986. Bulk deposition of ions in the Turkey Lakes watershed. *Water Poll Res J Can* 21: 474–485.

Somers, Keith M., and Harold H. Harvey. 1984. Alteration of fish communities in lakes stressed by acid deposition and heavy metals near Wawa, Ontario. *Can J Fish Aquat Sci* 41: 20–29.

Stephenson, M., and G.L. Mackie. 1986. Lake acidification as a limiting factor in the distribution of the freshwater amphipod *Hyalella azteca. Can J Fish Aquat Sci* 43: 288–292.

Summers, P.W., V.C. Bowersox, and G.J. Stensland. 1986. The geographical distribution and temporal variations of acidic deposition in eastern North American. *Water Air Soil Pollut* 31: 523–535.

Summers, P.W., and D.M. Whelpdale. 1976. Acid precipitation in Canada. *Water Air Soil Pollut* 6: 447–455.

Tang, A.J.S., A. Ahmed, and M.A.Lusis. 1986. *Summary: Some results from the APIOS atmospheric deposition monitoring program (1981–1984).* Ontario Ministry of the Environment, ARB-110-86, APIOS-011-86: 33 pp.

Trail Smelter Arbitral Tribunal. 1941. Trail smelter arbitral tribunal decision. *Amer. J. Int. Law* 35: 684–737.

Underwood, J.K. 1985. *Acid deposition in Nova Scotia: Seven years of record.* Presented at the International Symposium on Acid Precipitation, Muskoka, September 15–20.

Venkatram, A., and P.K. Misra. 1988. Testing a comprehensive acid deposition model. *Atmos Environ* 22: 737–747.

Watt, W.D., C.D. Scott, and W.J. White. 1983. Evidence of acidification of some Nova Scotia rivers and its impact on Atlantic salmon *Salmo salar. Can J Fish Aquat Sci* 40: 462–473.

Whitby, L.M., P.M. Stokes, T.C. Hutchinson, and G. Myslik. 1976. Ecological consequence of acidic and heavy-metal discharges from the Sudbury smelters. *Can Mineralogist* 14: 47–57.

Wright, R.F. 1983. *Predicting acidification of North American lakes.* Norwegian Institute for Water Research, Tech. Rep. 4/183: 165 p.

Yap, D., and J. Kurtz. 1986. Meteorological analyses of acidic precipitation in Ontario. *Water Air Soil Pollut* 30: 873–878.

Zakshek, E.M., and K.J. Puckett. 1986. Lichen sulphur and lead levels in relation to deposition patterns in eastern Canada. *Water Air Soil Pollut* 30: 161–169.

Acidic Precipitation in Japan

T. Katoh[*], T. Konno[**], I. Koyama[†], H. Tsuruta[‡]
and H. Makino[§]

Abstract

Serious interest in chemical pollution in precipitation began in Japan in 1930. However, the strong influence of the surrounding oceans on the atmospheric pollution environment in this island country was not at first fully realized. In the 1970s, people in the Kanto Plains—Tokyo and the surrounding coastal plain area—experienced a period of severe eye irritation. The major cause was at first thought to be industrial emission into the atmosphere of SO_x and NO_x. However, the severity of the pollution was found to be heavily influenced by the nature of the precipitation involved. In Japan approximately one-third of the days each year have some precipitation. In the summer there are many sudden rain squalls; from spring to the beginning of summer and then in the fall there is "weather-front-induced" precipitation. This is a lighter, steady precipitation but with small, concentrated areas of very heavy rainfall. The periods of most intense eye irritation are those with mist or very light rainfall. At these times, there is a stationary weather front extending east-west along the southern portion of the Kanto Plains area. The major sources of eye irritation are considered to be the strong acids H_2SO_4 and HNO_3, as well as formaldehyde, acrolein, formic acid, and H_2O_2. In addition to the human eye irritation there was pollution damage to eggplants, cucumbers, string beans, and other types of vegetables. Then, beginning in 1985,

[*]Hokkaido Research Institute for Environmental Pollution, Nisi-12cyoume, 19zyo, Kita-ku, Sapporo, 060, Japan.
[**]National Agriculture Research Center, Kannodai, Tukuba-shi, Ibaraki, 305, Japan.
[†]Tokyo Metropolitan Research Institute for Environmental Protection, Shinsuna, I-7-5, Kota-ku, Tokyo, 136, Japan.
[‡]Yokohama Environmental Research Institute, Takigashira, I-2-15, Isogo-ku, Yokohama-shi, Kanagawa, 233, Japan.
[§]Kanagawa Prefectural Environmental Center, 2-14 Toyohar-mati, Hiratsuka-si, Kanagawa, 245, Japan.
All authors are members of the Japan Acid Rain Society.

pollution damage weakening Japanese cedars in the northwest portion of the Kanto Plains became evident. The source was thought at first to be pollutant gases but was found to include acidic precipitation also.

Beginning in 1973 the Department of Air Pollution, Environmental Protection Measures Promotion Headquarters, Kanto District Governors Association (EPMPH), and the Environmental Agency began the systematic collection of pollution data. Their survey revealed that eye irritation was also related to the same meteorological conditions. The investigation showed that, with low pH (3.0 and less, during very light precipitation), the SO_4^{2-} and NO_3^- ion concentrations were high. Also, formaldehyde and other substances were found involved in the eye irritation attacks. The studies in the Kanto Plains showed that the NO_3^- concentration predominated over that of SO_4^{2-}. However, outside the Kanto Plains the situation was reversed, with the SO_4^{2-} concentration being highest. In the Tokyo vicinity, the NO_x and SO_x concentrations are high due to auto emissions and the extensive industrial complex. Even though the Kanto Plains are surrounded to the west and north by 1000- to 2000-m mountains, there is the possibility of the resultant pollutant concentrations being carried as far as 150 to 200 km distant. However, the phenomenon of a changing and then inverting NO_3^- to SO_4^{2-} ratio in regions further out is a subject of continuing study. The Environmental Agency conducted a long-term survey of pollutant concentrations from 1984 to 1986. The membrane-filter method was employed, taking weekly samples of collected precipitation. The pH was found to vary from 4.0 to 7.1 with an annual average of 4.5 to 5.2. The SO_4^{2-}–S was 780–1640 mg m^{-2} yr^{-1} and the excess (over ocean sources of S) SO_4^{2-}–S was 840–1270 mg m^{-2} yr^{-1} The NO_3^-–N was 170–760 mg m^{-2} yr^{-1}.

From 1950 to 1960, weakening of trees in the central Tokyo area occurred. The Japanese cedar was particularly hard hit and many trees had to be cut down. Now this problem has extended to the surrounding areas. Even the tops of large Japanese cedars have been severely damaged, and the problem worsens year by year. But the precise cause has not yet been determined.

I. Introduction

Japan has had environmental disasters such as "Minamata disease," "Itaiitai disease," "Yokkaichi disease"[1] such that the stigma of "the pollution country" has been attached to it. Since these incidents, through the

---

[1]"Minamata disease" began in 1953, centered in the city of Minamata in Kumamoto prefecture (Kyūshū), and then, starting in 1963, in the effluent region of the Agano River in Niigata prefecture. The disease affected the sensory nerves,

cooperation of government, industry, and researchers, pollution prevention measures have been introduced, so that now there is little danger of acute pollution. At present, countermeasures to reduce the levels of sulfur oxide and nitrogen oxide in exhaust gases, which remains a problem in Europe and the United States, have been instituted in Japan.

The analysis of the constituents of rain in Japan began in the 1930s (Miyake, 1939). The influence of sea salt on the rain in the inland areas was analyzed occasionally, but long-term continuous research surveys were not conducted (Sugawara, 1948).

Organized large-area surveys of the constituents of the rain have been carried on since 1975. The impetus for these studies was a large number of complaints from 1973 to 1975 of eye irritation from mist and drizzle, by people living in Shizuoka and Yamanashi prefectures and in the entire Kanto area (EPMPH , 1975, 1976; Japan Environmental Agency, 1979). But since then there have been virtually no reports of health hazards.

The long-range transport of atmospheric pollutants and the impact of acidic deposition on the ecosystems in remote areas are well known and constitute serious problems in Europe and in eastern North America. In Japan, mesoscale (intermediate range) acidic deposition has been a serious problem especially in the Kanto district. As described in Section II D, long-term trends in acidic deposition had not been assessable earlier due to lack of data.

However, there have been no seasonal or year-to-year trends in pH of the rain, except for the variations accompanying the rainfall pattern, so the cessation of health hazard problems is not yet fully explained (Koyama, 1986). In earlier years reports of acidification of lakes and the discoloration of forests in Europe and North America seemed unrelated

---

resulting in impaired speech, narrowing of the field of vision, and impairment of the motor functions. Many people died or were in serious condition, suffering paralyses due to damaged brain tissue, including many affected prenatally. A total of 1896 victims had been counted by the end of 1981. The cause was the discharge of mercury used in an acetaldehyde-producing process, over an extended period.

"Itaiitai disease" appeared beginning in 1910 in the middle and lower reaches of the Jintsu River in Toyama prefecture. The victims experienced pain in the lower back (lumbago), and the arms, hands, legs, and feet. Following this, protein appeared in the urine, and finally the bones became brittle, breaking easily, and severe aching was experienced. The patients died screaming "itaiitai" (it *hurts,* it *hurts!*). The cause was cadmium released from factories into the river. Through 1986 there were 123 victims, 102 of whom died because of the disease (Japan Environmental Agency, 1987b).

"Yokkaichi asthma": Beginning in 1955 there were large quantities of stack gases released into the air from the petrochemical complex in Yokkaichi, Mie prefecture. Beginning in 1971 there was a rapid spread of bronchial asthma among the residents in the vicinity of the plants. The seriousness of the disease was related to the level of SO_2 atmospheric pollution. Those who moved away recovered, so the disease was called "Yokkaichi asthma."

to Japan, but in recent years the frequent reports of such events caused the magnitude of the problem to be recognized.[2]

Reports of damage to foliage in the Kanto area was recently noted as a long-term effect of acidic rain (Sekiguchi, Hara, and Ujiiye, 1985). Japan has meteorological conditions and fertile soils favorable to plant growth. Since there has been considerable intrusion of human activity into the forest areas, assessing the influence of acidic rain and other pollution has become a complicated and difficult problem.

Also, whereas in Japan rainfall contours can be drawn, it is virtually impossible, compared with a number of other countries, to draw contours of equal rainfall pH (EPMPH, 1987a, 1987b, 1987c; Japan Environmental Agency, 1987a). The reason for this is that the Japan archipelago is long and narrow so that the influence of sea salt is felt inland, and the atmosphere is greatly affected by pollution from roads and residential development of the land areas.

As the reader will understand from the preceding discussion, the acidic rain problem in Japan is far from simple. It will be expeditious to consider first some of the geographical features of the country.

II. Geographical Characteristics of the Japanese Islands

A. Land Patterns

Japan is an archipelago, shaped something like a backward *S,* on the eastern edge of the Asian continent, extending from the northwestern edge of the Pacific Ocean in a southwesterly direction. It is approximately 3000 km long with an area of 372,000 km². Honshū is the largest island, located centrally in Japan, 1200 km long and with a maximum width of 450 km and an area of approximately 228,000 km². Next in size is Hokkaidō, to the north of Honshū, followed by Kyūshū and Shikoku to the south of Honshū, and a large number of very small islands (see Figure 2–15 on p. 00). The archipelago is narrow with a large number of relatively high mountains. Especially in central Honshū there are ranges of 3000 m high peaks. Most of these land areas are forests,[3] rivers, lakes, and barren land, so that 120 million people live on 90,000 km² of land, or only 24.4% of the total (Geographical Survey Institute).

---

[2]In the two-year period from 1984 to 1985 at least 10 news articles and wire service releases about acidic precipitation damage to forests in northern Europe and North America were carried by the four major Japanese newspapers—the *Asahi,* the *Mainichi,* the *Yomiuri,* and the *Nihon Keizai.* This was noted in studies by the Tokyo Metropolitan Research Institute for Environmental Protection.

In addition, see the S. Oden et al., 1976 entry in the References for a listing of research papers, books, and reports, a large number of which are well known in Japan.

[3]Apart from Hokkaidō, the forested areas are largely on the mountain slopes.

In the Kanto area acidic precipitation is a particular problem. As seen in the maps of Figures 2–6 and 2–12, the Kanto area is bounded on the south and the east by the Pacific Ocean. It extends approximately 100 to 130 km east-west, and 180 km north-south. To the west and to the north mountain ranges of 1000 to 2000 m border the Kanto Plains. The land pattern of the Kanto region is similar to that of the Los Angeles area in its retention of atmospheric pollution during summer weather conditions.

B. Climate and Meteorological Phenomena

The complex weather of Japan results from its being a long and narrow island chain, coupled with the effects of cold currents coming in from the Sea of Okhotsk and the warm coastal currents of the western Pacific. The climatic zones vary from the (sub) frigid zone to the subtropical (Yoshino, 1978). The coastal zones have some temperate zone characteristics and the mountains inland some subfrigid zone characteristics as well. As Japan is a volcanic archipelago, with 67 active volcanoes and 77 extinct ones, there are a great many places of geothermal activity and the emission of gases (Tokyo Astronomical Observatory, 1987).

The influence of atmospheric pollutants on urban areas due to natural phenomena other than lava flows from large volcanic eruptions, except for the effects of the Sakurajima volcanic activity, is not yet known (Hourai, Ootu, Takeyama, Minamizono, and Yamagawa, 1987). According to one estimate, the amount of SO_x released into the atmosphere by volcanic action is approximately the same as that from human sources (Fujita, 1987).

Because of the narrow island configuration, and the 2000 to 3000 m high mountains in the central regions of the islands, there is heavy rainfall, the average over the country being approximately 2000 mm. There are two parts to the summer season in the Kanto district. The first half is the rainy season called "Baiu," when the Baiu frontal zone lies along the Japan Islands east-west. The latter half is the period in which the Japan Islands are covered with a maritime tropical air mass, and experience the highest temperatures, resulting in frequent showers from convective clouds. Thus, from the Kanfo area to the south, rainfall is heavy due to the combination of the June-July Baiu resulting from a stationary front, the summer-to-fall typhoon season, and the September-October rainy season. The winter precipitation in the northwest coastal regions along the Sea of Japan is heavy due to the northwest cold winds from Siberia picking up heavy moisture content in passing over the Sea of Japan, and resulting in heavy snowfall on the coastal side of the high mountain ranges. The average annual precipitation in the Kanto area is 1600 mm. The precipitation in the Kanto region is relatively much greater (about twice) that of the east coast of North America and of central and northern

Europe. Also, the southern coast of the Kanto area has 300 mm of rainfall in the mid-June to mid-July rainy season due to a stationary front (Hukui et al., 1985). Because of the high mountains and narrow island pattern of Japan, the rainfall rapidly runs off in short rivers to the seas. Even in the longer rivers, the mountain rainfall runs off to the sea in 3 to 4 days. Almost all the winter snowfall runs off to the sea with the spring thaw.

C. Soil and Vegetation

The multitude of soil types in Japan results from five different sources: the original material of the stratum, biological material, topographical influences, climatological effects, and the effects of time. Particularly in Japan, with its complex geological structure, topography, and meteorological conditions, there are rapid changes in going from one narrow soil-type band to another.

Consequently there are a great number of soil types, approximately 320, which can be grouped in some 16 categories, among them gray lowland soil, brown forest soils, red soils, yellow soils, gley soil, and so on (Miyoshi, Shimada, Ishikawa, and Date, 1983). See Figure 2–1 which shows soil types in the Kanto Plains area (Koyama, 1980). Also, there are large regions of "kuroboku" soil, composed largely of volcanic ash.[4] In the Kanto area where much of this already acid soil exists, in depths of 10 cm to 1 m, there is much concern by some researchers as to the possible effects of acidic precipitation on the forested areas. Although the plains areas have been cultivated from ancient times, there is a wide variety of virgin soils. There are frequent cases where in the same soil type cellulose-decomposition tests show variations of several times to more than ten times in soil composition in locations only a short distance (centimeters to meters) apart (Tokyo Metropolitan Government, 1985). Thus with the influence of changes in soil types there are great variations in the plant ecology. Beyond this there are also great variations in the growth rates, and so on, of the flora on the various mountain slopes. The temperature changes frequently, and there is much rainfall. The influence of frequent changes in rainfall creates a complicated plant ecology pattern. In the Pacific coastal regions, the cyclic changes from spring to summer to the fall rains, coupled with the variations in the quantity of precipitation, produce a variety of growth rate and germination patterns. On the Sea of Japan side and in the northern regions the winters are cold, but the thick winter snow layer protects the trees against the cold. Because of these climatic and regional effects, the variations are great in the plant ecology and floral growth rates in only slightly separated regions. In investigating

---

[4]"Kuroboku" soils: One type of volcanic ash soil piled up over the Kanto plains and neighboring areas. Soil that has turned black.

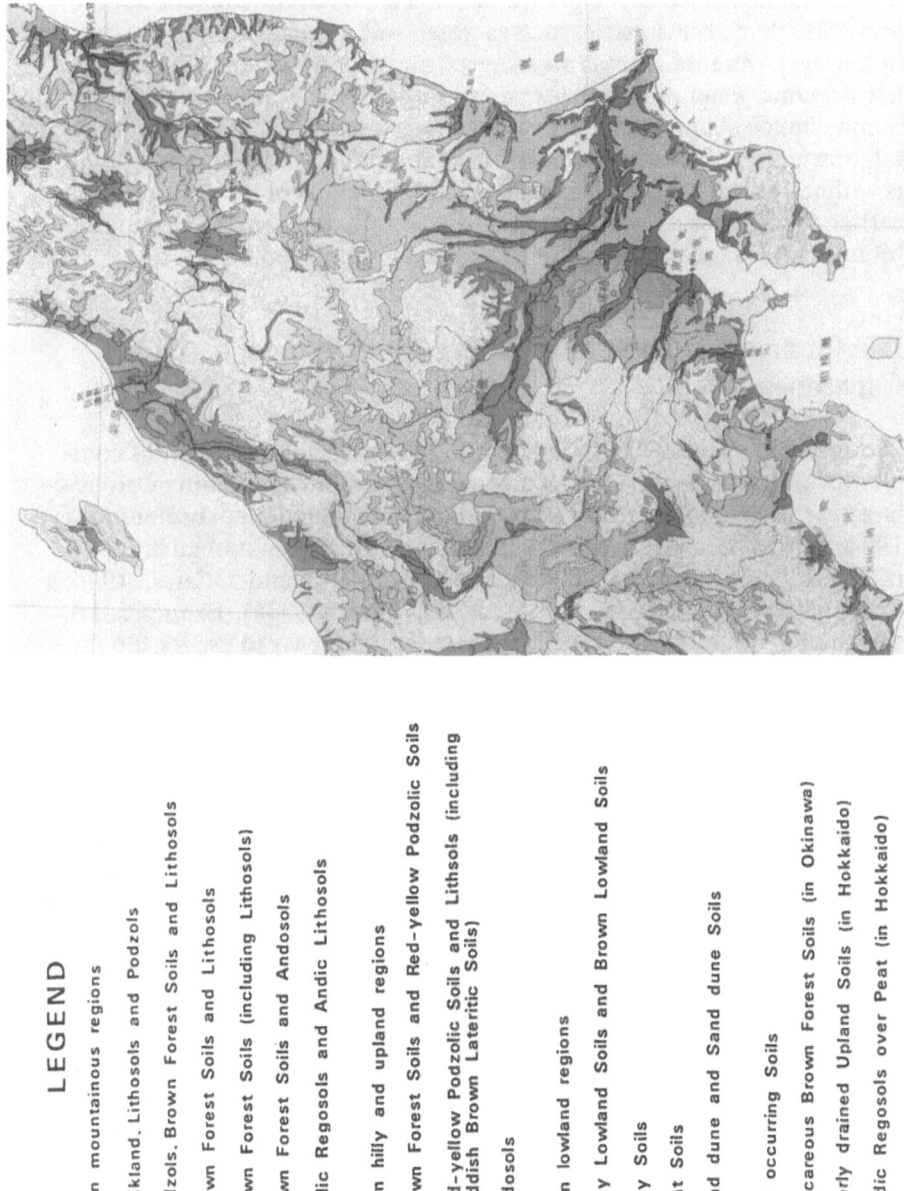

LEGEND

Soils on mountainous regions

Rockland, Lithosols and Podzols

Podzols, Brown Forest Soils and Lithosols

Brown Forest Soils and Lithosols

Brown Forest Soils (including Lithosols)

Brown Forest Soils and Andosols

Andic Regosols and Andic Lithosols

Soils on hilly and upland regions

Brown Forest Soils and Red-yellow Podzolic Soils

Red-yellow Podzolic Soils and Lithsols (including Reddish Brown Lateritic Soils)

Andosols

Soils on lowland regions

Gray Lowland Soils and Brown Lowland Soils

Gley Soils

Peat Soils

Sand dune and Sand dune Soils

Locally occurring Soils

Calcareous Brown Forest Soils (in Okinawa)

Poorly drained Upland Soils (in Hokkaido)

Andic Regosols over Peat (in Hokkaido)

Figure 2–1. Kanto area soil map.

the effects of air pollution on plant life, there is as yet no simple explanation of the influence of the pollution. This is due to the existence of factors other than atmospheric, such as water, soil, sunlight, and so on. Even
in the steep mountain areas and forests, many of which are reforested or
have second-generation forests, there is hardly a place untouched by
human hands. At present the greatest concern regarding the environmental destruction of forests and other vegetation due to acidic precipitation
is with respect to the Kanto area in central Honshū. For the reasons given
earlier, the evaluation of the cause-and-effect relationships in chronic debilitating influences on trees is extremely complicated.

III. Changes in Surveys of Acidic Precipitation Constituents

Systematic surveys of acidic precipitation began with the onset of complaints of eye irritation due to pollution. Since the complaints were related to the beginning of very light rainfall, the method used, beginning in
1984, was that of analyzing the pollution constituents contained in the initial 1 mm of precipitation (Fukuoka, Komeiji, and Odaira, 1976;
Komeiji, Sawada, Odaira, Hirosawa, and Kadoi, 1975). Examples are
shown in Figures 2–2a and b. Since this method is easy to use for the elu-
cidation of pollution mechanisms, it is still in use today by many researchers such as EPMPH. Since that time the purpose has shifted, and
surveys have been conducted to determine pollution mechanisms and the
long-term effects of pollution. For these purposes the total amount of pollutants in a single precipitation period is desired (EPMPH, 1983). Since
1984, rather than use the methods we have described, a filter method has
been developed to collect precipitation over an extended period of time
(Komeiji, Koyama, and Kadoi, 1983; Koyama, Komeiji, Onozuka,
Ohhoasi, and Ise, 1983). According to the researches of M. Tamaki, the
use of collector equipment of 10 cm. diameter or greater gives consistent
results (Tamaki and Havano, 1986). Examples are shown in Figures 2–3a
and b. With this method precipitation is collected over a period of one or
two weeks, a month, or longer without the need of a human attendant.
Also, methods of collecting snow precipitation have been developed (Japanese Environmental Agency, 1987a). For example, see Figure 2–4.

IV. Health Effects Episodes

The acidic rain problem in Japan began dramatically with a large number
of episodes of pollution irritation to people from 1973 to 1975, followed
by a period of more sporadic occurrence of the episodes. In particular, on
July 3 and 4, 1974, more than 30,000 people complained of eye, throat,

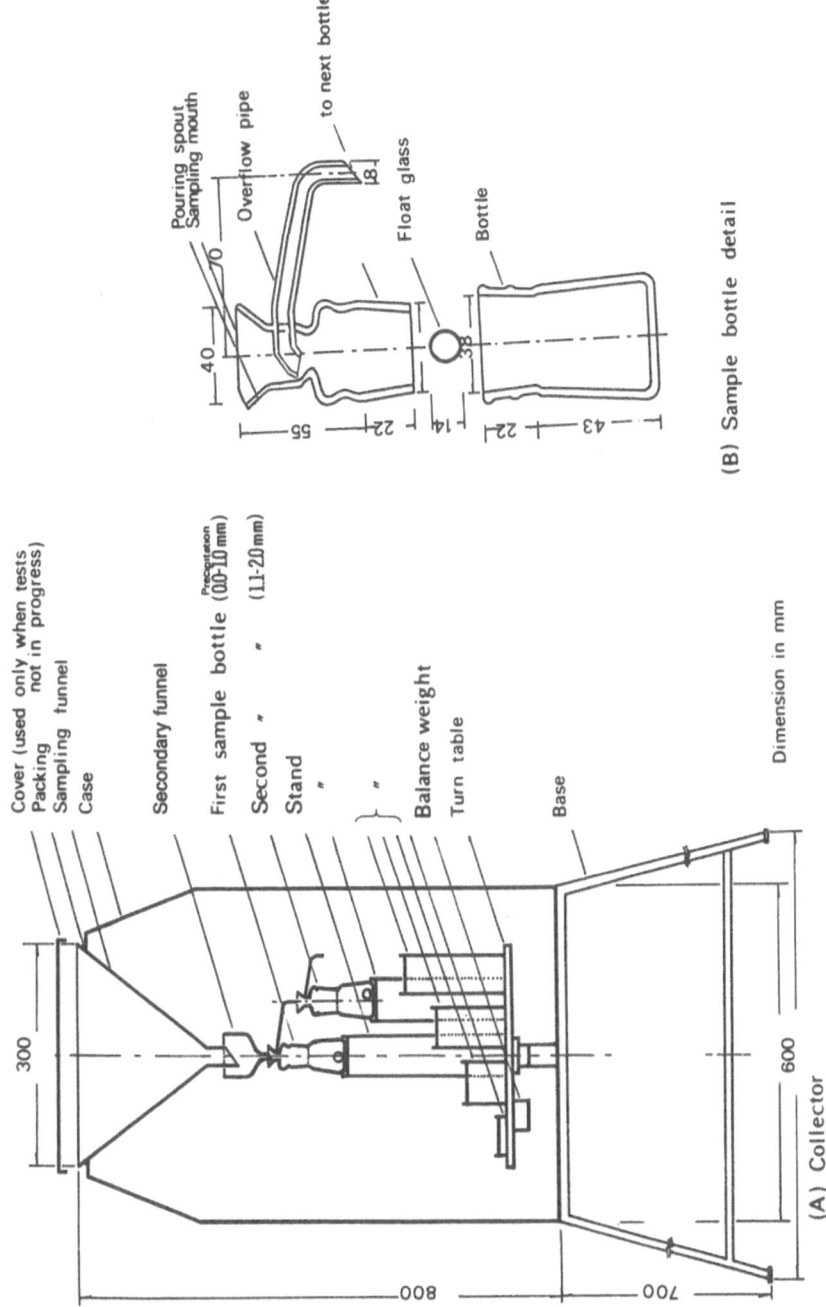

Figure 2–2a. One-millimeter precipitation collector (manual).

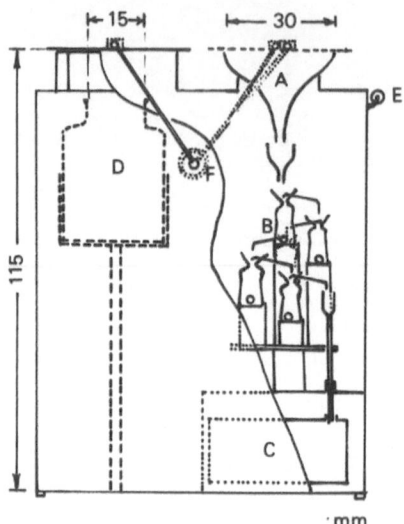

Figure 2–2b. Typical automatic precipitation sampler (dry and wet) A: wet precipitation collector; B: sampling bottle (first 5 mm); C: sampling bottle (over 5 mm); D: dry deposit collector; E: rain sensor; F: motor.

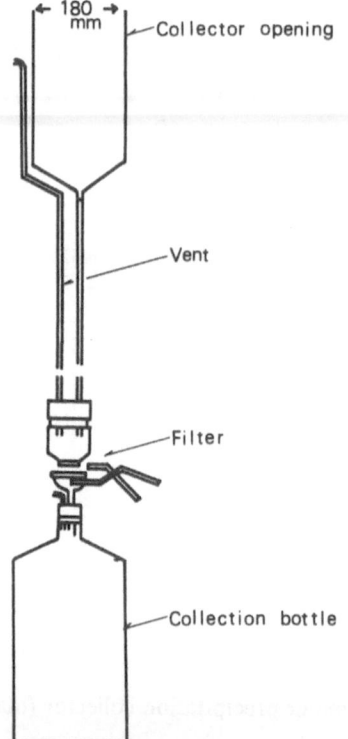

Figure 2–3a. Typical filter-type precipitation collector.

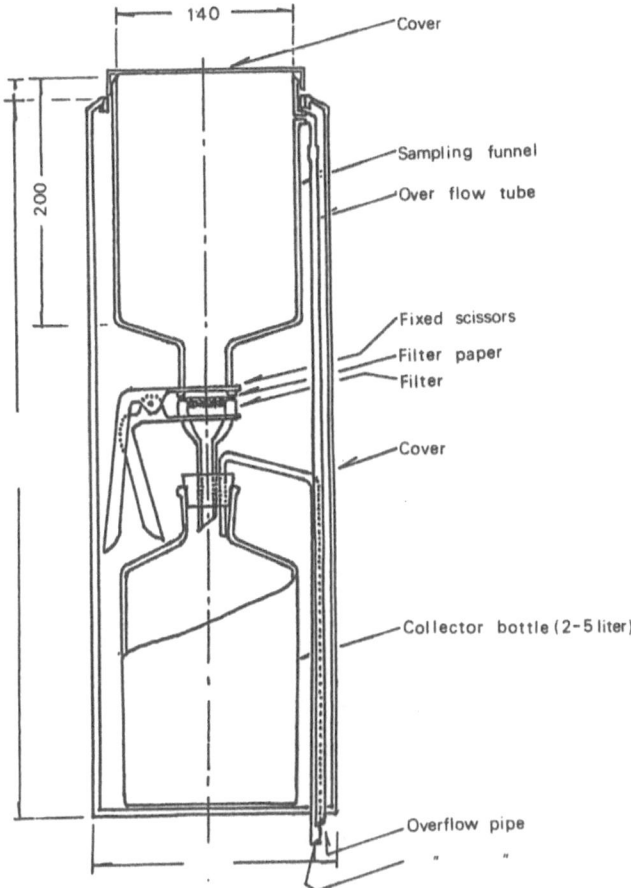

Figure 2–3b. Typical easily portable filter-type precipitation collector.

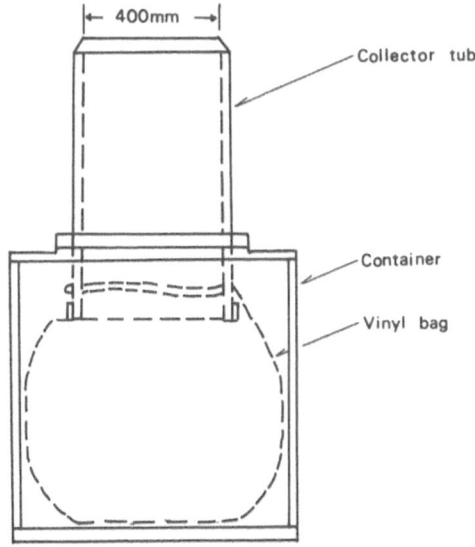

Figure 2–4. Snow collector.

and skin symptoms. Some public officials surmised that the actual number experiencing symptoms was much larger. The symptoms included eye pain, throat burning, and skin irritation due to contaminated fog and drizzle. The problem area was the Kanto Plains and its eastern neighboring areas. Initially the probable cause was felt to be such irritants as sulfuric acid, nitric acid, and formaldehyde. Surveys to investigate the causes of the problem were conducted in the various municipalities of the EPMPH, and the result of acidic rain surveys were given in a summary report on "wet air pollution"[5] (EPMPH, 1975, 1976; Japanese Environmental Agency, 1979). Following are major case outlines of pollution generation mechanisms and pollution irritation experiments.

A. Major Cases

1. Cases in 1973

In 1973 there were 547 complaints of pollution irritation. These are shown in Table 2–1.

The most serious cases were in Shizuoka and Yamanashi prefectures on June 28 and 29, 1973, in which there were attacks to eyes, shedding of tears, throat irritation, and coughing. The diagram showing the pollution complaints reveals 540 in limited areas 100 km apart. See Figure 2–5. Also, in Shizuoka prefecture damage to the crown of welsh onions and to tobacco, cucumbers, eggplants, and other plants was observed. On both days a stationary front was located on the southern coast of Honshū, and Shizuoka prefecture was cloudy with occasional rain. The hourly rainfall over most of the area was 0.0 mm, with a very slight drizzle. The Shizuoka Meteorological Observatory gave a poor visibility of 1.5 km at

---

[5]"Wet air pollution": an expression used in Japan to describe eye irritation producing pollution in drizzle known as "kirisame" and "bisame"—precipitation bordering on fog to extremely light rain.

Table 2–1. Pollution irritation incidents, 1973.

Date	Location (Pref.)	Compliants	Time of irritation	Type of irritation
6/28	Shizuoka	30	13:00–18:00	Eye irritation
	Yamanashi	151	18:00–19:00	
6/29	Shizuoka	359	13:00–18:00	Eye irritation, tears, throat irritation, coughing
9/13	Shizuoka	7	14:00–15:00	Eye irritation
	Total	547		

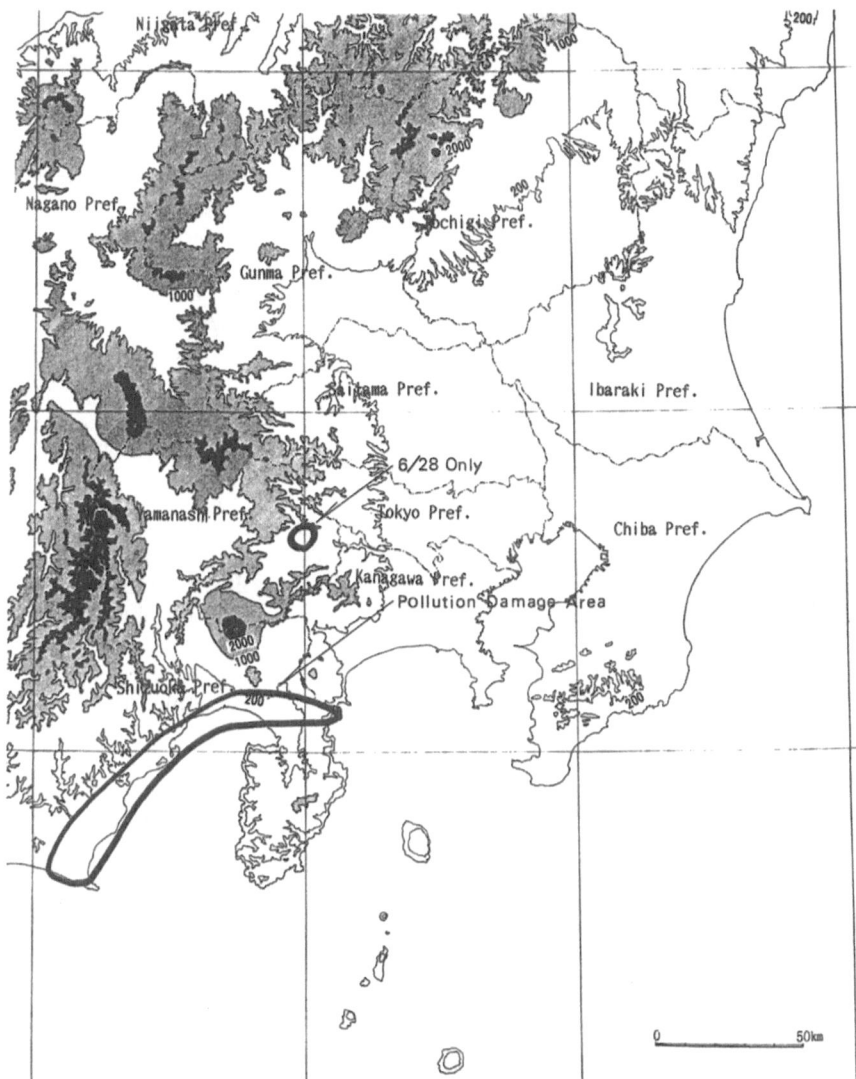

Figure 2–5. Area of acidic precipitation damage from very light rainfall—June 28 and 29, 1973.

3 P.M. on June 28, 1973. Precipitation analysis in Shizuoka prefecture generally showed a strong acidic content, with a pH from 2.7 to 3.5.

2. Cases in 1974

In 1974 complaints of pollution irritation numbered 33,181 (see Table 2–2).

Table 2-2. Pollution irritation incidents, 1974.

Date	Location (Pref.)	Compliants	Time of irritation	Type of irritation
7/3	Ibaraki	1,793	15:30–20:00	Eye irritation, arms
	Tochigi	28,762	14:00–18:00	irritation, throat
	Gunma	140	17:30–18:00	irritation
	Saitama	1,120	15:30–17:00	
	Subtotal	31,815		
7/4	Ibaraki	321	15:00–20:00	Eye irritation, eye
	Chiba	20	14:00–15:00	stimulus, arms
	Tokyo	203	11:00–20:00	irritation
	Kanagawa	187	15:00–18:00	
	Subtotal	731		
7/5	Ibaraki	3		Eye stimulus
	Chiba	9	11:00	
	Subtotal	12		
7/6	Chiba	9	8:50– 9:00	Eye stimulus
7/12	Shizuoka	13	13:00–14:00	Eye stimulus
7/13	Saitama	3	19:30–21:00	Eye stimulus
7/14	Saitama	1	11:00	Eye stimulus
7/16	Shizuoka	23	12:00–22:00	Eye irritation
7/17	Tochigi	71	10:30–15:15	Eye stimulus
7/18	Tochigi	225	10:00–16:00	Eye stimulus
	Saitama	281	10:00–16:00	Bloodshot
	Subtotal	506		
7/20	Saitama	1	21:00	Eye stimulus
7/24	Shizuoka	1	15:00	Eye irritation
Total		33,181		

Beginning July 3, the pollution irritation continued for four days, and then only sporadically until July 24. In Tochigi prefecture the data show that on July 3 there were 28,762 complaints, of which 28,285 were elementary school children and 477 others. On July 4 there was a total of 731 complaints. As seen in the pollution distribution shown in Figures 2–6 and 2–7, the complaints on July 3 were heavy in the northern Kanto Plains, centering in southern Tochigi, Saitama, Gunma, and Ibaraki prefectures. On July 4 the pollution had drifted to Ibaraki prefecture and the Tokyo Bay coastal areas. The irritation included combinations of eye pain and bloodshot eyes, and in Tochigi, Ibaraki, and Kanagawa prefectures additional complaints of skin irritation and pain with smarting of the arms. In southern Tochigi prefecture damage to cucumbers, eggplants, kidney beans, taro plants, peanuts, and so on, was observed. The weather conditions on July 3 involved rain and light drizzle in the northern Kanto Plains area resulting from a stationary front over Tokyo and neighboring Chiba areas. Surface winds to the north of the front were

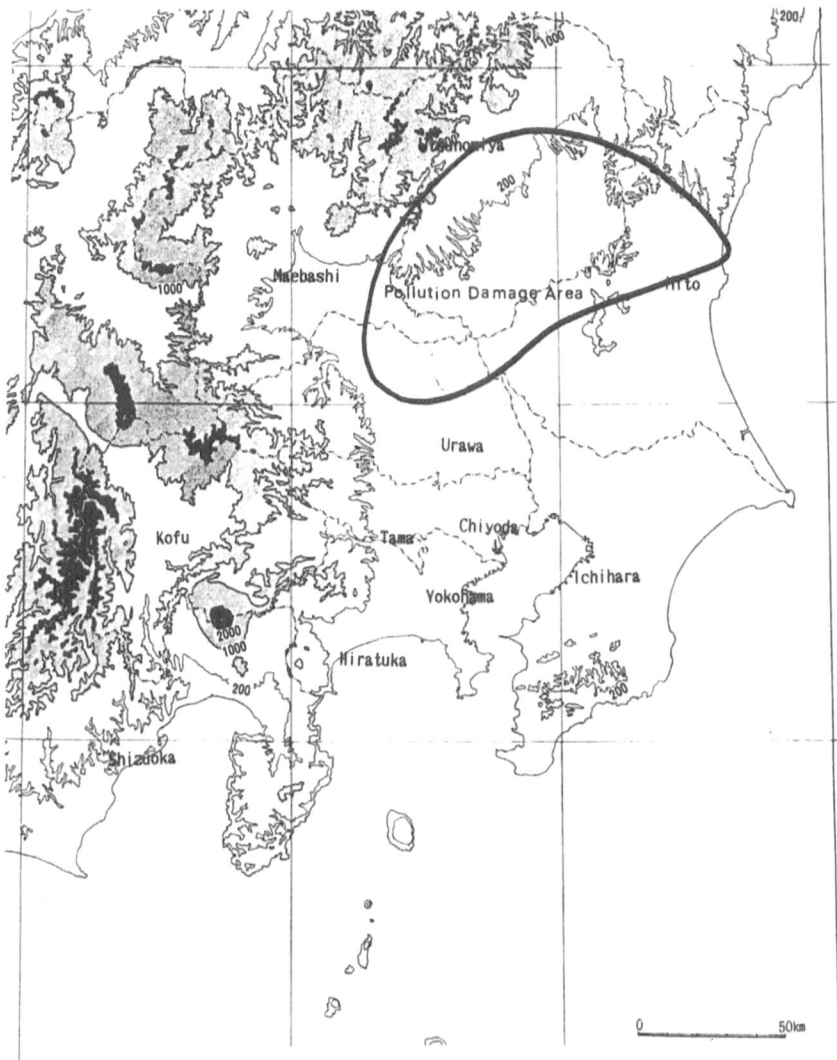

Figure 2–6. Area of acidic precipitation damage from very light rainfall—July 3, 1974.

northeast, with strong winds of the order of 10 m/s from the south on the south side of the front. In the air strata a few hundred meters above the ground, winds were all from the south. On July 4 the front had moved to the vicinity of southern Kanagawa prefecture, and northeasterly flow covered the entire Kanto area, with clouds and occasional rain or light drizzle. Most irritation complaints came from Tochigi prefecture with northeasterly winds of 0.8 to 2 m/s, the base of a temperature inversion at

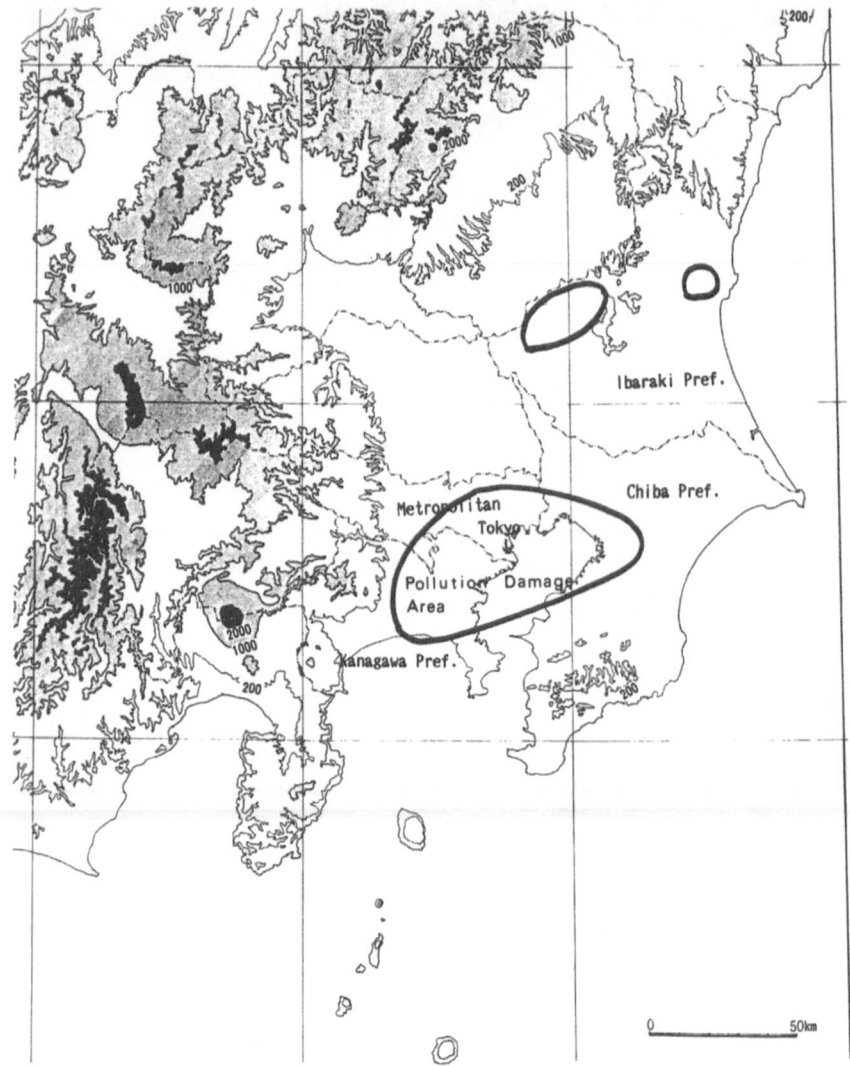

Figure 2-7. Area of acidic precipitation damage from very light rainfall—July 4, 1974.

500 m, and heavy fog in the afternoon with visibility reduced to only 500 m. The weather of July 4 is shown in the weather map of Figure 2-8.

In the period from July 2 to 5 the highest acidity in precipitation samples in the Kanto area gave a pH of 3.0. Again on July 18, 506 people suffered from eye irritation and bloodshot eyes, over the area comprising west-central and southwest Toch·gi prefecture and northern and southern Saitama prefecture. The weather conditions on the previous day involved intermittent precipitation. Whereas the amount of precipitation in the

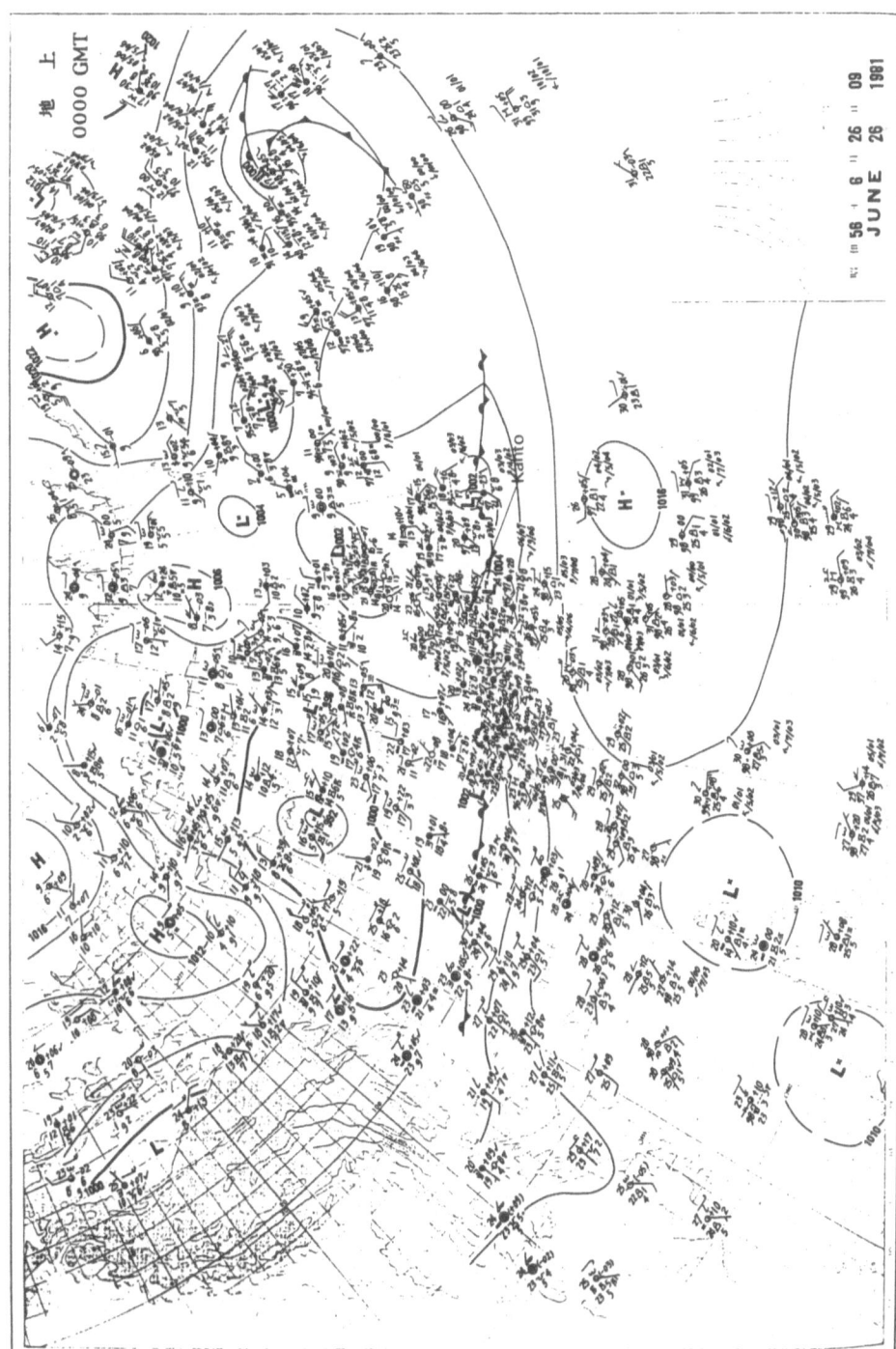

Figure 2–8. East Asia weather at time of drop in precipitation pH in the Kanto area of Japan.

southern Kanto area was relatively heavy, that in the northern Kanto was so slight as to register 0 mm. This was due to the front extending from the southern Kanto coast to the vicinity of the southern Izu peninsula, which remained stationary for a time and then moved somewhat to the north. The acidity of the precipitation observed in the area with a lot of irritation gave a pH of 3.0 in west-central Tochigi prefecture and a pH of 3.2 in the northern portion of Saitama prefecture.

3. Cases in 1975

The 1975 pollution irritation resulting from wet air pollution is shown in Table 2–3. In the period from May 3 to July 10 the total number of irritation complaints was 244. On June 25 there were 143, more than half of which were in central Tochigi prefecture and northern Saitama prefecture as well as Tokyo and environs. The majority were eye-related complaints involving pain and tears, with skin irritation in the Kanagawa prefecture cases. The weather conditions involved a front along the southern Honshū coast, with much cloudiness throughout the morning and precipitation of 0.5 mm in the afternoon. In the areas where most complaints were received, acidity was highest and the pH in northern Saitama prefecture was 3.1.

4. Cases in 1976 and After

Table 2–4 gives the number of complainants by area. On August 16, 1976, there was one complaint in Tokyo, and on June 25, 1981, there were four in Gumma prefecture. In 1981 an extremely strong acidic pH of 2.86 was recorded in Gumma prefecture. As reported by Sekiguchi and others, the cause was a high concentration of NO_3^- rather than SO_2^{2-} (Sekiguchi, Kano, and Ujiiye, 1983).

Table 2–3. Pollution irritation incidents, 1975.

Date	Location (Pref.)	Compliants	Time of irritation	Type of irritation
5/3	Saitama	1	7:10– 7:30	Eye irritation, tears
5/19	Ibaraki	72	7:00–14:10	Eye irritation, tears
6/24	Saitama	9	19:05–19:50	Eye irritation
6/25	Tochigi	90	8:00–17:00	Eye irritation
	Saitama	43	11:20–12:30	
	Tokyo	9	16:00–23:00	Eye irritation, tears
	Subtotal	143		
6/26	Gunma	18	17:00–	Eye irritation
7/10	Saitama	1	19:30–21:00	Eye irritation
Total		244		

Table 2–4. Pollution irritation incidents, 1976 and after.

Year	Date	Location (Pref.)	Compliants	Time of irritation	Type of irritation
1976	8/16	Tokyo	1	about 18:00	Eye irritation
1981	6/26	Gunma	4	19:00–	Eye stimulus, tears

B. Causal Mechanisms

It was thought that these episodes of pollution irritation were caused by irritants contained in fog or drizzle attacking the eyes and skin. The mechanism causing this irritant precipitation may be divided into two parts: (1) meteorological conditions resulting in formation of fog or drizzle and (2) the absorption of irritants in the precipitation.

1. Meteorological Conditions

Examining the meteorological conditions at the time of pollution irritation, we found a warm front between a northeast cold air current from across the Sea of Okhotsk and a south-southwest warm air current from the Ogasawara high pressure area. The surface of this front became one form of thermal inversion, with the pollutants in the lower stratum being held there and prevented from rising and dispersing. In the warm front light warmer rain fell from higher clouds producing fog or very light rain as it fell through the lower cold air mass. At the time of the episodes there was very light rain resulting from a warm air front. The episodic region was approximately 40 to 100 km to the north of the front.

2. Generation of Pollution Irritants, Their Absorption into the Precipitation, and the Resulting Irritant Precipitation

In addition to the described conditions, there was the possibility of the presence of irritants such as hydrogen ions, formaldehyde, acetaldehyde, acrolein and such aldehydes, formic acid, acetic acid, organic peroxides, and hydrogen peroxide in the episodic regions. It is thought that most of the hydrogen ions are present in the atmosphere in the form of sulphuric acid, hydrochloric acid, and nitric acid. A portion of the formaldehyde, sulphuric acid, and hydrochloric acid in these pollutant irritants comes directly from pollution sources. The major portion of pollutants other than these are generated in the atmosphere. That is, sulphuric acid results from the oxidation of sulphur oxides, nitric acid from the oxidation of oxides of nitrogen, hydrochloric acid from the reaction of atmospheric sea salt particles with sulphuric and nitric acids. Formaldehydes and acroleins result from photochemical reactions in the atmosphere. Sulphuric acid results from the oxidation of sulphur oxides in precipitation droplets. The mechanism by which pollutant irritants are incorporated into precipitation droplets is as follows: The larger particles contained in sulphuric acid mists enter droplets through inertial collisions. The smaller particles enter precipitation droplets by diffusion. Also, these particles may serve to form the nucleus of mist or fog droplets. Precipitation droplets may serve to bring gases into solution.

3. Eye Irritation Due to Pollution

Kurokawa has reported from tests that eye irritation occurs due to the presence of formaldehyde in very light precipitation (Kurokawa et al., 1975). The tests involved using eyedrops, made by adding formaldehyde and acrolein to sulphuric acid solutions of pH 2 to pH 5, in the edge of the eye. The results showed frequent eye irritation with the use of solutions of approximately pH 4, together with 3 mg L^{-1} or more of formaldehyde and 3.0 mg L^{-1} or more of acrolein.

V. Kanto Area Survey of Pollution Mechanisms

In the period from 1973 to 1975 there were large numbers of people in the central Kanto area experiencing pollution irritation at the time of very light precipitation. Recently a problem of injury to the crowns of Japanese cedars has developed in the Kanto area. It may be seen from these two effects that the wet air pollution problem has somewhat expanded.

A. Major Air Pollution Sources

The Kanto area, including the capital city, has approximately 33% of Japan's population—that is, some 37 million people (Japanese Statistical Yearbook, 1987). Of this number, one-third are concentrated in the central (23 wards) Tokyo population of approximately 8.4 million, the Yokohama population of 3 million, and the Kawasaki population of 1.1 million. To supply this high-density population area there is a huge industrial complex in the coastal industrial zone in the area on both sides of Tokyo Bay and the east-central Kanto Plains area.

In the course of production, and in the industrial complex as a whole, there are stationary sources of tremendous amounts of air pollution. There is a veritable forest of 200-m smokestacks associated with power generating stations, refineries, iron works, and petrochemical plants.

In recent years mining and manufacturing industries in Japan have stopped growing, as manufacturing operations in other countries have increased dramatically. However, there is concern that the pollution in the coastal industrial zone surrounding Tokyo Bay will increase due to the concentration of plants in the area.

Figure 2–9 shows major pollution sources. To satisfy the demands of the capital city residents there is an extremely high level of commerce, transportation of goods, and leisure activities. In addition, there is the moving source of pollution due to the network of roads and freeways over which 100,000 to 200,000 vehicles travel daily (Metropolitan Police Traffic Yearbook, 1986). A great quantity of pollutants has resulted from these sources of movement. The web of roads radiating from Tokyo in the

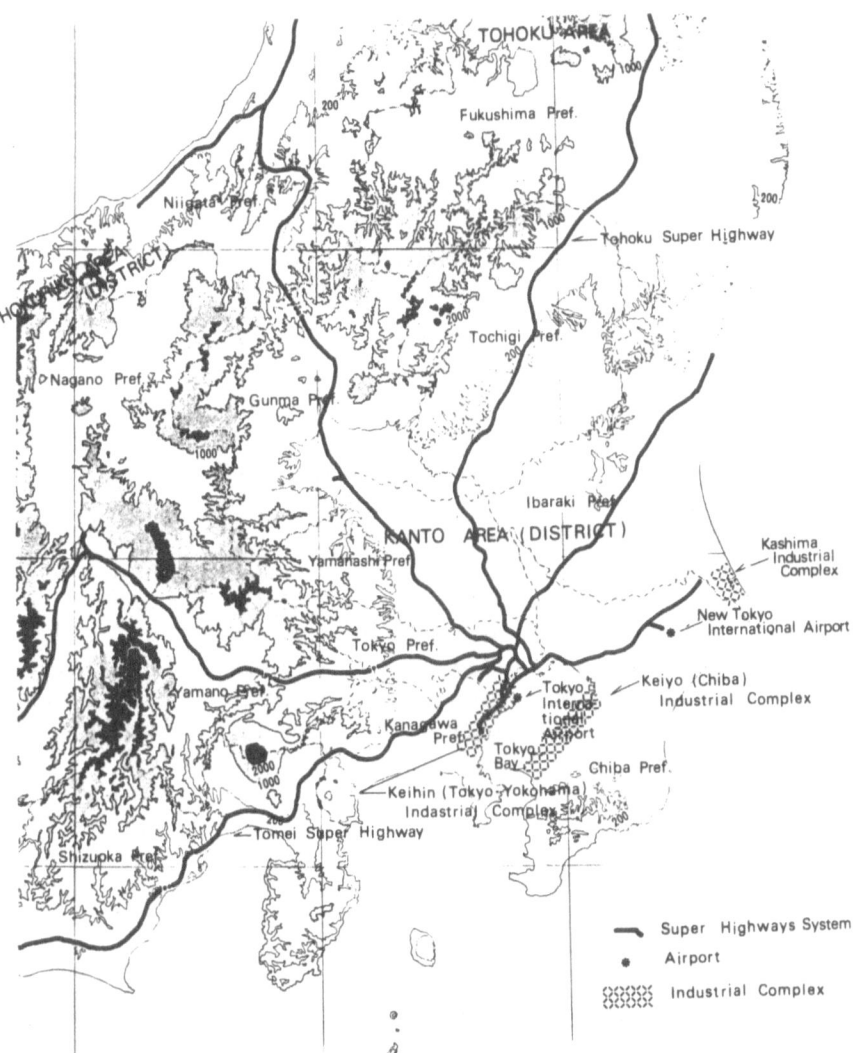

Figure 2–9. Major air pollution sources, Kanto coastal area.

center to the surrounding plains continues to develop, such that the expressways have a traffic volume of hundreds of thousands of vehicles a day, increasing year by year. The bulk of pollution from aircraft results from the concentration of air traffic in the three major airport areas of Narita, Haneda, and Yokota. In 1985 there were 78,000 takeoffs and lands at the New Tokyo International Airport at Narita, and 157,000 at the Tokyo International Airport at Haneda (Air Transport Statistics Yearbook, 1985). Also there is a yearly increase both in large size and in numbers of aircraft in use.

The air pollution from ships is concentrated in the harbors around Tokyo Bay, such as Yokohama, Tokyo, Chiba, and harbors associated with industrial complex groupings. In 1985, ship arrivals and departures in Tokyo harbor totaled 55,000, with a total gross tonnage of 95 million (Tokyo Metropolitan Government, 1987). Year by year the number of arrivals and departures is decreasing, but with an increase in gross tonnage due to the use of larger ships.

The area-by-area discharge of NO_x and SO_x pollutants in the Kanto area is shown in Figures 2–10 and 2–11 (Tanooka, 1987). At all locations in the plains area there are these sources of gas pollution, especially in the industrial complex and the area of the major highways.

B. Survey of Pollution Mechanisms

From the time of irritation to people by air pollution, the EPMPH has conducted surveys to determine wet air pollution mechanisms and their explanation (EPMPH, 1975, 1976, 1983, 1987a, 1987b, and 1987c). For these surveys of wet air pollution mechanisms the precipitation collection method involved collecting the first 3 mm of precipitation in 1 mm increments (see Figure 2–2). Through this 1 mm-increment method the change in concentration in the precipitation became well understood. However, since the collector also collected fallout constituents, the later surveys involved collection of the total precipitation for a precipitation period. The survey period was two weeks in the latter half of June and at the beginning of the rainy season. Initially, the analyses were for pH, SO_4^{2-}, NO_3^-, Cl^-, NH_4^+, HCHO, and for electrical conductivity (EC). Several years later measurements were made of Na^+, K^+, Mg^{2+}, Ca^{2+}, and so on, a check being made by ion equivalent balance. In addition data were always taken on SO_x, NO, NO_2, dust, along with meteorological observations and analyses conducted using these data. The major results of these analyses are as follows:

1. There are many cases of lowered pH of the precipitation even after irritation to personnel had ceased to be observed. Measurements of the pH in the initial precipitations at various stations in the Kanto area are shown in Table 2–5 (Koyama, 1986). It was learned that the low pH was in no way confined to the same areas from year to year.

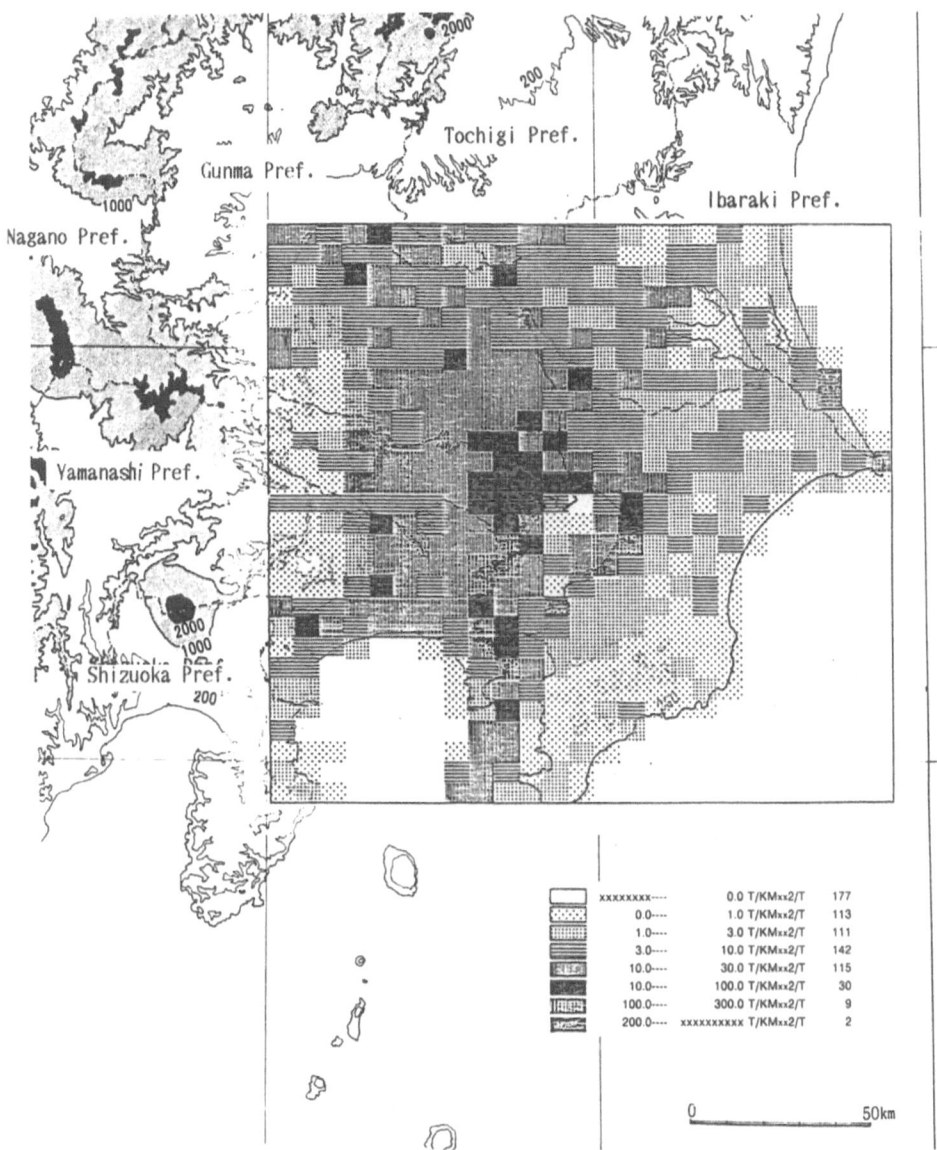

Figure 2–10. NO$_x$ pollutant sources map, southern half of Kanto area.

2. The irritation effects on personnel of concentrations of SO$_4^{2-}$, NO$_3^-$, and other pollutants are found now to be only one-tenth what they were at the outset some years ago. However, the phenomenon of the spread of concentrations of pollutants over broad polluted areas has not been determined. This is because the diffusion of pollutants by tall smokestacks,

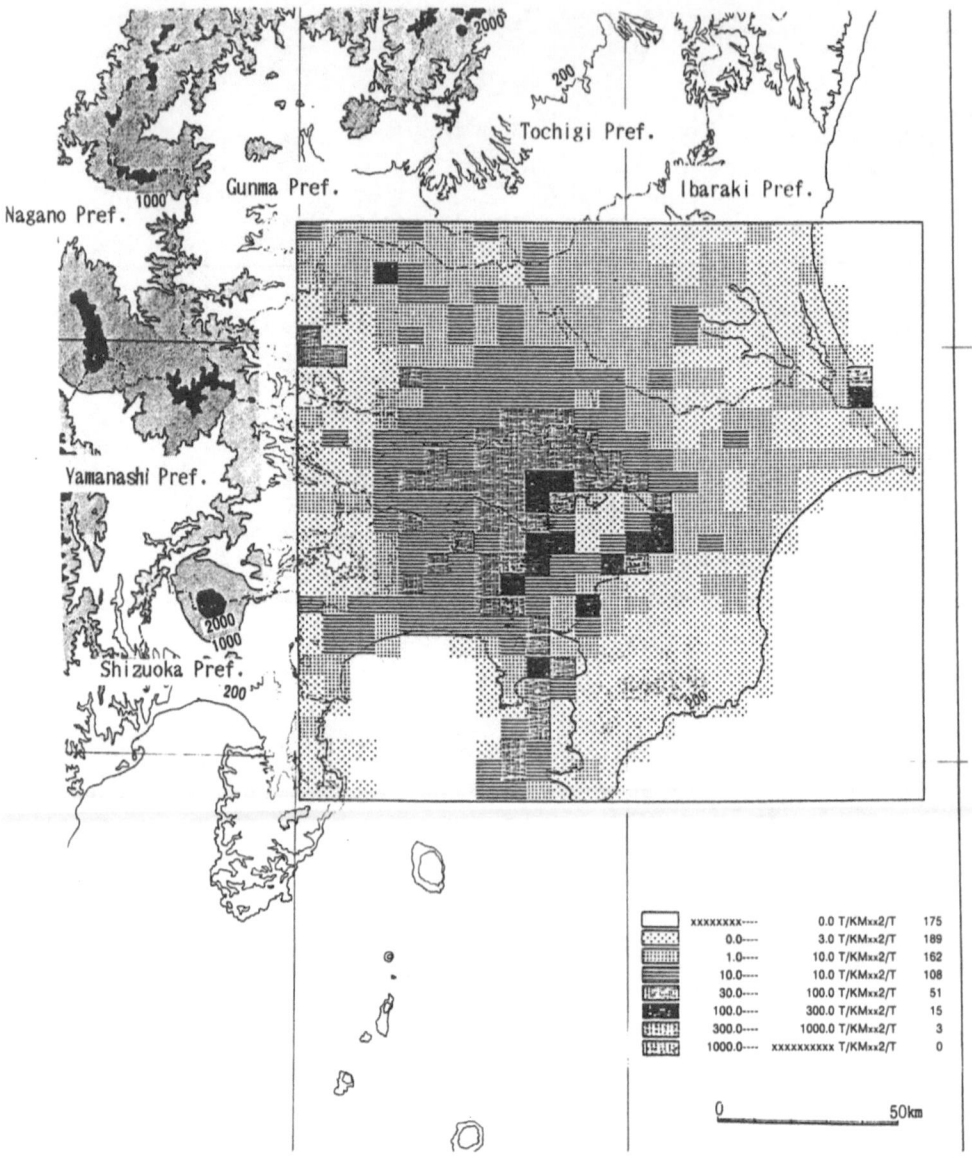

Figure 2-11. SO$_x$ pollutant sources map, southern half of Kanto area.

highway networks, and so on, has not been investigated by surveys at various stations.

3. The cause of low pH regions has been found to be related to the location of stationary fronts along the southern coast of Tokyo Bay. Even though the precipitation is the same, the pH is seen to vary with stationary fronts, the location and velocity of low pressure areas, resulting from

Table 2-5. Values of pH in initial 1 mm of precipitation as measured in prefectural and metropolitan surveys.

	Survey site								
Year[b]	Mito	Utsunomiya	Maebashi	Urawa	Ichihara[a]	Chiyoda	Tama	Yokohama	Hiratu
1975	3.3	3.2	4.4	3.1	3.7	3.9	3.5	3.3	3.4
1976	4.0	3.3	3.9	3.6	4.4	4.1	4.7	3.3	3.7
1977	4.6	4.1	4.6	3.7	3.9	4.0	3.5	3.0	3.8
1978	4.2	3.3	3.6	3.5	3.6	3.5	3.4	3.4	3.1
1979	4.3	3.9	4.3	4.1	3.8	4.0	4.2	4.0	4.5
1980	4.0	3.8	—	4.1	5.7	4.4	4.0	4.4	4.0
1981	3.9	3.3	3.1	3.3	4.0	3.8	3.9	3.5	3.6
1982	5.5	4.2	3.6	3.4	3.8	4.2	3.5	3.7	3.5

a In a precipitation of only 0.28 mm at Ichihara, the pH was 3.7.

b The survey periods were; June 25–July 8, 1978; June 26–July 7, 1981; June 22–July 3, 1976; June 28–July 9, 1979; June 25–July 6, 1982; June 21–July 2, 1977; June 27–July 8, 1980; June 23–July 4, 1975.

a variation in the concentration of precipitation constituents. The cause is thought to be due to the change in quantity of discharged air pollutants. Sudden changes in discharged pollutants result from the response to photochemical smog warnings. But apart from times of such governmental pollution control, it appears that no such reductions take place. Monitoring of large-scale stationary sources of SO_x pollutants are now conducted on the basis of law and company self-governing agreements. These data have been made public only in summary fashion, and details are not available.

4. The pH of precipitation drops sharply with meteorological conditions, as seen in precipitation from a stationary front in the southern coast of the Kanto area, in a very weak pressure gradient, and in the occurrence of a weak low pressure system. In accord with these meteorological situations a low pH region developed from the western Kanto plains area of Hiratsuka, Tama, and Kumagaya to the northern area of Maebashi, Utsunomiya, and to Shimodate. Especially from Tama to Maebashi there were many low pH regions. This is seen in Figure 2–12. Photo A shows the clouds at this time.

5. Concurrently with the appearance of low pH, the concentrations of SO_4^{2-} and NO_3^- went up sharply above the average. In this area the ratio of increase of NO_3^- tended to be greater than that of SO_4^{2-}. However, outside the Kanto area there was the reverse tendency (Makino and Kaneko, 1984). See Figure 2–13.

6. In precipitation heavily influenced by sea salt, the precipitation was largely alkaline, but in precipitation resulting from thunderstorms, especially in the surrounding areas with low precipitation, the NO_3^- concentration rose, and the pH was low.

7. In large urban areas with higher suspended particulate matter concentration and dustfall, the pH was neutral.

8. In the southern coastal region of the Kanto area the drop in pH at the time of a stationary front would be caused by the prevention of the escape of pollutants from the area (see Figure 12). Along the northern side of the front, with surface winds from the east but not from the north, the northerly movement of the south coast front is stopped by the mountains. A large region of air stagnation developed along the base of the western mountains. The wind on the southern side of the front drove the clouds all along the front slowly upward with a slope of 1/100. At the same time all along the northern face of the front even the winds from the north turned and rose. The bulk of the pollutant-carrying precipitation, being unable to rise over the northern and western mountains, caused a high concentration of pollutants to fall in the northern and western areas of the plains short of the mountains.

9. The minimum pH of fog at Mt. Tsukuba (at the 870 m level) and Ohyama (at the 720 m level) has been recorded by environmental agency surveys at 2.8 and 3.05 (Ohta, Okita, and Kato, 1981). This value is rela-

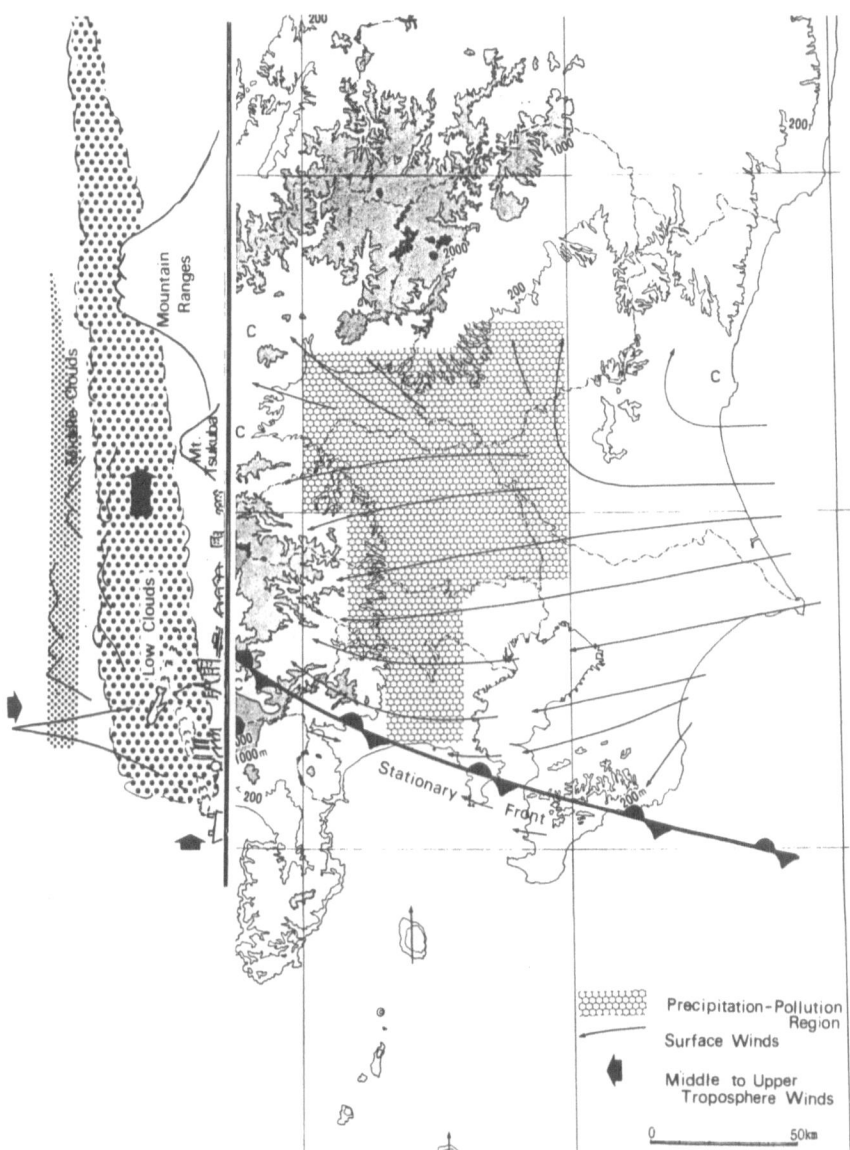

Figure 2–12. Meteorological conditions in the Kanto region, which is highly susceptible to precipitation pollution.

tively low compared with that for surface level precipitation. Also, the pollutant concentrations were several times higher (see Table 2–6).

10. Since it became apparent that the pH of precipitation at the various stations in the western and northern Kanto plains area could easily drop to low values, it is important to examine the ratios of pollutant

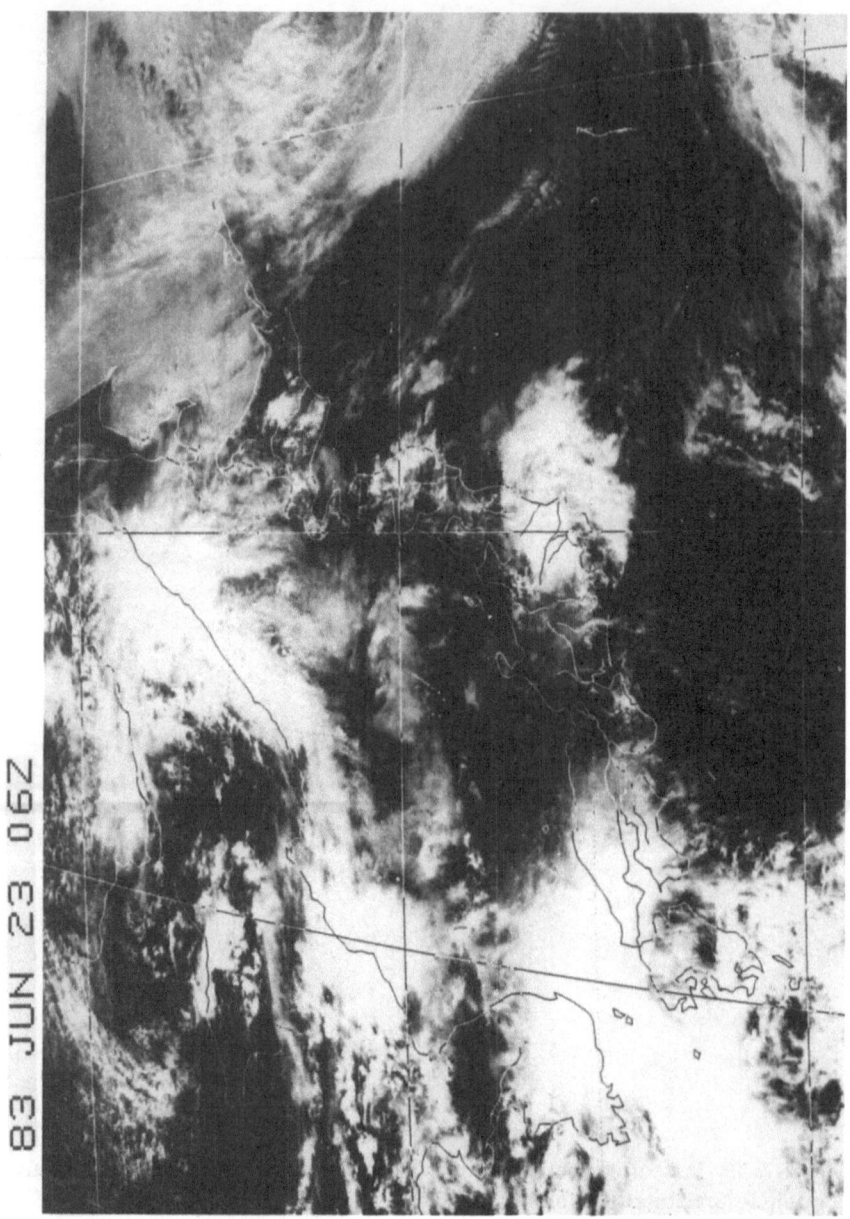

A. **Meteorological Satellite Photo**

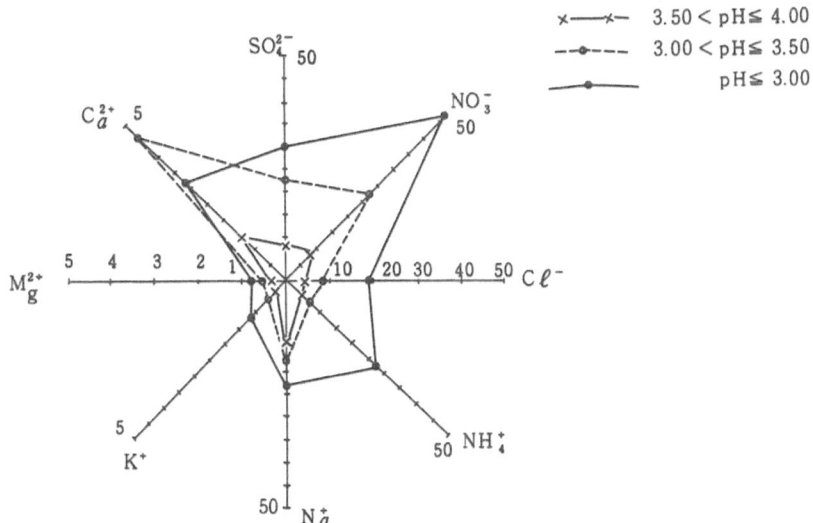

Figure 2–13. "Spiderweb" graph of the concentrations of precipitation components. Regions of pollutant concentrations resulting average pH values 4.00. (initial 1 mm of precipitation) 1983 example.

concentrations at these various stations. For example, statistical cluster analyses were run on the fallout particulates collected at the various test locations in the two-week surveys of June 1986.[6] The results are shown in Figure 2–14. Six cluster areas—a central Tokyo cluster, a Kanto region from southwest to northwest cluster, a Kanto region central-northern sector, a Pacific coastal region from Shizuoka to Koriyama cluster, a crescent from Niigata to Yamanashi cluster, a Hamamatsu cluster, and an Isezaki cluster—may be seen.

C. Subjects for Further Investigation

The EPMPH has conducted pollution mechanism surveys for the past 10 years, and in this period there have been great advances in the sensitivity and accuracy of the analytical instruments. In the past the object was to facilitate the easier and cheaper conduct of surveys. However, in the future the object is to investigate the movement of pollutants in the atmosphere, from their production to their elimination, and to prevent the accumulation of these pollutants over large areas. In the Kanto area NO_3^- is the predominant cause of low pH; in other areas it is SO_4^{2-}. It remains a

---

[6]"Cluster analysis": according to the ordinary method of multivariate analysis. It involves the statistical analysis of the observed values of the multirandom variables in a matrix array.

Table 2–6. pH and pollution concentrations in precipitation and fog water droplets at Mt. Tukuba, Oyama, and nearby areas.

Survey site	Date	Time	pH	NO_3^-	SO_4^{2-}	NH_4^+	Cl^-	HCHO	Type of precipitation
Tsukubasan	1977.7. 2	—	2.8	85.0	68.3	20.0	6.5	0.38	Fog
Oyama	1978.6.27	14:50–16	3.05	38.	69.8	30.9	25.4	0.70	Fog
"	"	9:50–16	—	12.3	—	1.52	—	0.15	Precipitation 0.2mm
Isehara	"	10–16h	4.5	3.0	—	0.95	1.5	0.2	Precipitation 0.35mm
Hiratsuka	"	–15h	3.36	9.8	—	2.74	3.1	0.39	Precipitation 0.15mm

Unit: μ g ml^{-1}

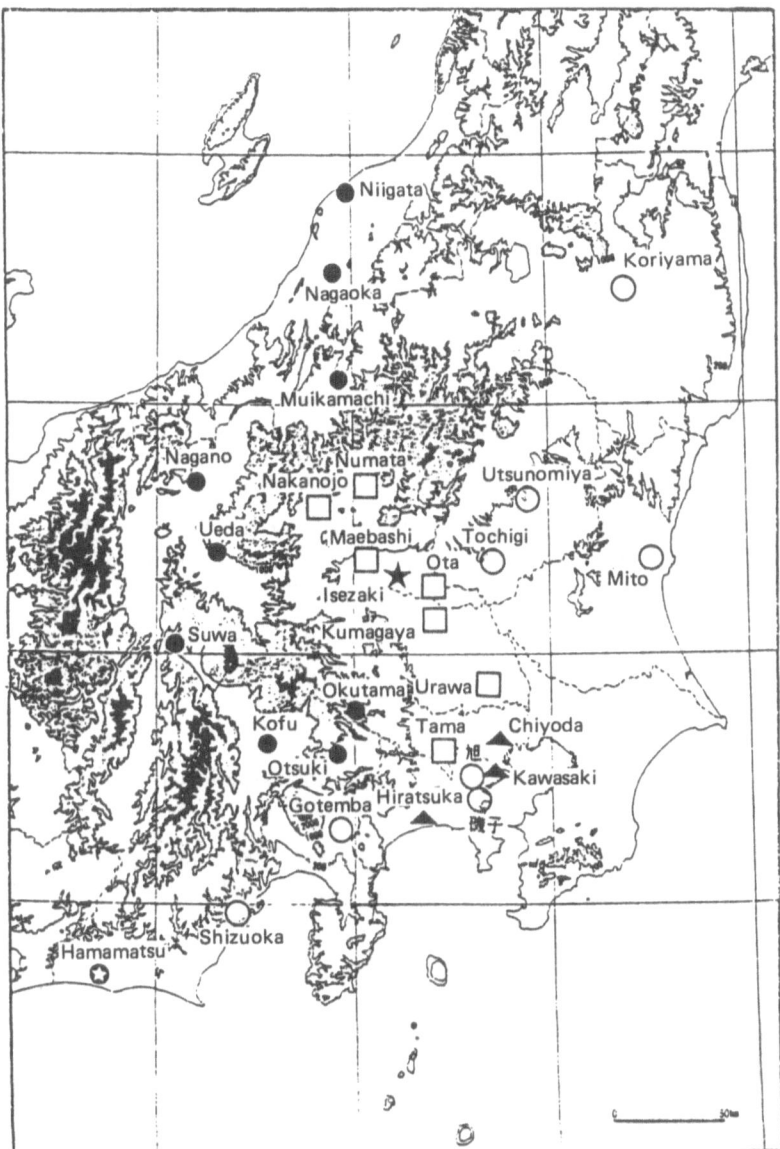

Figure 2–14. 1986 example of "cluster analysis" regional distribution (initial precipitation).

question as to the reason for this. Apart from the above, the production of SO_x pollution in Japan due to human activity has tended to decrease. However, the pH has not changed significantly, so much of this problem remains unsolved.

Agencies other than the EPMPH, such as the Environmental Laboratories Association, Hokkaido-Tohoku Branch and Chugoku-Shikoku Branch, began in 1986 to conduct long-term wide-area surveys so it is now possible to coordinate measurements in a team effort.

VI. Wide-Area Surveys of Fallout of Acidic Materials

In this section we describe the principal results of surveys already detailed in other reports (Japan Environmental Agency, 1987a). The Environmental Agency has summarized acidic rain survey data covering a two-year period from April 1984 to March 1986. These surveys were made to cover the entire country over a long period. First, the methods of calculating the average pH and the "excess SO_4^{2-} are given here. For calculating the average pH the following equation was used:

$$pH = \log_{10} [\{The sum of (Q) \times (H^+)\}/(Q)]^{-1}$$
$$Q = \text{quantity of water collected per week or month}$$
$$H^+ = \text{hydrogen ion concentration}$$

The symbol "excess SO_4^{2-}" designates the excess of the concentration of that ion over the normal seawater concentration of that ion. This is calculated by the following method:

$$\text{Seawater concentration} = (SO_4^{2-}/Na^+)\text{sea salt} \times (Na^+)\text{fallout (g)}$$
$$(SO_4^{2-}/Na^+)\text{sea salt (g/g)} = 0.251$$
$$\text{"excess } SO_4^{2-}\text{"} = (SO_4^{2-}) - (\text{seawater})$$

Since the ratios of the various acidic fallout pollutants is shown by Environmental Agency surveys to be different in the Sea of Japan area from that in the Pacific coastal area (Kanto area), this matter is treated briefly here.

The annual mean pH in bulk sampling was higher than that obtained in wet-only sampling by no more than approximately 0.2. Comparing the data obtained by those two sampling methods (bulk and wet-only), the highest acidity in precipitation alone was estimated to be pH 4.4. The lowest value of pH in the Kanto district is slightly higher than that of 4.0 to 4.2 in eastern North America and Europe.

A. National Surveys

Surveys were conducted countrywide in seven prefectures or cities as shown in Figure 2–15. In each prefecture two samplings were conducted,

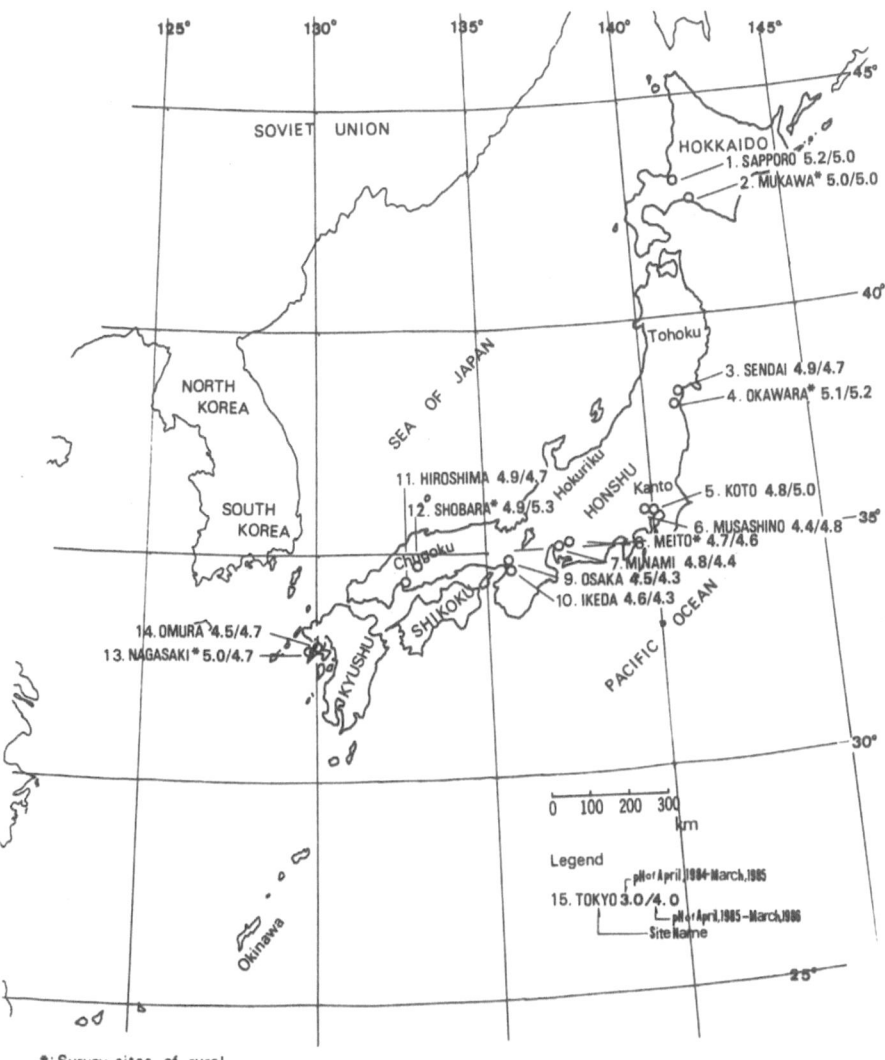

*: Survey sites of rural

Figure 2–15. Large area survey of precipitation pH in Japan.

one in an urban area and one in a rural area. The precipitation collection apparatus involved the methods described earlier. Analyses were made of electrical conductivity (EC), pH, and precipitation constituents every week, using one-week samples. In Hokkaidō, because of low temperatures and heavy snowfall from December to March, and for the same reason in Miyagi prefecture for January to March, snow collection apparatus was used. Snow samples were melted at room temperature and immediately analyzed.

1. pH

The results of pH measurements are shown in Table 2–7 and Figure 2–15. As shown in the table, in all the areas the average pH falls in the 4.4 to 5.3 range. Using as a standard an acidic precipitation with a pH of 5.6 resulting from the saturation of atmospheric CO_2 in pure water, it can be said that precipitation in Japan is definitely acidic. Also, by comparing the results in the first and second years it is seen that changes ranging from −0.3 (Nagasaki 5.0 to 4.7) to +0.4 (Musashino 4.4 to 4.8) occurred. But when the monthly pH average varies from area to area much more than the annual averages—such as at Sapporo, Sendai, and Tokyo Koto Ku— the difference was as large as 2.6.

A reason for the great difference, with high pH occurring for the most part from the beginning of winter to early spring, is urban activity causing the generation of dust particles (e.g., in Tokyo), while in the snow country studded tires produce dust by running on the surface of asphalt-paved roads.

2. Soluble Components in Fallout

Tables 2–8a and 2–8b give the results of measurements of the quantities of soluble components in fallout materials. The quantity of constituents in fallout materials varies greatly with the amount of precipitation and with the area. Comparing the maximum-minimum ratios, it is seen that although the differences in excess SO_4^{2-}(3.3) and SO_4^{2-}(3.5) at different

Table 2–7. The results of pH measurements.

		\multicolumn{3}{c}{1984.4~1985.3}			\multicolumn{3}{c}{1985.3~1986.3}		
		Mean	Max.	Min.	Mean	Max.	Min.
Hokkaido	1	5.2	6.9	4.9	5.0	6.7	4.3
	2	5.0	5.9	4.7	5.0	5.4	4.7
Miyagi	3	4.9	6.8	4.2	4.7	6.8	4.4
	4	5.1	6.1	4.2	5.2	6.2	4.3
Tokyo	5	4.8	6.5	4.5	5.0	7.1	4.6
	6	4.4	5.4	4.2	4.8	6.8	4.5
Nagoya	7	4.8	5.7	4.4	4.8	6.3	4.4
	8	4.7	5.3	4.2	4.6	6.1	4.2
Osaka	9	4.5	5.1	4.3	4.7	6.7	4.4
	10	4.6	4.9	4.3	4.7	5.5	4.4
Hiroshima	11	4.9	6.0	4.5	4.7	5.7	4.2
	12	4.9	6.1	4.2	5.3	6.2	4.7
Nagasaki	13	5.0	5.8	4.4	4.7	5.5	4.4
	14	4.5	5.7	4.0	4.7	5.0	4.0

Max., Min.; Monthly Average

sampling locations is quite small, there is a great difference in the sea salt constituents and H^+ (i.e., H^+ 11.6; Na^+ 10.8; Mg^{2+} 7.6; K^+ 7.6; Cl^- 7.7, etc.) between only slightly separated areas. The mean fractions of sea salt ions (Cl^-, Na^+) in the total ions were Cl^- 26% and Na^+ 12%. Also, SO_4 was highest at 31%, and excess SO_4^{2-} was 28%. Thus the three constituents, Cl^-, Na^+, SO_4^{2-}, comprised 69% of the total ions. However, there were differences at the several stations—for example, high NO_3—occurring at Musashino and high proportions of Cl^- at Sapporo, Mukawa, and Nagasaki.

a. Fallout of SO_4^{2-}–S and Excess SO_4^{2-}–S

The fallout of SO_4^{2-}–S ranged from 470 mg m^{-2} yr^{-1} at Mukawa to 1640 mg m^{-2} yr^{-1} at Sapporo; and of excess SO_4^{2-}–S, from 380 mg m^{-2} yr^{-1} at Mukawa to 1270 mg m^{-2} yr^{-1} at Sapporo. In Sapporo and Nagasaki the excess SO_4^{2-}–S was 77% of the total SO_4^{2-}–S; in Osaka and Hiroshima it was 96%. Generally, the high proportion of the SO_4^{2-} S is excess SO_4^{2-}–S, so the high sea salt proportion at Sapporo makes the Nagasaki excess SO_4^{2-}–S seem relatively small.

b. Fallout of NO_3^- N and NH_4^+ N

The fallout of NO_3^- N ranged from 150 mg m^{-2} yr^{-1} at Mukawa to 760 mg m^{-2} yr^{-1} at Musashino. The NH_4^+ N fallout ranged from 120 mg m^{-2} yr^{-1} at Mukawa to 1640 mg m^{-2} yr^{-1} at Shobara, the total nitrogen fallout ranging from 2470 mg m^{-2} yr^{-1} at Mukawa to 1300 mg m^{-2} yr^{-1} at Musashino. It is thought that the NO_3^- N fallout, which is heavy in and around the large cities, is caused in large part by the NO_x produced by cars. Also, the ratio of NH_4^+ N to NO_3^- N ranges from 0.67 at Musashino to 2.8 at Shobara. But since the NO_3^- N exceeds the NH_4^+ N only at the Musashino and Nagasaki stations, it is clear that the total nitrogen results primarily from the NH_4^+ N.

c. Fallout of Cl^-

The Cl^- fallout ranges from 1040 mg m^{-2} yr^{-1} at the Nagoya Tomei station to 8040 mg m^{-2} yr^{-1} at Sapporo, comprising 12% to 38% of the total measured fallout constituents. The major sources are sea salt and waste disposal plants, but since the contribution of the latter is expected to be virtually eliminated the correlation between Cl^- and Na^+, K^+, Mg^{2+}, which is within a level of significance of 1%, is felt to be determined almost entirely by sea salt. This is surmised from the fact that Japan is an island country, surrounded by seas. Also, in this case the Cl^- of sea salt origin is conjectured to suppress rather than enhance the acidity of precipitation.

Table 2–8a. The quantity of fallout (1984.4~1985.3).

Site	Rain (mm)	H⁺	SO_4^{2-}-S	Excess	NO_3^--N N	Cl⁻	NH_4^+-N N	Ca²⁺	Mg²⁺	K⁺	Na⁺
1	765	5	1170	900	180	5690	210	2340	360	260	3110
2	603	8	470	380	150	1750	120	380	140	120	1020
3	992	21	800	710	290	2490	470	1050	150	190	970
4	930	16	1000	930	300	2690		1390	190	450	910
5	1135	18	1230	1140	520	2260	540	920	200	140	1080
6	1143	50	940	890	690	1870	610	790	120	110	590
7	992	16	1080	1000	330	1190	370	800	96	73	610
8	1084	25	940	870	380	1150	390	400	76	74	620
9	1276	44	1180	1130	400	2420	680	1120	98	150	620
10	1404	35	790	730	300	1470	340	570	98	91	710
11	1251	15	1200	1110	330	1580	320	940	110	99	730
12	1245	16	1100	1000	330	2310	650	700	130	110	910
13	1495	16	1380	1060	220	6470	450	570	490	380	3630
14	1748	58	1320	1130	330	3700	440	710	310	230	2070

Unit: mg m⁻² yr⁻¹

Table 2–8b. The quantity of fallout (1985.4~1986.3).

Site	Rain (mm)	H^+	SO_4^{2-}-S		NO_3^--N	Cl^-	NH_4^+-N	Ca^{2+}	Mg^{2+}	K^+	Na^+
				Excess	N		N				
1	1198	19	1640	1270	210	8040	330	2160	530	400	4420
2	973	9	620	530	170	2010	200	360	170	91	1100
3	1092	26	850	790	330	2280	580	1070	130	120	700
4	827	12	560	520	210	1130		580	77	85	410
5	1333	13	1100	1020	350	2830	480	2120	280	92	900
6	1398	23	950	900	760	2170	510	870	110	80	570
7	1387	23	1200	1130	400	1440	530	860	110	79	610
8	1427	31	870	810	350	1040	370	390	70	100	520
9	1105	23	1070	1020	310	2050	510	1060	98	280	640
10	1405	28	910	840	340	1410	370	550	110	120	740
11	1522	32	1320	1270	340	1450	290	1200	110	59	590
12	1751	9	1240	1150	310	2550	860	500	140	120	1030
13	1891	34	1460	1180	210	5740	350	550	440	320	3300
14	2177	45	1420	1210	230	4310	270	610	300	200	2140

Unit: mg m^{-2} yr^{-1}

d. Fallout of Ca^{2+} and Mg^{2+}

The Ca^{2+} fallout ranged from 360 mg m^{-2} yr^{-1} at Mukawa to 2340 mg m^{-2} yr^{-1} at Sapporo, while the Mg^{2+} fallout ranged from 70 mg m^{-2} yr^{-1} in the Meito-ku of Nagoya to 530 mg m^{-2} yr^{-1} at Sapporo. The sources generating this include yellow soil, sea salt, and particles produced by soil and urban activity. In addition, in Hokkaidō and Miyagi prefectures there is a great deal of filler related to slake lime contained in particles generated by studded tires on asphalt pavement. Also, in spite of a relatively large amount of SO_4^{2-}, NO_3^- fallout the average monthly pH not falling below 4 is thought to be due to Ca^{2+} and Mg^{2+} precipitation.

e. Fallout of Na^+ and K^+

The Na^+ fallout ranged from 410 mg m^{-2} yr^{-1} at Okawara to 4420 mg m^{-2} yr^{-1} at Sapporo, and the fallout of K^+ ranged from 59 mg m^{-2} yr^{-1} at Hiroshima to 450 mg m^{-2} yr^{-1} at Okawara. The primary source of this fallout was sea salt.

3. Summary

At present, because the survey summary covers only two years, only the relative highs and lows of this period show in the results. The environmental agency will be continuing this survey until March 1988, after which the survey results should facilitate the development of a model to estimate and monitor the effects of acidic precipitation on soils, vegetation, and so on.

B. Kanto Area Survey of Acidic Fallout Constituents—Mesoscale Acidic Deposition Around Kanto District

According to Kanto area surveys by the EPMPH, one of the conclusions is that there are many cases where the pH of initial precipitation falls below 4.0.

An intensive field program was carried out to investigate the mesoscale precipitation chemistry in and around the Kanto district for the year September 1984 to August 1985. Bulk precipitation collectors were operated with a sampling period of half a month. In the sampling network as shown in Figure 2–16, 20 sites were located in the Kanto district, including the Tokyo metropolitan area, facing the Pacific Ocean. Three sites were set up in the Hokuriku district facing the Sea of Japan. The climate of the Hokuriku district in winter contrasts greatly with that of the Kanto district.

The spatial distribution and the seasonal variation in pH and the concentrations of the major ions will be shown in this section. The annual mean values for pH and ion concentrations were volume-weighted by the monthly amount of precipitation.

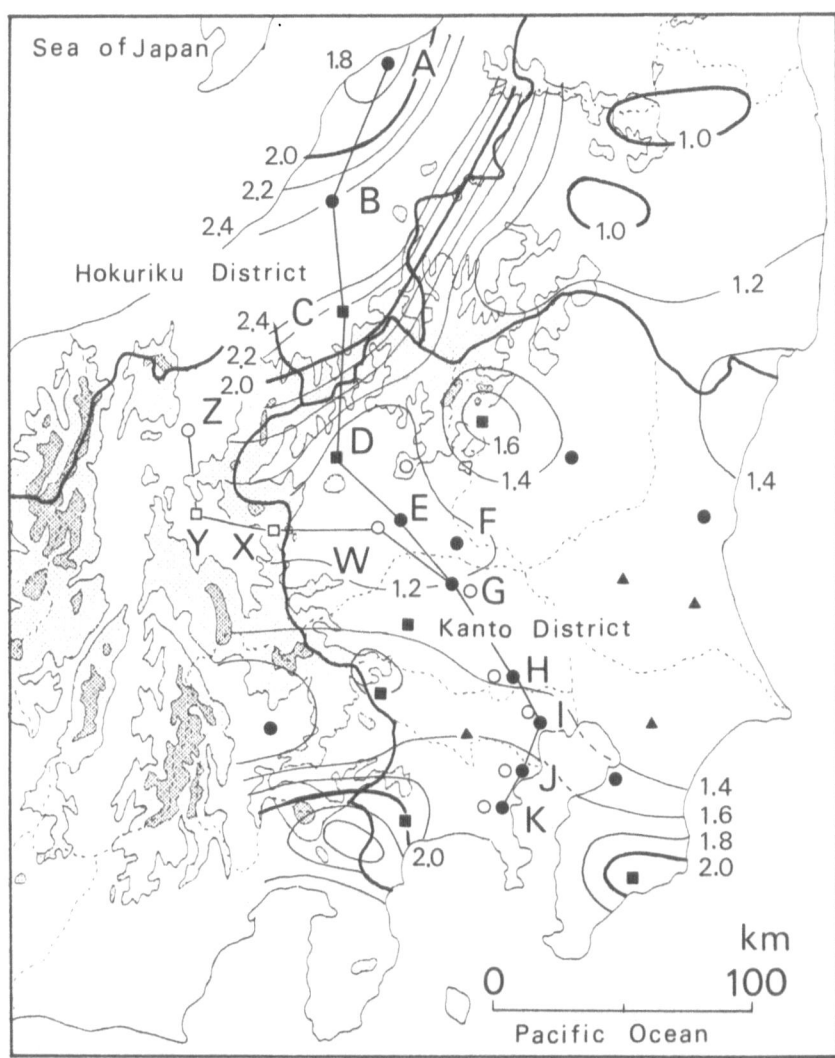

Figure 2-16. Sampling sites of monitoring network for precipitation in Kanto and Hokurikui districts (●, ▲, ■) between September 1984 and August 1985, and for secondary atmospheric pollutants in the summer of 1983 (○, □).

1. Annual Deposition

a. Precipitation

The spatial distribution of the amount of annual precipitation is also shown in Figure 2–16, for the year September 1984 to August 1985, from the data obtained with the Automated Meteorological Data Acquisition System (AMeDAS) operated by the Japan Meteorological Agency (JMA).

A minimum value of 1000 to 1200 mm was found in the central part of the Kanto district, and the annual precipitation was found to increase southward to the coast of the Pacific Ocean and northward to the mountains. In the Hokuriku district along the Sea of Japan the annual precipitation also increased southward from the coast, and showed a maximum of about 2500 mm near the mountainous area forming the climatic boundary between the Kanto district and the Hokuriku district.

This spatial pattern of annual precipitation was similar to that of the latest 30-year climatological mean rainfall for the period 1951 to 1980, and the anomaly of the year 1984–1985 relative to it was within +/− 15%. The annual precipitation in the Kanto district is greater than that in eastern North America and in central Europe.

b. pH

The stations of highest acidity, about 4.6 in annual mean pH, were located in the west side of the central Kanto district as shown in Figure 2–17, corresponding to the northwest of the area of the heavy industrialization and urbanization along Tokyo Bay, where anthropogenic emissions of SO_x and NO_x are high, and the annual mean pH was between 4.8 and 5.2.

The annual mean pH in bulk sampling was higher than that in wet-only sampling by approximately 0.2, as seen in the comparison of these two sampling methods. Therefore, the highest acidity in the precipitation-only sampling was estimated to be 4.4 in pH, slightly higher than that of 4.0 to 4.2 in eastern North America for 1982 (Finkelstein, 1984) and Europe from 1978 to 1982 (EMEP/CCC, 1984).

The value of pH in rural sites of the eastern Kanto district was greater than 5.0. The most unpolluted precipitation was obtained at Nikko, located in the mountainous area about 100 km north of Tokyo, where the annual mean EC was 9.0 μS cm^{-1} and the pH was 5.1. In the Hokuriku district, the annual mean pH was about 5.0, higher than that in the urban sites of the Kanto district, except for Nagaoka where the cause of low pH has not been determined.

c. SO_4^{2-}

Although Japan is surrounded by the sea, the amount of sulfate derived from the sea salt particles was calculated to be less than 10% of the total deposition of sulfate in the Kanto district, except for Kiyosumi, immediately facing the Pacific Ocean. As shown in Figure 2–18, the annual deposition of excess (i.e., nonmarine) sulfate was high, about 3.6 to 4.6 g m^{-2} yr^{-1} in the coastal area of Tokyo Bay and 3.4 to 3.8 g m^{-2} yr^{-1} in the urban sites of the central Kanto district. These values are almost the same as the annual deposition in eastern North America.

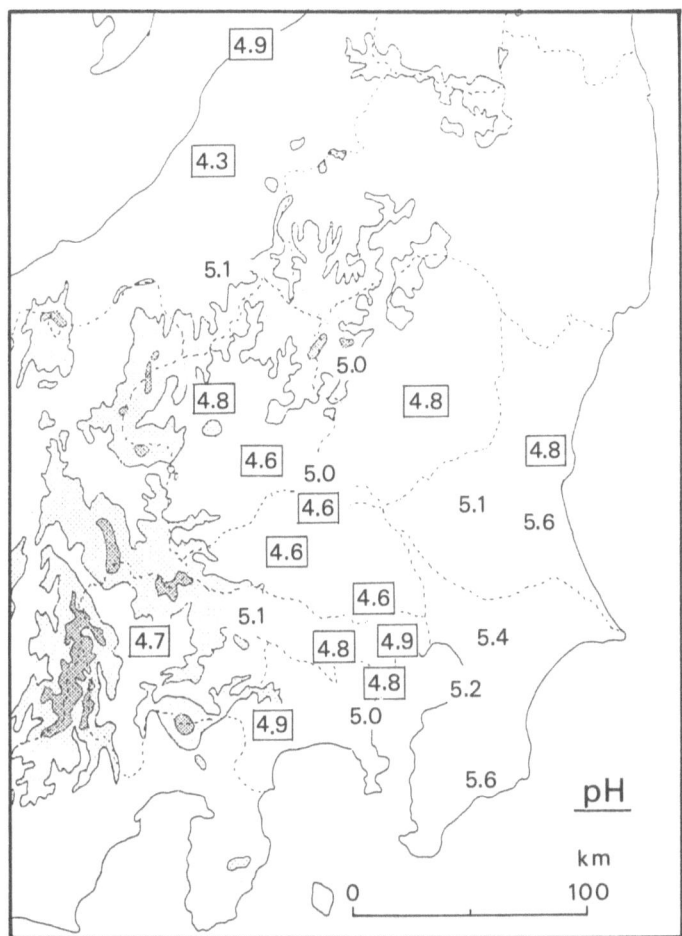

Figure 2-17. Annual mean volume-weighted pH in precipitation around Kanto district (Sept. 1984 to Aug. 1985).

In the Hokuriku district, the contribution of sea salt particles was 13% to 30% of the total deposition of sulfate, much higher than in the Kanto district. This was due to the fact that more than half of the annual precipitation was observed in the winter, when the continental air mass is transported from the northwest over the Sea of Japan.

At two stations in the Hokuriku district, the annual deposition of excess sulfate was higher than that in the urban and industrial area of the Kanto district. But the annual mean concentration of excess sulfate in the Hokuriku district was lower than in the Kanto district. This was caused by the larger amount of annual precipitation in the Hokuriku district. In

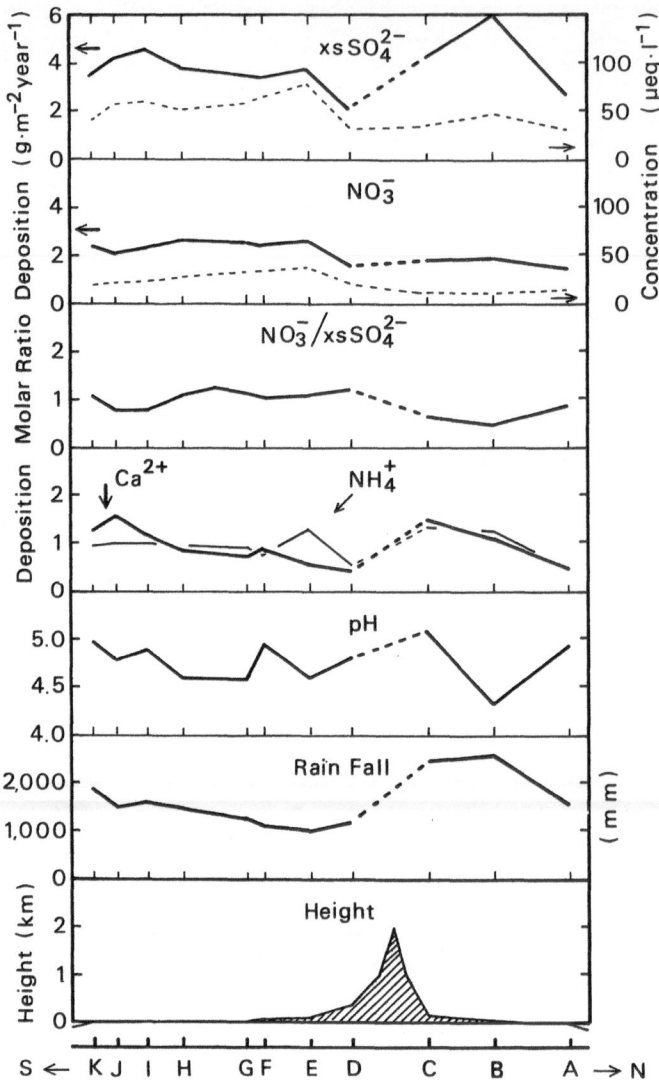

Figure 2–18. Annual depositions of major ions in precipitation, and annual mean values of $NO_3^-/xsSO_4^{2-}$ and pH from the south to the north in Kanto and Hokuriku districts.

the inland areas of the Kanto district, the annual deposition was minimal, half of that in the coastal area of the Kanto district.

d. NO_3^-

The annual deposition of nitrate was high, about 2.1 to 3.1 g m⁻² yr⁻¹, in the west side of the central Kanto district as shown in Figure 2–19, corre-

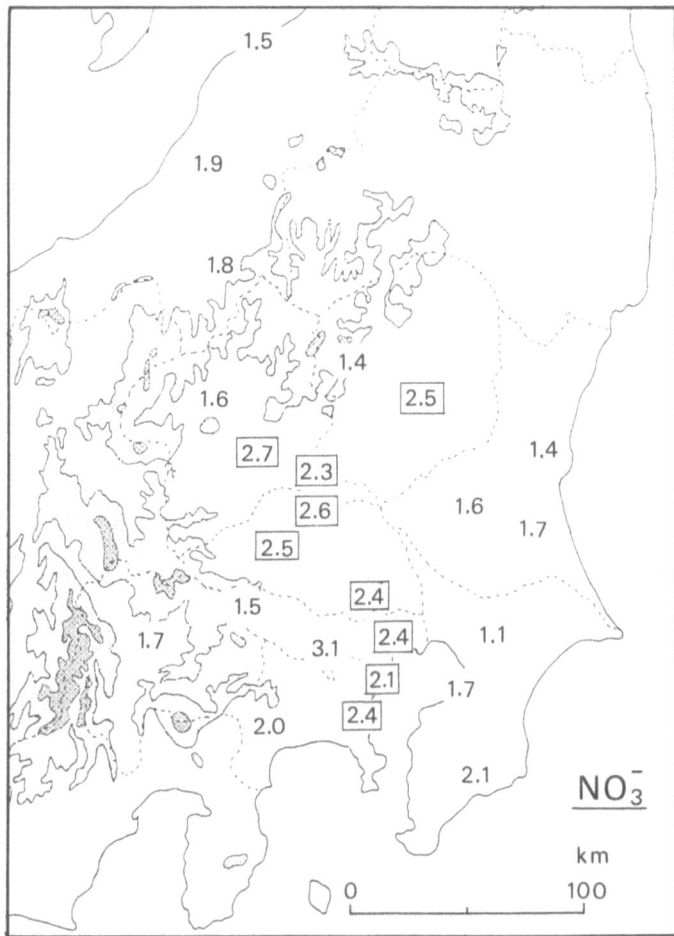

Figure 2–19. Annual deposition of NO_3^- in precipitation around Kanto district (unit: $g\ m^{-2}\ yr^{-1}$).

sponding to the northwest of the coast along Tokyo Bay where sulfate deposition was maximum. The maximum value for nitrates in the Kanto district is nearly equal to that along the Ohio River in eastern North America, and higher than that in Europe.

The annual deposition as well as the concentration of nitrate in the Hokuriku district was smaller than that in the Kanto district, mainly resulting from the amount of NO_x emissions, less than in the Kanto district. At the Higashi-Chichibu station, 850 m above sea level, the annual deposition of $2.6\ g\ m^{-2}\ yr^{-1}$ was nearly equal to the maximum value in the urban area. But the concentration of NO_3^- was lower than that in the urban areas of the Kanto district because of the large amount of precipitation.

e. NO_3^-/SO_4^{2-}

The ratio of nitrate to sulfate in precipitation increased northward from the southern coastal site of Kanto district corresponding to a decrease in the pH value, as shown in Figure 2–18. The value of this ratio was higher than that in eastern North America and Europe. In the Hokuriku district, however, the ratio of nitrate to sulfate was much smaller than that in the Kanto district.

f. NH_4^+

The annual deposition of NH_4^+, about 0.8 to 1.3 g m^{-2} yr^{-1}, showed a maximum in industrial sites of the Kanto district, and a low value of 0.4 to 0.7 g m^{-2} yr^{-1} was observed in rural and mountainous areas. This pattern suggests that the anthropogenic emissions of NH_3 are high enough to increase the concentration and the deposition in urban and industrial areas.

g. Ca^{2+}

As Ca^{2+} in bulk sampling is mainly derived from dry fallout rather than precipitation, it is not a good index for wet-only deposition. The annual deposition of Ca^{2+} however, showed a marked maximum in the urban and industrial areas, and the minimum in mountainous areas was less than half of that. This is the same feature as for NH_3^+. One of the main sources of Ca^{2+} is the dust from roads paved with asphalt including $CaCO_3$ or CaO in the industrial and urban areas.

2. Seasonal Variation

Seasonal values of pH, as shown in Figure 2–20, decreased in spring and summer and increased in fall and winter in the Kanto district. This is a common feature in the areas facing the Pacific Ocean. A decrease of pH in the Hokuriku district, however, was found in fall and winter, which is also a common variation in the areas facing the Sea of Japan.

Seasonal variations in excess SO_4^{2-}, NO_3^-, and NH_4^+ in both districts showed the same pattern as that in the monthly amount of precipitation. The ratio of NO_3^- to SO_4^{2-} was a maximum in summer and a minimum in winter in both districts.

These data suggest that nitrates rather than sulfates are related to the decrease of pH in the Kanto district, whereas the sulfates are related to it in the Hokuriku district.

C. Mechanism of Acidic Deposition in Summer

The molar ratio of NO_x to SO_x emissions in the Kanto district is about 1.9, twice as much as that in eastern North America and Europe. This is

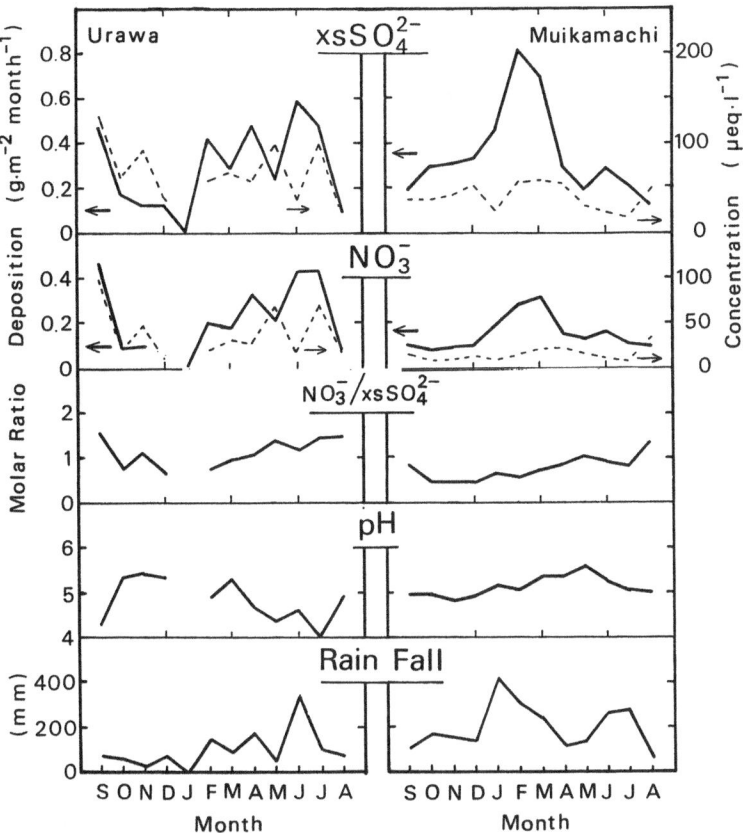

Figure 2-20. Seasonal variation of monthly deposition for xsSO$_4$$^{2-}$ and NO$_{3-}$, and of NO$_3$$^-$/xsSO$_4$$^{2-}$ pH and rainfall, at Urawa (station H in Fig. 2-16) and Muikamachi (station C in Fig. 2-16), September 1984 to August 1985.

the reason why the ratio of NO$_3$$^-$ to SO$_4$$^{2-}$ in precipitation in the Kanto district was higher, as described in Section VA. As shown in Figures 2-10 and 2-11, anthropogenic sources of atmospheric pollutants are concentrated in the southern part of the Kanto district. Tokyo has the broad area sources of NO$_x$ from automobiles. On the contrary, more than half of SO$_x$ and NO$_x$ in three prefectures is emitted from a large industrial zone located at the coast. Therefore, the NO$_x$ concentration in ambient air is higher than that of SO$_x$ in the Kanto district. In the Hokuriku district, however, the amount of NO$_x$ is smaller than that of SO$_x$ and NO$_x$ in the Kanto district.

In Los Angeles, as in the Kanto district, where the amount of NO$_x$ emissions is higher than that of SO$_x$, the ratio of NO$_3$$^-$ to SO$_4$$^{2-}$ in precipitation is also higher than in eastern North America (Stuart, 1984).

In July in the Kanto district, the lowest value of pH was observed as approximately 4.1 among all the seasons, and the ratio of nitrate to sulfate was also a maximum. This suggests that the oxidation of NO_x to HNO_3 rather than that of SO_x to H_2SO_4 is a more important chemical reaction in the acidic precipitation in summer. In eastern North America, however, the ratio of nitrate to sulfate was smallest in summer when the pH decreased, (Calvert et al., 1985) which means the importance of oxidation of SO_x rather than that of NO_x. This difference between eastern North America and the Kanto district of Japan seems to be due to the difference in the amount of SO_x and NO_x emissions.

The summer season in the Kanto district is divided into two periods. The first half is the rainy season called the "Baiu" period, when the Baiu frontal zone lies along the Japan Islands from west to east. The latter half is the period when the Japan Islands are covered with a maritime tropical air mass, and have the highest temperatures and frequent showers from the convective clouds.

In the Baiu period, the value of pH in precipitation is frequently below 4.0 and the ratio of nitrate to sulfate is high, as described in Section IV. Because the intensity of solar radiation is small due to the dense clouds formed over the Baiu frontal zone, oxidation of atmospheric pollutants seems to occur under the aqueous phase rather than with the photochemical reaction in the gaseous phase. Therefore, HNO_3 may be produced in the aqueous phase from the oxidation of NO_x, as well as H_2SO_4 being formed by the reaction of SO_2 with H_2O_2 or O_3 in the aqueous phase.

In the second period of summer, however, the concentrations of photochemically produced pollutants, such as nitric acid gas, sulfate aerosol, and ozone from the industrial and urban areas, are higher than at other seasons. In the daytime, these secondary pollutants are transported northward from the coastal region of the Tokyo metropolitan area by extended sea breezes, as shown in Figure 2–21 (Tsuruta, 1986). In the western area of the central Kanto district, the molar ratio of total nitrate ($NO_3^- + HNO_3$) to sulfate in ambient air is about 1.5, as shown in Figure 2–22 (Tsuruta, 1987). This value was almost the same as that in precipitation of summer in the Kanto district, as shown in Figure 2–23.

The rate of production of HNO_3 gas in the photochemical reaction is greater than that of H_2SO_4, and the dry deposition velocity of nitric acid gas is also greater than that of fine particles in which sulfate is predominant. Therefore, the dominant mechanism in summer is that HNO_3 gas, as well as sulfate aerosol, generated in the daytime through photochemical reaction, is scavenged by the cloud droplets or raindrops and is deposited through precipitation or in dry deposition. As the lifetime of nitric acid gas seems to be shorter than that of sulfate aerosol, the deposition of total nitrate is supposed to be limited to a region near the source area, in contrast to the case of sulfate aerosol deposition.

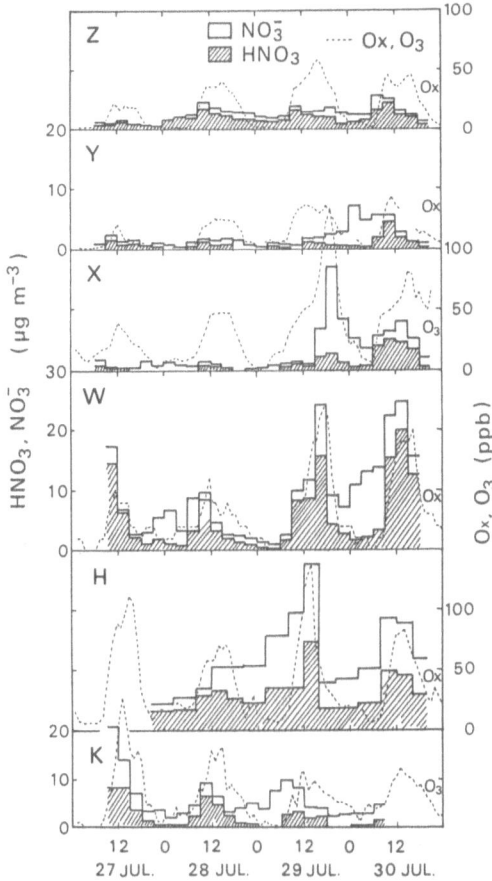

Figure 2-21. Diurnal variation of O_3, O_x, HNO_3, and NO_3^- aerosol at six stations shown in Figure 2-16 (July 27 to 30, 1983).

In conclusion, nitric acid in the gas and aqueous phase is the main contributor to the decrease of pH in summer in the Kanto district, as described in Section V B.

In the Hokuriku district, however, the mechanism of decrease of pH in fall and winter is different from that in the summer of the Kanto district. The increased SO_4^{2-} production from SO_2 in the aqueous phase is primarily responsible for the decrease in pH. The sources of SO_4^{2-} in precipitation are not only SO_2 in the local sources of the Hokuriku district but also that from remote areas of the Asian continent. The air mass moving from the northwest is drastically humidified in crossing the Sea of Japan, favoring oxidation of SO_2 to SO_4^{2-} and the coastal area along the Sea of Japan has the heavy precipitation, largely snow, in fall and winter. Of course other factors, such as the concentration of metallic catalysts from

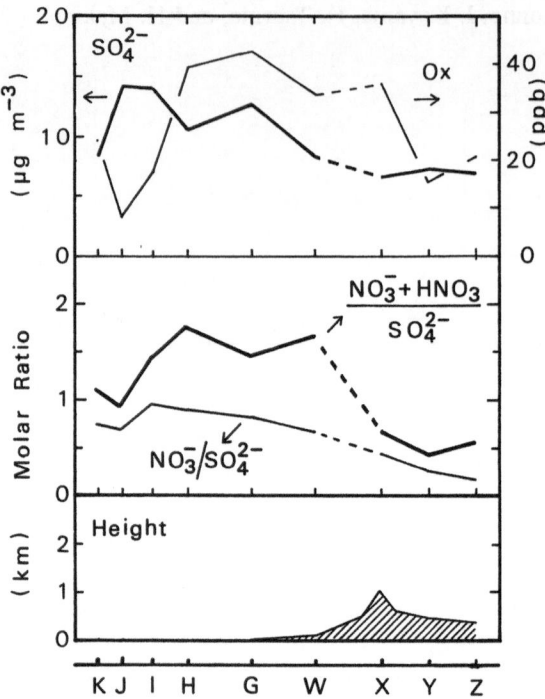

Figure 2–22. Molar ratio of total nitrate ($NO_3^- + HNO_3$) to sulfate and the concentration of sulfate aerosol and O_x in ambient at nine stations shown in Figure 2–16 (July 27 to 30, 1983).

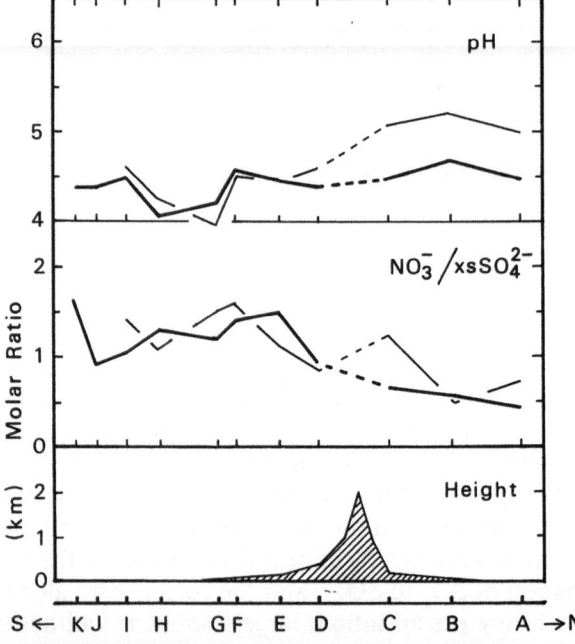

Figure 2–23. Molar ratio of nitrate to excess sulfate and pH in precipitation in July 1985 from the south to the north in Kanto and Hokuriku districts.

automobile and stack gases and solar radiation also affect the rate of sulfate production.

As the study of the mechanism of the oxidation process, as well as of the transport and scavenging of atmospheric pollutants has just begun for both these districts of Japan, a detailed understanding should be available in the near future.

VII. Effect on Vegetation (Japanese Red Cedar)

In Japan there is a strong tradition for the protection of Japanese red cedar (*Cryptomeria japonica* D. Don) and pine, where the trees are virtually regarded as sacred. From a few hundred to more than a thousand years ago the original mountain forests and the forests within the precincts of the shrines and temples were left with large trees. Now it is said that within a short time these trees are weakening, for various reasons.

In central Tokyo there has been concern about the weakening of trees by factory pollution since 1940. In a brief time, the damage to trees spread greatly, but there has been a partial recovery of the "keyaki" (*Zelkova serrata* Makino) and other trees in the park areas of central Tokyo (Oohashi, 1987). However, recently in the areas over and surrounding the Kanto Plains the advance in the deleterious effects on the cedar and other trees has come to be understood (Government Forest Experiment Station, 1986; Sekiguchi, Hara, and Ujiiye, 1985). Beginning in 1950, and becoming acute in 1960, in the Kansai and areas surrounding the Inland Sea, the wide-area deleterious effects on the red pine have advanced. More recently this problem has spread to the Tohoku region. Clearly understood as the primary cause is insect damage due to the "matsunozaisen-chu" (*Bursphelenchus xylophilus*), carried by the long-horned beetle (*Monochamus, alternatus; Cerambycidae*) (Uemura, 1985). The reasons are not fully understood. Even in the Fukui Plains of central Honshū's Sea of Japan area there has been noticeable damage to the cedar and other trees from 1970.

A. Fifty-Year Changes of Vegetation in Tokyo

Natural flora in Tokyo were "suda-shii" (*Shiia sieboldii* Nakai), white oak (*Quercus myrsimaefolia* Blume), and laurel forests. After human cultivation changed the arable land to fields for rice and vegetables, the remaining hills became second-generation forests of "kunugi" (Japanese chestnut oak) (*Quercus acutissima* Carruth) and "konara" (*Quercus serrata* Thunb). Also in various areas temple groves were left. In these groups the Japanese red cedar and pine, which are a species requiring good sunlight, were planted and preserved. However, many changes occurred with the rapid expansion of factories in dense urban areas.

In the 1940s with the emphasis on national development and a strong military, and the ensuing world war followed by difficult conditions, there was not great concern for damage to vegetation. There did not seem to be a consensus regarding the problem of damage to areas surrounding factories. So at that time a large number of Japanese red cedars in the central Tokyo area were wiped out.

In the 1950s there was great injury done to the cherry trees in the Sumida River area and to the weeping willows in the Ginza area. The air pollution spread, but this problem continued in an atmosphere of uncertainty. In Tokyo for some years after 1969 great irritation episodes from photochemical smog arose. Although there were some adults affected, the majority of the complaints were from middle school and high school girls. There was no understanding of the cause, but these were sudden attacks of trembling and gasping for breath. At this time there was great damage to vegetation in the various sections of Tokyo. There was also severe damage to the "keyaki" in the deciduous broad-leaved forest. In the summer days with little wind and a sky that sparkled as if with dancing particles of aluminum, the glittering weather brought a rapid falling of leaves. The species most severely affected were the Himalaya cedar (*Cedrus deodora* Loud), white oak, cypress "keyaki," "mukunoki" (*Aphananthe aspera* Planchoy), and the cherry (National Institute of Resources, Science and Technology Agency, 1970). After this event, the great damage to vegetation from air pollution decreased markedly. Recently attention has been directed to the damage to the Japanese red cedar in the western district.

B. Weakening of the Japanese Red Cedar in the Northern Kanto Plains

Surveys of the weakening effects on the Japanese red cedar were quickly taken by the forestry experimental station in the central northern Kanto Plains. Included here is an overall explanation of the results gleaned from a reading of those reports. The reports were limited to surveys of actual deleterious effects on the Japanese red cedars in the northern Kanto Plains. The area distribution of these effects as of 1973 is shown in Figure 2–24. No significant change was seen in 1985 from the 12-year earlier survey of air pollution weakening of the Japanese red cedars, as shown in Figures 2–24 and 2–25 (Government Forest Experiment Station, 1986; Yambe, 1978). However, the area of damage extended significantly to the northwestern Kanto Plains. Also, the zones of great, moderate, or light damage were notably confined to narrower boundaries.

C. Growth Conditions for Japanese Red Cedar in Tokyo

Natural changes in the trees in the Nature Educational Park in Meguro, Tokyo, are shown in Figure 2–26 (National Institute of Resources, Science and Technology Agency, 1971). The bad withering of the pines, and the even greater deterioration of the Japanese red cedar is shown by the fact that only one greater than 40 cm in diameter remains. Even this one

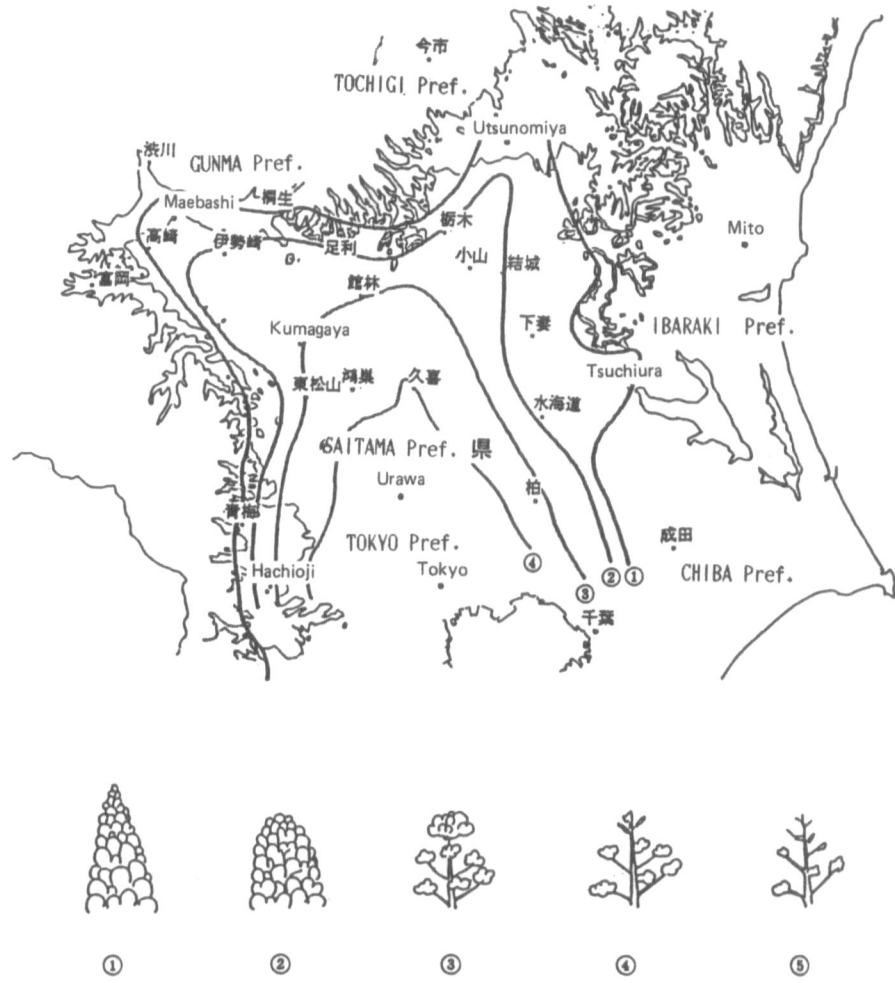

Figure 2–24. Regional distribution of damage to Japanese red cedar (1985).

is defoliated to a point approximately 3 m below the top. This is all that remains of some 60 trees there from the 1950 to 1960 period.

Figure 2–27 shows the degree of weakening of Japanese red cedars in the parks in Tokyo and temple and shrine grounds (Koyama and Oohashi, 1988). The standard for the degree of deforestation is patterned according to the Yambe method. This is a summary of the information from the fall 1986 preparatory surveys. In general, the Japanese red cedar weaken markedly with increased age and size, and the crowns die. The result is that there are almost none remaining in central Tokyo with a diameter greater than 60 cm. Within the capital city the damage factor to young Japanese red cedars was 2 to 3 on this scale (see Photo B).

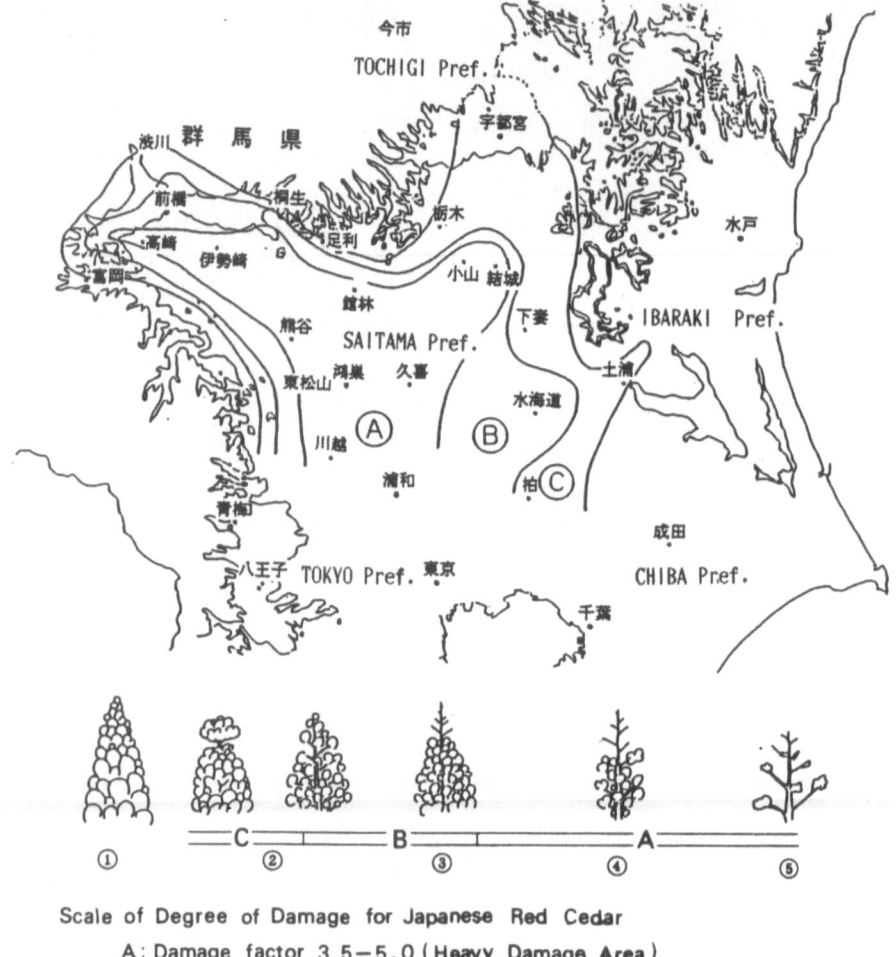

Scale of Degree of Damage for Japanese Red Cedar
    A : Damage factor 3.5−5.0 (Heavy Damage Area)
    B :   "   "   2.5−3.0 (Medium   "   ")
    C :   "   "   1.5−2.0 (Light   "   ")

Note: In comparing this 1985 survey with the 1973 survey (see Fig. 24),
      similarities in Japanese Red Cedar damage are seen. However, the
      damage factor scales are slightly different due to the progressive
      tree damage.

Figure 2–25. Regional distribution of damage to Japanese red cedar (1987).

In the central plains there was a large amount of cutting of damaged trees in temple and shrine groves in the 1950s and 1960s. In this region the damage factor to trees 40 cm and over in diameter was 3 (see Photo C). In the western plains along the mountains is an area with large Japanese cedars 80 cm in diameter and over. The damage to these trees was

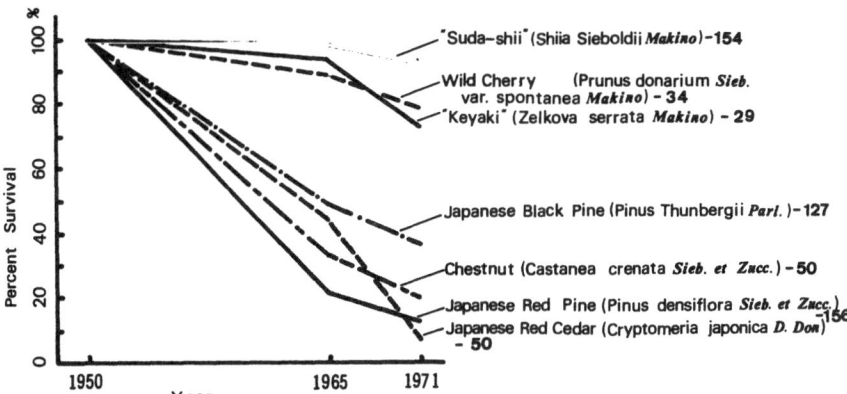

Figure 2-26. Survival study on various tree species over a 21-year period—The Institute for Nature Study, Tokyo (Meguro). Note: The total number of trees included in the survey is indicated by the number to the right of the name of that species.

great, with a factor of 4 to 5 on the scale (see Photo D). On Mt. Mitake at 900 m. elevation there was notable damage to these trees in the shrine grounds, with a damage factor of 4 to 5, while in its vicinity similar large trees showed a damage factor of 3 to 4 (see Photo E).

The source of the weakening of these Japanese red cedars is the altering of the area by humans, such as a rise or fall in the water table due to concrete walls or cuts in banks. Also, the hardening of the surface due to crowds walking in the temple grounds has had a bad effect. Gley soil, tree disease, pests, and climatic changes must all be considered. At present the source of tree decline is not clearly understood.

D. Weakening of Japanese Red Cedar in the Fukui Plain

There has been notable weakening of the Japanese red cedar and "keyaki" in the Fukui Plains, along the Sea of Japan coast in central Honshū. The Fukui Prefectural Pollution Experts Council and Awara Cho are conducting surveys to determine the causes of severe tree weakening in the Fukui Plains (Awara City Government, 1984; Fukui Prefectural Government, 1986). Here these results are touched on only briefly. It was found that in the same area with the same survey methods the sources of pollution are different. The Fukui Plains comprise an area 20 km east-west and 40 km north-south, with a coastal side extending 20 km south-southwest from the northwest corner. The prefectural survey analysis included a survey of the flora and growth activity, dendritic rings, exhumed root structure, chemical analysis of leaf content, soil analysis, and so on. In addition, the Awara Cho survey included a health survey of students. Skillful use was made of pollutant particulate data, results of prefectural measurements,

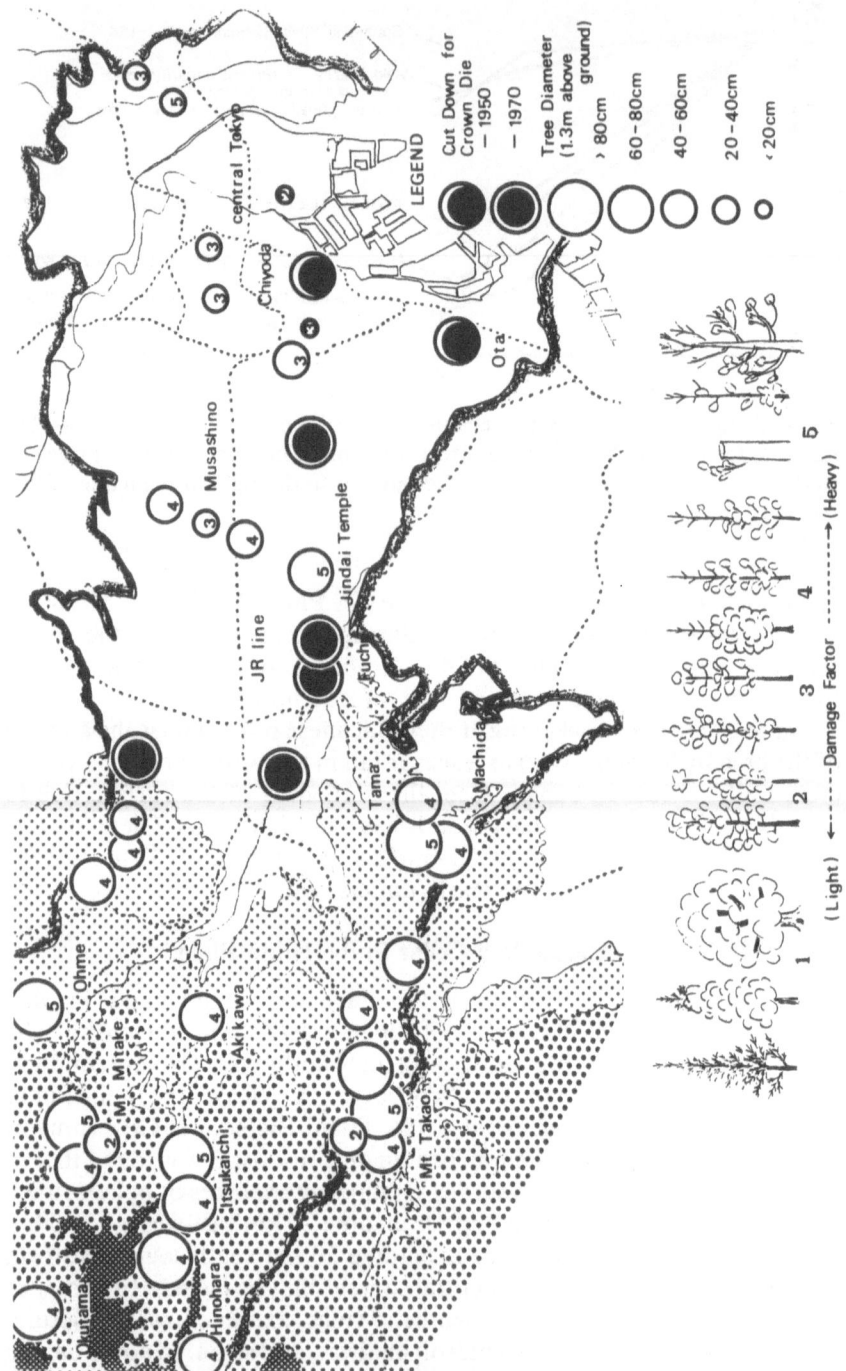

Figure 2–27. Degree of damage to Japanese red cedar in the Tokyo area as of October 1987.

B. Japanese Cedar in Central Tokyo (Katsushika-ku)

meteorological data, and a variety of data from public agencies. The prefectural report states, "The weakening of the Japanese red cedar in the center of the lower central region of the Fukui Plains can be seen to have begun in the 1960s as a result of changes in floral growth environment factors such as soil and water conditions." On the other hand the Awara Cho survey report states, "There is no doubt as to the fact that the increase in damage to vegetation is due to atmospheric pollutants including SO_x." Both reports recognize the presence of the damage sources given by the other, but claim the sources given in their own to be paramount. Future records will prove which analysis is most correct.

E. Summary

Deleterious effects on vegetation were first noted in many portions of the Kanto area. Notable damage is found also in areas where man-made

C. Japanese Cedar in West-central Tokyo (Jindai Temple)

pollution is not thought to have advanced, such as the Hokkaido wilderness areas and the Lake Akan area. Opinions as to whether these effects are the result of human activity or the result of natural phenomena vary with the expert commenting.

A Japanese red cedar standing alone is said to be particularly vulnerable. Present surveys in the northern Kanto area bear this out. However, are there not many examples of trees hundreds or even thousands of years old that stand alone without any restrictions? A splendid example is the growing and very large Japanese red cedar in Taiwa Cho, Miyagi prefecture (see Photo F). On small islands, Japanese red cedar experiencing strong sea winds do not grow branches on the windward side but have

D. Japanese Cedar in Western Tokyo (Tama area)

healthy deep-green needles. Many trees have recovered from great damage caused by typhoons or lightning. In Japan's southern Kyushu on Yaku Island there is a Japanese red cedar thought to be approximately 7000 years old.

The terrible damage to vegetation from man-made pollution including acidic precipitation should be recognized as a fact. The world's tallest tree (112.1 m), a California redwood, also shows damage to its crown (Japan Broadcast Publishing Co., Ltd., 1987).

A veritable mountain of not-yet-understood problems of man-made pollution, including those of acidic precipitation, remain to be solved. If problems such as long-distance transport of pollutants and other long-range problems are not quickly solved, great damage to vegetation will take place. Damage to vegetation is not simply that of loss of valuable lumber. The great benefit of oxygen production and other essentials to our peace of mind are the unseen hidden factors.

Surveys and the best possible attempts at corrective action must be made soon to alleviate the effects of acidic precipitation on soils, trees, agricultural products, surface water, and aquatic life (Kon'no, 1986).

E. Japanese Cedar in Okutama Area

F. Japanese Cedar in Miyagi Prefecture (Strong, Old Cedar)

VIII. Summary

In Japan, surveys of pollution damage began with the dramatic problem of injury to individuals. For this reason surveys began primarily in the Kanto region.

Starting in 1975 precipitation sampling methods shifted to the measurement of quite small amounts of pollution components to evaluate the damage to the human body. For these measurements automatic and/or manual collection equipment is used involving a 30 cm or greater diameter funnel capable of collecting the initial 3 to 5 mm of precipitation in 1 mm increments. With this method of collection it became easy to solve the problem of analyzing the pollution mechanism by noting the great change in the constituent components of successive 1 mm precipitation samples. This method is still used by a large number of researchers even though there are now almost no reports of pollution injury to people.

Since 1980 the method of collecting the total precipitation for each precipitation period is used by many researchers. The effects on flora are being studied through surveys of long-term and large-area long-distance transport of pollutants in areas all over the country. Apart from these methods, an acidic precipitation collection device has been developed using a filter method especially for the purpose of studying the long-term effects of acidic precipitation pollutants on trees. With this method it is possible to collect the atmospheric precipitants such as rainfall and particulate fallout over periods of one or two weeks or a month. A reason for applying this filter method acidic precipitant collection device is to simultaneously measure precipitation quantity, prevent, for the most part, insects, and so on, from falling into and decomposing in the device, and thereby stabilize the concentration of the collected constituent pollutants. In northern Europe and North America many are deeply concerned about the spread of acidic rain damage to forests, lakes and marshes, and in Japan there are those who are similarly concerned. However, there are others with a different view who emphasize the fact that since Japan has a lot of volcanic ash and the trees are strong in the presence of acidic soils, there should be little concern for the low pH resulting from precipitation. However, there have been research reports that there is decay caused in the Japanese red cedar by large amounts of acidic precipitants, so there is a rapidly developing identification with the North European type of acidic rain pollution problem. Because of this, methods of using acidic particulate collection devices for the purpose of monitoring the long-term effects have begun to attract attention in Japan as well. At present, long-term wide-area acidic rain surveys have been undertaken by the National Environmental Agency and the EPMPH. In Japan, systematic surveys of acidic rain began because of the problems of eye irritation and bodily injury.

A. Cases of Personal Bodily Injury

From 1973 to 1975 there were cases of eye irritation and skin irritation at times of a very light drizzle or extremely light rainfall in Shizuoka and Yamanashi prefectures and throughout the Kanto area. At that time there were some tens of thousands of people coming to the city halls and area health centers with such complaints. Along with this, damage to cucumbers, eggplant, onions, tobacco, beans, taro, and other plants was observed. Studies were begun by the Wet Air Pollution Research Group under the environmental protection air pollution department, established by the EPMPH, to find the source of this pollution. It was impossible to investigate the rainfall just at the time and location where this pollution unexpectedly appeared, but examining the rainwater at nearby locations showed a low pH with high concentrations of SO_4^{2-} and NO_3^-. However, without having solved the problem, the pollution damage disappeared in five or six years.

B. Survey of Pollution Mechanism

A survey of wet air pollution was begun by the environmental agency in the Kanto area as a result of these many cases of bodily irritation due to pollution. A five-year survey begun in 1975 by the environmental agency was mainly centered on gaining information on the concentrations of pollutants in the area. The survey measured concentrations of the pollutant constituents in rainfall samples collected in 1 mm increments. The survey covered a 12-day period at the end of June and beginning of July each year. The target area of the survey included Tokyo and all prefectures of the Kanto area. Starting in 1980, the Kanto EPM picked up the task of conducting the surveys. Following this, surveys have been conducted in the central Honshū district comprising Yamanashi, Nagano, Shizuoka prefectures, and Yokohama as the designated city in Kanagawa prefecture, as well as Niigata prefecture, and Fukushima prefecture in the Tohoku district. This results in a survey coverage involving 11 prefectures plus the Tokyo metropolis and the major city of Yokohama. The surveys involve an area of approximately 300 km from north to south and 200 km east to west.

C. Yearly Surveys of Acidic Precipitation and Other Particle Fallout Pollutants

A cooperative effort combines the survey results on acidic fallout by the EPMPH and those by the Environmental Agency, although the former involves the yearly June surveys only and the latter a comprehensive five-year plan. The first extended-area yearly surveys of acidic fallout by the environmental agency began in 1983. Now the EPMPH is also conducting year-round surveys in the Kanto district. Both agencies are using the filter collection method. In experiments reported by Tamaki the amounts of SO_4^{2-}, NO_3^-, and so on, in one total precipitation are the same without regard to the diameter of the collection funnel, presence or absence of a filter, or initial presence or absence of water in the case where a dust jar is used. But the amounts of NH_4^+ and of insoluble components are different. Acidic fallout surveys conducted by the Environmental Agency have comprised a five-year plan from 1983 in 14 areas over the entire country, large area surveys in 34 areas over the country, and from the fall of 1984 to the summer of 1985, surveys in 24 areas of the Kanto, Niigata, and Yamanashi districts. Expanding and continuing from the EPMPH Kanto area surveys, acidic precipitation surveys have since 1987 been conducted from the Hokkaido-Tohoku block to the Chugoku-Shikoku block. In all these, the filter method of precipitation collection is used. The collector developed by Tamaki et al. is used to collect the precipitation. This experiment involves using vessels of different diameters, filter or no filter, particulate fallout collector, and containers with and without initial water. Using this method the NH_4^+ concentration and that of insoluble components was shown to vary with the method of collection. But the SO_4^{2-} and NO_3^- concentrations were consistent regardless of method. The pH measured as an average in a series of single precipitation measurements was less by no more than 0.5 than that measured using the filter method over the same period.

D. Additional Surveys by Prefectural and City Agencies

Since 1987 many acidic fallout surveys have been conducted by a large number of prefectures and major cities as well as by the Chugoku-Shikoku and Hokkaido-Tohoku Groups of the "Zenkooken." The purpose of these surveys, as with those by the EPMPH, was originally to investigate the bodily irritation from pollution, but now included are the evident effects on trees and plant life. The results of these surveys are published for the most part in the annual pollution research reports and information publications of the various prefectures and major cities.

E. Surveys of Damage to Plant Life

Withering of the crowns of the Japanese red cedars in the Kanto Plains area was observed and reported in 1985 by Sekiguchi. Following this, the Forestry Agency Forest Preservation Research Center initiated a detailed survey over the entire Kanto area. In 1986, an intermediate report was issued revealing the debilitating effects of this pollution. In a separate survey conducted by metropolitan Tokyo, preliminary observations showed virtual annihilation in the central area of a fir species (*Abies mill*) similar to that abundant in northern Europe, and very deleterious effects on the

Japanese cedar in the western section. The order Fir (Cnififerales), family Cedar, Japanese red cedar is showing conspicuous weakening in the western area of Tokyo.

IX. Further Topics for Investigation

Research on acidic precipitation in Japan is only in its first stages. The explanation of mechanisms resulting in low pH precipitation and the causes of the long-range effects on forests must yet be found. Along with this, the long-distance transport of pollutants must be studied and the means for solving the worldwide pollution problem must be sought. The monitoring and modeling of pollution phenomena must extend beyond national boundaries through the cooperation of many researchers. The checking of pollution sources and the public dissemination of information on the subject is a major concern.

Acknowledgments

We acknowledge with deep thanks the EPA reports and materials concerning the work and data of local government agencies and researchers from all over Japan, from which we have drawn and which we have attempted to summarize. The permission to use these reports is greatly appreciated.

Special thanks to Dr. T. Okita, formerly of the National Environmental Research Institute and now of Oberlin University, Japan, for his careful review and pertinent advice.

Also, thanks to Mr. D. Lyon and Mrs. Y. Mukaihira for their help in translating the manuscript.

References

Air Transport Statistics Yearbook. 1985. All-Japan Air Transport and Service Association.

Awara City Government (Fukui Prefecture). 1984. *Effects of air pollution on the environment and residents in Awara Machi.*

Calvert, J.G., A. Lazrus, G.L. Kok, B.G. Heikes, J.G. Walega, J. Lind, and C.A. Cantrell. Chemical mechanisms of acid generation in the troposphere. *Nature* 317: 27–35.

EMEP/CCC. 1984. Summary report from the chemical coordinating center for the second phase of EMEP, EMEP/CCC—report 2/84. (E)*

EPMPH. 1975. *Report on what is called "acid rain" (wet air pollution), 1974.*

EPMPH. 1976. *Report on what is called "acid rain" (wet air pollution), 1975.*

EPMPH. 1983. *Report on wet air pollution, 1981.*

EPMPH. 1987a. *Report on wet air pollution, 1984.*
EPMPH. 1987b. *Report on wet air pollution, 1985.*
EPMPH. 1987c. *Report on wet air pollution, 1986.*
Finkelstein, P.F. 1984. The spatial analysis of acid precipitation data. *J Climate and Applied Meteorology* 23: 52–62. (E)*
Fujita, S. 1987. *Japanese acid rain.* Meteorological study technical note no. 158: 437.
Fukui Prefectural Government. 1986. *Studies of the cause of damage to Japanese cedar* (Cryptomeria Japonica *D. Don) in the Fukui Plains.*
Fukuoka, S., T. Komeiji, and T. Oadira. 1976. Development of the automatic measurement apparatus for rainwater components and its results. *J Japan Society Air Pollut.* 10: 367.
Government Forest Experiment Station. 1986. *Survey report on the effects of acid precipitation in causing Japanese red cedar* (Cryptomeria Japonica *D. Don) grove decline in the Kanto area in 1985.*
Hourai, S., Ootu, M., Takeyama, E., Minamizono, H., and Yamagawa, T. 1987. Specific high pollution of rainwater from eruptions of Sakurajima volcano. *The Proceedings of the 28th Annual Meeting of the Japan Society of Air Pollution,* Tokyo.
Hukui, E., et al (ed.). 1985. Map of Japanese and world climates. Tokyodo Shuppan Co., Ltd., Tokyo.
Japan Broadcast Publishing Co., Ltd. (NHK). 1987. *The miracle planet,* no. 3, 130–137.
Japan Environmental Agency. 1979. Summary report on wet air pollution by the Photochemical Products Investigating Committee (Wet Air Pollution Subcommittee).
Japan Environmental Agency. 1987a. Interim report on acid rain survey by investigative survey group (Acid Rain Countermeasures Committee, Air Pollution Subcommittee).
Japan Environmental Agency. 1987b. Environmental white paper.
Japan Statistical Yearbook. 1987. Bureau of Statistics, Management and Coordination.
Komeiji, T., I. Koyama, and M. Kadoi. 1983. *The Proceedings of the 28th Annual Meeting of the Japan Society of Air Pollution,* Mie Prefecture, 549.
Komeiji, T., M. Sawada, T. Odaira, K. Hirosawa, and M. Kadoi, 1975. *Research on precipitation.* Annual report of the Tokyo Metropolitan Research Institute for Environmental Protection, 104–112.
Korokawa, M., T. Asoh, H. Mimura, K. Aihara, M. Kaneko, Y. Saiki, N. Nishiyama, K. Himi, and S. Kanno. 1975. Concentration of formaldehyde in rainwater and the eye irritation due to aldehyde group. *J Japan Society Air Pollut* 10: 628.
Koyama, I. 1986a. *Analysis by areas of precipitation constituents in the Kanto region.* The National Institute for Environmental Studies. Report of First National Institute on Environmental Protection Research, 12–13.
Koyama, I., T. Komeiji, H. Onozuka, T. Ohhoasi, and H. Ise. 1983. Improvement of dust fall apparatus for reference area. *The Proceedings of the 28th Annual Meeting of the Japan Society of Air Pollution,* Mie Prefecture.
Koyama, I., and T. Oohashi. 1988. The dieback of "sugi" (Japanese red cedar) in Tokyo (photograph). Annual report of the Tokyo Metropolitan Research Insti-

tute for Environmental Protection, 69–74.

Koyama, M. (ed.). 1980. Soil map of Japan, Encyclopedia of Plant Nutrition and Soil Science. Tokyo: Yokendo Co., Ltd. (E + J)*

Makino, H., and M. Kaneko. 1984. Studies on acid rain (part 2)—contribution of atmospheric H_2So_4, HNO_3, and HCl to acid rain (pH <4.0), *Bulletin of Kanagawa Prefectural Environmental Center*, 1–7.

Metropolitan Police Traffic Yearbook. 1986. Japan Metropolitan Police Department Traffic Bureau.

Miyake, Y. 1939. Chemistry of rain. *J Meteorological Society of Japan*, 11, 17, 20–38.

Miyoski, H., N. Shimada, M. Ishikawa, and N. Date (eds.). 1983. Glossary of terms used in soil science and manure studies. Tokyo: Nousanson Bunka Kyokai.

National Institute of Resources, Science and Technology Agency. 1970. *Survey of Tokyo wooded areas.*

National Institute of Resources, Science and Technology Agency. 1971. *Twenty-one year comparison of plants and animals.* Institute for Nature Study, National Science Museum, Tokyo.

Oden, S., et al. 1976. The acidity problem—an outline of concepts. *Water Air Soil Pollut.* 6: 137–166. (E)* Also see Tomlinson, G.H. 1983. Air pollutants and forest decline. *Environmental Science & Technology* 17: 146–256. (E)*; *Scientific American* 241: 4, 3–11 (Oct. 1979). (E)*; *Chemical & Engineering News,* Nov. 22, 1976 (E)*; The OECD program on long-range transport of air pollutants. 1979. OECD. (E)*; Swedish Ministry of Agriculture. 1983. *Acidification.* (E)*; EMEP/CCC, 1984: *Summary Report from the Chemical Coordinating Center for the Second Phase of EMEP,* EMEP/CCC—report 2/84. (E)*

Ohta, S., T. Okita, and C. Kato. 1981. A numerical model of the acidification of cloud water. *J Meteorological Society of Japan* 59: 892–901.

Oohashi, T. 1987. *Studies on "Keyaki" vitality.* Japan Environmental Agency, Symposium no. 17.

Sekiguchi, K., K. Kano, and A. Ujiiye. 1983. Acid rain (pH 2.86) in Maebashi. *J Japan Society of Air Pollut.* 18: 1–7.

Sekigucki, K., Y. Hara, and A. Ujiiye. 1985. Dieback of Japanese cedar and distribution of acid deposition and oxidation in Kanto district of Japan. *Environmental Technology Letters* 7: 263–268. (E)*

Stuart, J.A. 1984. Acid deposition in the south coast air basin, South Coast Air Quality Management District, El Monte, CA. (E)*

Sugawara, K. 1948. Distribution of salt particles in air. *J Japanese Chemistry* 2(8): 9–13.

Tamaki, M., and T. Hirano. 1986. Collection methods for deposition of atmospheric pollutants.

Tokyo Astronomical Observatory (ed.). 1987. Chronological scientific tables. Tokyo: Maruzen.

(E)*: reference is in English

(E + J)*: reference has both English and Japanese versions

All other references are in Japanese only at this time.

Tokyo Metropolitan Government. 1985. *Report on acid precipitation in the Kanto region.*

Tokyo Metropolitan Government. 1987. *Functions of the Tokyo port and harbor bureau.*

Tonooka, Y. *Recent trends in NO$_x$ and NMHC emission technical meeting on photochemical air pollution modeling and its applications.* National Institute for Environmental Studies, November 11, 1987. (E)*

Tsuruta, H. 1986. *Transport and transformation of atmospheric pollutants to the inland from the Tokyo metropolitan area by extended sea breeze.* WMO symposium on air pollution modeling and its dispersion, held in Leningrad. (E)*

Tsuruta, H. 1987. *Acid generation mechanism in Japan.* The 28th annual meeting of the Air Pollution Society of Japan, held in Tokyo.

Uemura, S. 1985. *Pine wilting.* Consumers Union of Japan, 14–28.

Yambe, Y. 1978. *Declining of trees and maicrobial florae as the index of pollution in some urban areas.* Bulletin of the Forestry and Forest Products Research Institute, no. 301, 119–129.

Yoshino, M. 1978. *Climatology.* Tokyo: Taimeido.

Acidic Precipitation and Forest Damage Research in Austria

R. Orthofer* and K. Kienzl†

Abstract

A review of Austrian research projects and results of acidic precipitation and forest damage research is presented. Data are included for country-wide annual emission rates and the forest damage situation. Wet and bulk deposition data and air quality characteristics are reported for selected monitoring sites.

With 86% of pH recordings below 5.1, acidic precipitation in Austria is frequent. In characteristic open-field rural/background stations precipitation pH is 4.2 to 4.8; annual deposition rates are about 7 to 15 kg ha^{-1} yr^{-1} SO_4–S (as S) and 3.0 to 7.0 kg ha^{-1} yr^{-1} NO_3–N (as N). However throughfall deposition in forests is higher, especially for sulfur: Data from 11 research sites indicate 15 to 54 kg ha^{-1} yr^{-1} SO_4–S (as S), 5 to 24 kg ha^{-1} yr^{-1} NO_3–N (as N), and 12 to 49 kg ha^{-1} yr^{-1} total N. Because there are only a few sensitive regions, no acute damage to aquatic ecosystems was reported until 1987. Nationwide surveys from 1983 to 1986 showed the majority of rivers and streams were not acidified in terms of pH or alkalinity, but confirmed there is a considerable risk for sensitive alpine lakes. On the other hand, since the early 1980s vast visible forest damage has been reported. Annual nationwide surveys, which were established in 1984, indicated a peak damage intensity in 1986. In 1989 visible damage was noted for 25.4% of the trees. In terms of affected areas this is 18.5% of the total forests. A research program aimed at the problems of forest damage was set up, focused mainly on diagnosis and causal analysis. Most research work is concentrated on three locations, each with different air quality characteristics. Present results indicate that the damage is not uniform and cannot be attributed to any one cause. In general it is

*Environmental Planning Department, Austrian Research Centre, A-2444 Seibersdorf, Austria.

†Federal Environmental Agency (Umweltbundesamt), Radetzkystrasse 2, A-1031 Vienna, Austria.

agreed that air pollutants, soil acidification, or acidic depositions play an important role in the damage process. Of all the air pollutants, ozone (and photooxidants) are most important and seem to have reached damaging levels throughout the country. Tree physiological data indicate that because of possible latent damages, visible crown condition is not a reliable damage indicator.

I. Introduction: Air Pollution, Acidic Precipitation, and Forest Damage in Austria

With most of the soil and water sufficiently buffered, no acute effects of acidification, such as lake dying or acidification-related fish decline, were reported in Austria until 1989. But in recent years there has been serious public concern about widespread forest damage (forest dieback, *Waldsterben*), which has been mainly attributed to acidic precipitation and acidification effects. From 1981 to 1983, visible symptoms like crown thinning, needle loss, branching anomalies, or needle yellowing were increasingly reported. With the knowledge gained from long-term ecosystem studies (e.g., Solling, FRG: Ellenberg, Mayer, and Schauermann, 1986), it was suggested that acidic precipitation might be a primary cause of forest damage because of soil acidification or direct damage of leaves and needles (Ulrich and Matzner, 1983). Although there had been research on air pollution, acidification, and forest damage before, the Waldsterben discussion led to an immense surge of research.

Certainly it is not possible here to present all the Austrian research activities and results, but we give a comprehensive overview of the data regarding acidification and its effects (Section II), and we describe the current research program especially dedicated to the problems of the "new" forest damage (Section III). One aim of this chapter is to provide a basis for possible future international research cooperation. Thus we will facilitate contacts by providing a list of principal investigators and their addresses (see Appendix).

A. Air Pollution

Compared to other countries in Central and Western Europe, Austrian annual emission rates of air pollutants are moderate on a *per capita* basis (data for 1985: SO_2: 18 kg yr⁻¹; NO_x (as NO_2): 28 kg yr⁻¹; CO: 142 kg yr⁻¹; anthropogenic nonmethane volatile organic compounds (VOCs): 69 kg yr⁻¹) or on a *per hectare* basis (data for 1985: SO_2: 16 kg yr⁻¹; NO_x (as NO_2): 25 kg yr⁻¹; CO: 127 kg yr⁻¹; anthropogenic nonmethane VOCs: 62 kg yr⁻¹). But because of the unhomogenous population density and the differentiated topography, some very high regional emission densities occur. Local and regional air pollution problems may be even most serious in some alpine valleys, when under specific meteorological conditions rapid dispersion of pollutants is hindered. Total Austrian emission levels for 1980 and 1985 are given in Table 3–1. As may be seen, measures to reduce air pollution have been successful mainly in respect to SO_2, where an almost 60% reduction has been achieved. Measures to reduce NO_x, CO, and VOCs have been taken only recently and have not yet shown any considerable effect. The first country in Europe to do so, Austria has stipulated since October 1987 that all new cars with gasoline engines have to be equipped with a three-way catalytic exhaust converter. Thus certain reductions can be expected over the next few years.

It has to be emphasized that the problem of air pollution in Europe will not be solved on the basis of individual nations only. Austria is located in the center of Europe, and it was calculated that about 70% of sulfur depositions in Austria are imported (EMEP, 1984). Even if the calculations did not sufficiently consider local valley depositions, the impact of imported acidic depositions remains significant. Similarly, European synoptic data show that Austria and Switzerland have the highest O_3 levels from all stations (Grennfelt, Saltbones, and Schjoldager, 1988), so it must be concluded that O_3 pollution is a transboundary continental problem, too.

B. Acidic Precipitation

Acidic precipitation with a pH below 5.1 is frequent in Austria, but because most of the soil and water is very well buffered there are only a few sensitive regions where acute damages might occur. Nevertheless there is great concern about the possible long-term effects of acidic depositions on soils and vegetation. The deposition situation must be seen in the context of Austrian topography. The Alps form a European continental barrier for long-range transboundary fluxes of air pollutants, and frequent

Table 3–1. Annual emission rates of air pollutants in Austria (data from Energiebericht, 1986).

Pollutant	1980	1985
	(1000 tons yr⁻¹)	
SO_2	325	138
NO_x (as NO_2)	201	208
CO	1126	1068
VOCs	121[a]	517[b]
Partic.	50	53
Pb	0.9	0.3

[a] Emissions from combustion sources (stationary/traffic) only.

[b] Total anthropogenic nonmethane VOC emission for 1987 (Orthofer and Urban, 1989).

rain on the lower slopes of the Alps leads to high depositions of imported pollutants. There is also considerable local influence on deposition rates by local emission sources in some mountain valleys, and sometimes heavy local soil and vegetation damage is reported. However, the characterization of the countrywide acidic deposition situation is not possible with the existing 69 monitoring sites, because in a country as mountain-dominated as Austria all data can only be interpreted very locally.

The exact determination of solid, liquid, and gaseous acidic depositions is one of the more sophisticated tasks of environmental monitoring. Reliable equipment for the routine determination of the different deposition types and the many different substances has been introduced only recently. Thus much of the Austrian data refer to bulk depositions only, which allow reasonable estimates of the total deposition situation but do not sufficiently differentiate the various forms of pollutant input. A simple sampling device for the determination of wet-and-dry deposition is increasingly used for routine deposition monitoring. In 1987 about one-third of all monitoring stations were equipped with the new wet-and-dry sampler (see Section IIB).

C. Forest Damages

Much of Austria's surface is covered with forests (44.8%). The predominating tree species is spruce (*Picea abies,* 61.9%); followed by beech (*Fagus sylvatica,* 9.1%); pine (*Pinus sylvatica* et *nigra,* 8.0%); larch (*Laryx europaea,* 4.5%), and fir (*Abies alba,* 3.0%).

Forest ecosystems have always been exposed to natural and anthropogenous influences that cause damage. Air pollutants have been known for centuries to be a significant factor in tree and forest damage. The influence of flue gas SO_2 on forests close to emission sources has been subject to scientific studies by the Austrian Forest Research Station for more than 100 years. But all of this forest damage was limited to areas near emission sources, and there was no doubt about the causal relations. Only in the early 1980s was vast forest damage reported for areas where no emission sources were located. Air quality measurements rather indicated very low levels of SO_2 and/or HF in some affected areas so that these pollutants could be excluded as primary causes. On the other hand, it was demonstrated that there certainly is a significant relationship with respect to air pollution (Schöpfer and Hradetzky, 1984). This new forest damage (called *forest dieback*) rapidly became a matter of concern for the public and for science. Interestingly it was not the symptoms that were "new"; most of the obvious symptoms like needle thinning, needle yellowing, and branching anomalies had been observed for a long time. The really "new" facts rather were (1) the rapid yearly increases in visible damage, (2) the inability of establishing cause-effect relationships with available knowledge, and (3) the vast areas of affected forests. In 1986, when explainable "normal" forest damage (mechanical, parasites, altitude, frost episodes) were excluded, stands with so-called new forest damage made up 31% of the total Austrian forest area, which comes to about 1 million hectares. However, a substantial recovery of visible damage was noticed from 1986 to 1989.

Meanwhile many theories have been developed to explain the new forest damage in Central and Western Europe, and some of them have been supported by scientific evidence (for a summary see Cowling, Krahl-Urban, and Schimansky, 1986; McLaughlin, 1985), but general undoubted scientific evidence for causality of symptoms has not been demonstrated yet. Taking into consideration the complexity of forest ecosystems, it may be anticipated that there is not a common general cause for the many symptoms of forest damage at all.

II. Environmental Data Research: Air Quality, Deposition, and Forest Damage Monitoring

A. Air Quality Monitoring

Meteorological parameters (amount of precipitation, air temperature and humidity, wind direction and velocity, solar radiation) have been monitored in a very dense nationwide network by the Central Meteorological Service (*Zentralanstalt für Meteorologie und Geodynamik*) for many years. Some temperature and precipitation data may be traced back for about 300 years.

Chemical parameters of air quality have only been monitored since rather recently. Although reliable equipment is easily available for the registration of SO_2, NO_x, and O_3, measurements of other atmospheric components are difficult and demand sophisticated methods of sampling. At the Technical University in Vienna an annular diffusion denuder sampling system was developed for registration of gaseous HCl, HNO_3, SO_2, NH_3, CH_3COOH, HCOOH, as well as particulate chloride, nitrate, and sulphate. This system was successfully tested in Europe and in the United States (Puxbaum, Weber, and Pech, 1985; Puxbaum et al., 1988; Rosenberg et al., 1988).

In 1987 a total of 153 air quality monitoring stations were operated continuously for the measurement of mainly SO_2 (at 145 stations), NO/NO_2 (at 95 stations), and O_3 (at 37 stations) (see Figure 3–1). Most stations are part of provincial automated networks; a nationwide network is planned until 1995. Regular monthly or yearly reports are published by several provincial governments. The majority of stations are located in polluted or densely populated areas, but 32 stations are in rural areas or near forests (Smidt, 1988a). Characteristic data for the air pollution situation in some forest areas are shown in Table 3–7. However, it is not

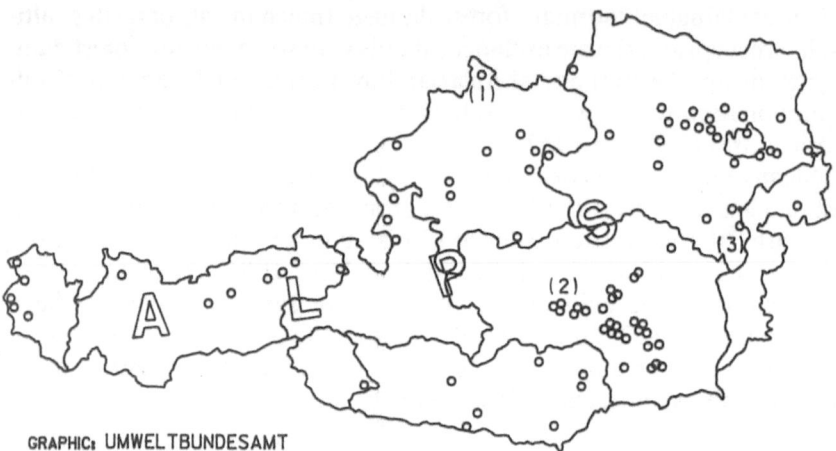

GRAPHIC: UMWELTBUNDESAMT

Figure 3–1. Air quality monitoring stations in Austria (Eber and Hojesky, 1987) and locations of the main forest damage research areas (see Section III A.2); (1) Schöneben, (2) Judenburg, (3) Rosalia.

possible to give a synoptic overview of the Austrian air quality situation. This is due to the very differentiated geographical situation (alpine valleys, mountains), which causes many problems of spatial data interpretation.

B. Bulk and Wet Deposition Monitoring

Standardization of deposition monitoring is guaranteed by guidelines specified by the Federal Environmental Agency (*Umweltbundesamt*). Recommendations are given for the determination of pH, conductivity, and "strong acids" as well as concentrations of sulphate, nitrate, ammonium, sodium, calcium, magnesium, and chloride. For the separate determination of wet-and-dry depositions, an effective wet-and-dry-only-sampling system (WADOS) is increasingly used (Puxbaum and Ober, 1987). The WADOS contains two parallel samplers with one automatic cover (controlled by a humidity sensor) alternatively open for wet deposition or for dry deposition only.

In 1987 a total of 69 deposition-monitoring stations were operated routinely throughout Austria, 22 of them of the WADOS-type, 38 of the bulk-type, and 19 special forest-monitoring stations. Provincial governments have the main responsibility for operation and analyses. The location of monitoring stations and some annual deposition rates (1984–1985) are shown in Figure 3–2 and Table 3–2. Deposition parameters change with topography and vegetation heterogeneity, thus deposition rates in alpine regions are very diverse, and spatial patterns of

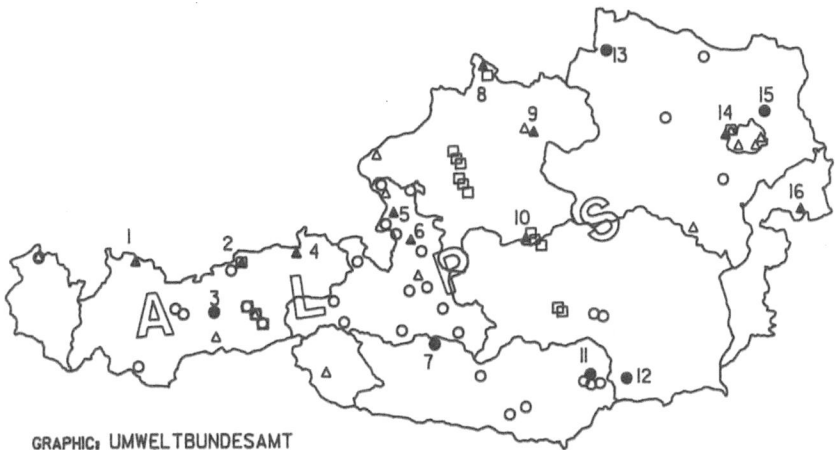

GRAPHIC: UMWELTBUNDESAMT

Figure 3–2. Wet deposition monitoring stations in Austria (data from Smidt, 1988b). Numbers refer to deposition data in Table 3–2 (dark symbols); △ Wet deposition (WADOS); ○ Bulk deposition; ☐ Monitoring station in forests below canopy.

Table 3–2. Annual wet-only and bulk deposition in Austria (SO_4-S is calculated as S; NO_3-N is calculated as N)[a].

	Altitude (m)	Wet/Bulk (W/B)	pH	SO_4-S	NO_3-N
				(kg ha^{-1}yr^{-1})	
Rural					
Reutte (1)	930	W	4.5	7.5	4.0
Achenkirch (2)	990	W	4.6	10.0	5.1
Nuβdorf-Haunsberg (5)	480	W	4.5	12.4	6.2
St.Koloman (6)	1000	W	4.2	13.2	7.0
Schöneben (8)	850	W	4.4	11.7	5.3
Wiel (12)	900	B	4.8	14.6	3.8
Illmitz (16)	120	B	n.a.	7.2	3.1
Mountain					
Patscherkofel (3)	1900	B	5.1	3.6	2.7
Malta (7)	1900	B	4.6	4.3	2.8
Wurzeralm (10)	1400	W	4.6	10.3	4.6
Klippitzthörl (11)	1310	B	4.9	5.6	3.0
Industrial Influence					
Kufstein (4)	700	W	4.3	13.2	7.4
Steyregg (9)	335	W	6.2	32.6	9.5
City Influence					
Exelberg (14)	490	W	4.7	5.7	2.7
Wolkersdorf (15)	200	B	5.6	9.6	6.9

[a] Numbers refer to site locations as indicated in Figure 3–2 (data from Smidt, 1988b).

depositions might be only rough estimates. A summary of deposition data for all Austrian stations was prepared by Smidt (1988b).

In winter there are serious problems determining atmospheric input effectively. In alpine regions snow makes up a great proportion of the total precipitation, and special equipment is needed for regular sample collection. In addition, the concentration of pollutants increases because deposition on snow surface is intensified. High and previously underestimated input of pollutants is also due to fog and hoar frost. Special equipment has been developed and tested by Glatzel, Kazda, and Markart (1986): Fog is collected with a passive fog-sensitive harp-type sampler, frost is collected with a similar passive thread sampler, and hoar frost with exposed fluorocarbon sheets. It is important to differentiate the various forms of winter pollutant deposition because fog, hoar frost, or dew with high loads of pollutants destroy epicuticular waxes of tree leaves and needles (Kazda and Glatzel, 1986).

Deposition monitoring in forests needs to be carefully evaluated: Because of pollutant interception in the canopy and needle washout or uptake effects, the actual pollutant inputs into forest ecosystems may be under- or overestimated. Sometimes correction factors calculated from the Cl^- budgets are used for estimation of atmospheric inputs, but sequential sampling methods have also proved to be a good way to determine depositions (Kazda, 1989). Glatzel et al. (1988) summarized depositions for 11 Austrian forest monitoring sites: Inputs are 15 to 54 kg ha^{-1} yr^{-1} SO_4-S (as S), 5 to 24 kg ha^{-1} yr^{-1} NO_3-N (as N), and 12 to 49 kg ha^{-1} yr^{-1} total N.

Unpolluted rain is expected to have a pH of 5.6 due to dissolved atmospheric CO_2. Eighty-six percent of recorded precipitation measurements in Austria have a pH below 5.1, which is more than half a unit below the "natural acidity" (Smidt, 1983). This indicates that acidic precipitation is frequent and serious throughout the country. However, it is important not to rely on the pH only as an acidification indicator, as pH is easily increased by other pollutants and the actual situation may be underestimated. Samples in the "natural" pH range (pH 5.1 to 6.1) often show high electrolytical conductivities, indicating that original acidity is neutralized by (mostly locally emitted) alkaline air pollutants or dusts. Synoptic views of the geographical distribution of deposition patterns are possible with selected sites only. Similarly to air quality monitoring (see Section IIA), most stations are influenced by local emission sources and very local deposition parameters. Data from background stations show a pH gradient at the northern and southern slopes of the Alps (see Figure 3–3). This would indicate a close relation of precipitation pH to the emission distribution, as most industrialized and populated areas are concentrated north and south of the Alps. On the other hand, it may be concluded that in fact the Alps form a continental barrier for long-range transported pollutants and that air pollutants are deposited with frequent rains on the

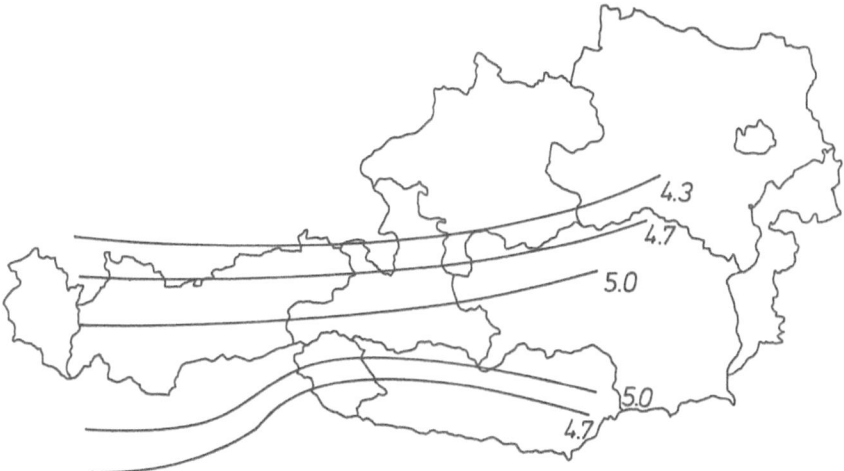

Figure 3–3. Precipitation pH from background stations, 1984 to 1985. (Reproduced with permission from Psenner, 1987.)

lower slopes of the Alps. Indeed some regions with no local emission sources in northern Austria have high deposition rates, most probably from long-range transport of air pollutants (see Table 3–2 and Section III B.1.). Data from other stations must be interpreted in respect to the local differentiated topography, and only a few stations might be directly compared. Annual wet-only and bulk deposition recordings for H^+, S, and N of all Austrian stations (including monitoring sites in mountain valleys, where local sources play an important role in valley acidification) are given in Figure 3–4 (Smidt, 1988b). Although there is a great variability, results show that in high altitudes the input of S and N is less than in valleys. But clearly the input of protones seems to increase in high altitudes. This reflects the typical pattern of depositions caused mainly by local emission sources in mountain valleys, as shown in Figure 3–5: Sulfur and nitrogen concentrations in rain and snow as well as electrolytical conductivity are much higher in the valley, but proton concentrations increase with altitude. This poses an important risk to sensitive high altitude forests (Smidt, 1986).

C. Aquatic Ecosystems Monitoring

Areas which are sensitive for acute acidification are located in northern Upper Austria (crystalline bedrock) and in some small areas of the Central Alps (alpine lakes on crystalline rock formations in the provinces of Tyrol and Carinthia).

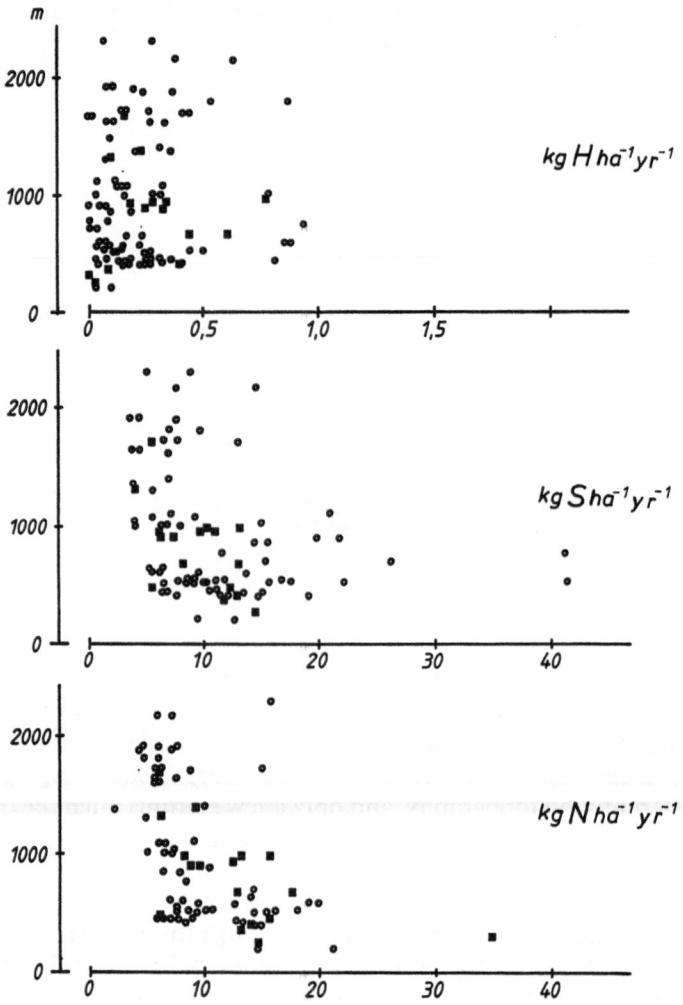

Figure 3–4. Deposition data of H⁺, S, and N (1984 to 1985) from all Austrian monitoring stations in respect to altitude (meters above sea level). (Graph reproduced with permission from Smidt, 1988b.)
○ Bulk   □ WADOS

Streams, rivers, and lakes, notably high altitude alpine lakes, were monitored nationwide for acidification from 1983 to 1986. Until the present all results indicated that there is no acute acidification of streams and rivers in terms of low pH or increased Al^{3+} ions. But evidence was given of exhausted buffer capacity in some rivers in sensitive areas which might lead to acidic episodes during snow melt (Janauer, 1986). The lowest recorded pH in rivers was 4.8 in northern Upper Austria (Butz and

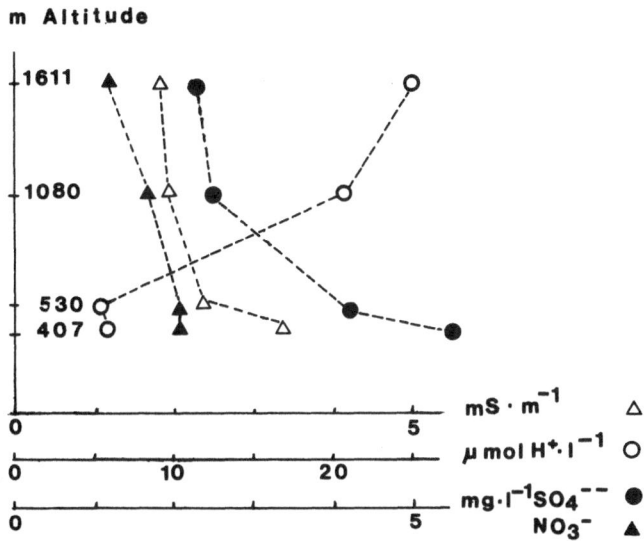

Figure 3–5. The influence of altitude on deposition (rain, snow) of H^+, S, and N and electrolytic conductivity in a mountain valley with a local emission source (Koralpe, Carinthia). Data from Smidt (1986).

Rydlo, 1987); some indications of acidification-injured benthos were determined in sensitive rivers.

Alpine lakes in crystalline rock formations are more strongly affected by acidification because of their lower buffer capacity and their very labile ecosystem. The majority of the lakes studied did not appear to have been acidified in terms of pH or alkalinity. But several sensitive alpine lakes in Carinthia and Tyrol were determined to be at high risk, and a clear tendency toward acidification was found for all high altitude small ponds (< 2000 m^2). When 71 small alpine lakes were surveyed from 1983 to 1986, 20% had a pH below 6.0 and in 10% of the lakes the carbonate buffer capacity was exhausted (Psenner et al., 1988). This is because precipitation frequently shows a pH of 4.3 to 4.9, and watershed runoffs in this region have low pH especially during snow melt (Honsig-Erlenburg and Psenner, 1986). Conclusions drawn from the nationwide monitoring projects indicate that there are considerable risks of future acidification particularly because of increasing nitrogen compound depositions (Janauer, 1986).

D. Forest Damage Monitoring

It is not yet clear what the actual causes of the forest damage are, but there is strong evidence that depositions of air pollutants play an important role (Schöpfer and Hradetzky, 1984; Ulrich and Matzner, 1983).

Forest trees act as filters for air pollutants. Thus deposition rates in or below the canopy are higher than in the open field. Air pollutants contribute to the destabilization of forest ecosystems in various ways. Acidic depositions decompose the protective wax layers on leaves or needles and injure the cuticles, eventually leading to increased leaching of nutrients. Long-term input of acidity also causes irreversible soil acidification with consequences to nutrient availability and washout. Gaseous pollutants (SO_2, NO_x, NH_3, O_3, and other oxidants) damage leaf surfaces and, if penetrating the stomata, disturb physiological processes of leaf cells (Bermadinger, Grill, and Pfeifhofer, 1987). Finally, continued inputs of nitrogen into forests lead to eutrophication, disturbed nutrient uptake, and an altered plant metabolism (Glatzel, Kazda, Grill, Halbwachs, and Katzensteiner, 1987); a latent deficiency of nutrients may thus become acute. Weakened trees generally have a higher predisposition to pathogen attack and abiotic stress factors. Although air pollution is a major stress factor, it certainly is not the only cause of forest damage. The actual damage to forests in central Europe makes clear the complex processes and living conditions of forest ecosystems. They are prestressed by many other anthropogeneous agents, such as forest management, as well as by natural factors. Causal analysis of damages thus must include all possible influences on the ecosystem.

There is no practicable method of differentiating air pollution effects on forests and other stress factors which could be applied for nationwide surveys. Only mechanical damages may be clearly diagnosed. Visible damage symptoms (needle loss, needle yellowing) may be caused by many biotic or abiotic factors and attributing certain symptoms to air pollution only is not possible. On the other hand, if just visible damages are considered for an assessment of forest damages, latent but serious physiological damages may not be detected.

In Austria there are two different, but complementary, nationwide monitoring systems: the Bioindicator Grid (*Bioindikatornetz,* BIN) and the Forest Condition Inventory (*Waldzustandsinventur,* WZI). Both operations are run by the Federal Forest Research Station. In addition considerable effort is spent on the utilization of remote sensing techniques for forest damage monitoring.

1. Bioindicator Grid (BIN)

The BIN is an annual nationwide survey of sulfur, fluorine, and nutrient contents in tree leaves and needles; it was started in 1983. The concept was developed in response to the "traditional" forest damage caused by SO_2 (and, to a lesser extent, by HF) in industrial emission areas, damage which has been serious in some regions. Its purpose was mainly to identify problem regions and to help forest property owners with evidence when they wanted to sue industry for compensation. The method takes advantage of the fact that high concentrations of SO$_2$ and HF in forest areas lead to an accumulation of S and F in needles or leaves.

There are marked sample trees (spruce, pine, beech) in a 16 × 16 km grid; the basic grid consists of 317 sample plots, but it is intensified in industrial areas and in areas of special interest (mountain slopes). The total of sample points comes to about 1700 (Stefan, 1987). Needle and leaf samples are taken in summer and chemically analyzed, and the results are evaluated according to the standards set by the Austrian Forest Law. For spruce this means a maximum acceptable limit of 0.11% S (on a dry weight basis) in the youngest needles and 0.14% S in last year's needles, respectively. The extent of Austrian forest areas with high sulfur needles as indicated by the BIN data is shown in Figure 3–6.

The advantage of the BIN is that areas of SO$_2$ stress may be identified regardless of the visible appearance of the trees. But experiences since 1983 show that even if the regional distribution of damage was similar, there was an unexpected annual variability in the number of trees with sulfur levels above the tolerable limits (see Table 3–3). Another striking fact is that a SO$_2$ emission reduction of 60% (from 1980 to 1985) was not reflected in the BIN results.

The conclusion must be drawn that the large variations of the sulfur accumulation levels in the years from 1983 to 1989 were caused mainly by climatic factors. Climatic factors certainly influence the physiological

Figure 3–6. Distribution of areas with considerable air pollution effects on forests, referring to high sulfur needle samples from the Bioindicator Network, 1985 to 1986. Arrows show transboundary fluxes of air pollutants. Mapping from the Federal Forest Research Station.

Table 3–3. Bioindicator grid (4 × 4 km basic grid): results of sulfur determinations 1983 to 1989.[a]

	Youngest needles (% S d.w.)		Last year's needles (% S d.w.)		Sample trees above limits
	Range	Mean	Range	Mean	(%)
1983	0.056–0.166	0.093	0.058–0.260	0.099	11.3
1984	0.055–0.130	0.083	0.056–0.189	0.090	5.5
1985	0.058–0.165	0.096	0.062–0.237	0.102	21.4
1986	0.066–0.144	0.095	0.062–0.188	0.095	13.3
1987	0.063–0.167	0.099	0.062–0.220	0.105	19.1
1988	0.067–0.151	0.094	0.066-0.212	0.098	8.7
1989	0.065–0.160	0.099	0.068–0.184	0.102	18.8

[a] Note the annual variability of trees above the legal limits of 0.11% in the youngest needles and 0.14% in last year's needles (data from Stefan [1990]).

activity of trees and hence the uptake of sulfur dioxide, too. From this point of view the results of the year 1984 (and probably 1988) must be regarded as not representative for the actual situation.

2. Forest Condition Inventory (WZI)

The "new" types of forest damage are characterized mainly by visible damage symptoms (needle loss, needle yellowing, crown thinning, branching anomalies) in areas where SO_2 and HF have been excluded as a primary cause. Since the BIN monitoring system was concerned with the accumulated pollutants but not with visible damage symptoms, a second nationwide forest monitoring system was set up 1985 with the special aim of surveying the visible status of forest trees regardless of possible causes.

A 4 × 4 km grid system fitting into the BIN (see Section II D.1) was set up leading to about 2200 sample plots, each one with 30 to 50 marked reference trees (mainly spruce 60 years or older). Every summer, symptoms like loss of needles, leaves and twigs, discoloration of needles and leaves, and general visual appearance of the crown structure are assessed in five categories for each tree. From single-tree data the extent of affected forest areas is calculated with statistical models (Pollanschütz and Neumann, 1987).

In Table 3–4 the WZI results are presented for the years 1985 to 1989. The data indicate that although a slight damage peak was noticed in 1986, there was no general severe deterioration of visible forest conditions after 1985. This is remarkable, because earlier regional surveys had confirmed the immense deterioration of visible forest conditions from 1983 to 1985. Even a considerable improvement has been noticed since 1986. For some species (e.g., beech, oak), though damages still increases

Table 3–4. Results of the Austrian Forest Condition Inventory 1985–1989 (data from the Federal Forest Research Station).

Category	Crown condition	% of trees				
		1985	1986	1987	1988	1989
1	No symptoms	65.3	62.8	66.5	71.2	74.6
2	Weak symptoms	31.1	32.6	29.9	25.2	21.0
3	Moderate symptoms	3.2	4.0	3.1	3.0	3.6
4	Heavy symptoms	0.3	0.5	0.3	0.3	0.7
5	Dead crown	0.1[a]	0.1[a]	0.2[a]	0.3[a]	0.1[b]

[a] Accumulated since 1985.
[b] Accumulated since 1988.

sharply until 1989, due to their small share of total trees this does not reflect in the "total" statistics. However, it should be noted that latent damage of trees may not be estimated by this method.

3. Remote Sensing

Annual visual damage monitoring requires many skilled and trained personnel and is very costly. Even more problems arise because all estimates are influenced by the personal bias of the observers. Therefore an attempt has been made to use remote sensing techniques to replace terrestrial crown evaluations. Aerial color-infrared (CIR) photography has been used in Austria since the 1960s mainly for the inventory of vegetation vitality. In 1984 and 1985 the first extensive inventory of forest (crown) damages was made for all of the forests in the province of Vorarlberg (Zirm et al., 1985). The main criteria for judgment in the CIR photo are structural and color characteristics of the tree crowns. Symptoms which are used for the terrestrial estimation are correspondingly defined in the CIR picture. As several phenotypes of tree species have to be differentiated, it is necessary to define separate interpretation keys for every region to guarantee a valid judgment. The evaluation is based on a 50 m grid and a photo scale of 1:7000 to 1:10,000. Interpretation data are stored digitally and can be used for statistical evaluation in relation to climate, exposition data, altitude, and so on.

Beginning in 1989 the nationwide terrestrial WZI was partly replaced by CIR aerial photography survey. Every year another one-fifth of the country should be surveyed, so every five years all forests might be covered (Pollanschütz, 1987).

E. Forest Soil Acidification

Acidic depositions may be only one of many causes if forest soils are poor and very acid; it should not be overlooked that frequently inadequate silviculture and tree harvesting systems may be blamed, too. Anion uptake from the soil by plants is much less than the uptake of cations, and to maintain the charging balance protones are exudated into the soil. In the plants the balance of charging is maintained via organic acid anions (Nielsson, Miller, and Miller, 1982). When biomass is taken out of forests, nutrient cations are removed from the forest ecosystems. This loss is especially enhanced when stems are removed with all the thin branches, twigs, and leaves or needles. Although less in biomass those parts of the tree contain higher concentrations of most mineral elements than the stems. Whereas the removal of 1 ton of wood has an acidification potential of 0.1 kmol H^+, the loss of 1 ton of tree needles and small twigs would cause an acidification of 1.2 kmol H^+ (Glatzel, Englisch, and Kazda, 1985). Whole-tree harvesting therefore causes considerable nutrient depletion, leading to the dominance of anions in the soil and an accelerated soil acidification rate; in most soils this can be counterbalanced by bedrock weathering rates for only short time periods.

However, at present many forest ecosystems in Austria receive on average half or more of their H^+ input from atmospheric depositions which are also a main source of mobile anions that increase the cation leaching in the soil. This could be demonstrated with analyses of soils around beech trees on slopes: Atmospheric depositions are concentrated in stemflow and affect only downhill soils around the stems, so atmospheric effects may be compared to stand management effects. Results showed that downhill of beech trees, exchange-buffer capacity of the upper soils was exhausted and the aluminum buffer region was about to become effective (Blum and Rampazzo, 1987; Glatzel, Sonderegger, Kazda, and Puxbaum, 1983). Thus Al^{3+} ions are released having toxic effects on mycorrhizae and fine roots.

Long-term changes in soil acidity are difficult to prove because of a lack of proper reference measurements. In Austria there is evidence for acidification in the last 25 years. Stöhr (1984) compared soil profile sample data from 1958 to 1970 with soil profiles from 1983 in 66 areas. The results showed that the frequency of low pH recordings in the A–1 horizon was clearly increasing.

III. The Forest Damage Research Program

Acidification research in Austria has been mainly dominated by the serious problems of the "new" forest damage (*Waldsterben*) in wide areas of the country. Assessments of research results (e.g., BMFT, 1987; NAPAP, 1987) led to the conclusion that the new forest damage cannot be attributed to the effects of just one single cause such as soil acidification. Most results confirm that morphological symptoms of the once-thought uniform "new" forest damage are distinct and caused by many influence factors, regionally and stand specifically. Thus there is no single research field exclusively dedicated to *Waldsterben:* On the contrary, forest damage research includes almost all fields of terrestrial ecosystem research. A recent survey of the Austrian Research Center Seibersdorf showed that about 950 research projects were carried out in Austria from 1982 to 1988 in the fields of

atmospheric chemistry and meteorology,
acidification and deposition science,
tree physiology,
forest ecosystems,
forest management, and
emission reduction strategies.

About 210 projects are especially connected with forest damages.

Responsibility for research administration and funding is shared by three competent and involved ministries on the federal level (for the majority of projects) and by local governments on the provincial level. Basic research projects are mostly funded by the National Science Fund (*Fonds zur Förderung der wissenschaftlichen Forschung*). Although the Federal Ministry of Agriculture and Forestry (with its research facility, the Federal Forest Research Station, *Forstliche Bundesversuchsanstalt*) is emphasizing practical themes like deposition and damage monitoring (see Sections II B and D) or management practices of damaged forests, the Federal Ministry of Environment, Family and Youth Affairs (and its research facility, the Federal Environmental Agency, *Umweltbundesamt*) accepts responsibility mainly for air quality monitoring. The Federal Ministry of Science and Research leads the special scientific program aimed at the *Waldsterben.* The program was planned for five years (1984 to 1989); it is about to be prolonged for another five years. This coordinated program (*Forschungsinitiative gegen das Waldsterben,* FIW) is based on research contracts; it includes many research fields but focuses mainly on the diagnosis and the causal analysis of forest damage (see Table 3–5). Scientific head is E. Führer from the Vienna University of Agriculture, the university that was awarded most of the research contracts; the coordination office for the program was established at the Austrian Research Center Seibersdorf. Until 1988, 100 projects (with 1 to 2 years duration each) have been supported with 72.5 million Austrian schillings (about U.S. $6 million). The distribution of projects and funding for the various research fields is shown in Table 3–6.

As the FIW is the scientific core of Austrian forest damage research, we will briefly describe its research strategy, field research locations, some tree physiology results, and an example of an integrated research project.

Table 3–5. Research fields of the Austrian Research Initiative Against the Forest Dieback (FIW).

A: *Basic data* (at research sites): Meteorology, climatology, air quality, monitoring, deposition monitoring

B: *Morphology:* Ultrastructure, anatomy

C: *Physiology:* Photosynthesis, water budgets, nutrients, ion balance, pigments, enzymes, secondary metabolites (lipids, phenols)

D: *General Forest Ecology:* Production parameters, tree-parasite interactions, forests and game

E: *Soil/Rhizosphere:* Soil biology, soil chemistry, biology of rhizosphere, mycorrhizae

F: *Revitalization:* Soil revitalization, In vitro fertilization of trees

G: *System Analysis/Coordination*

H: *Emission Control Strategies:* Emission inventories, cost-benefit assessment

J: *Remote Sensing:* CIR aerial photographs, satellite data

Table 3–6. Distribution of projects and funds of the Austrian Research Initiative Against the Forest Dieback (until December 1988).

Research fields	Projects number = %	Funding millions Austrian schillings	Funding %
A: Basic data	15	11.3	15.6
B: Morphology	8	3.5	4.8
C: Physiology	22	9.9	13.7
D: General forest ecology	14	12.6	17.4
E: Soil/Rhizosphere	7	2.7	3.7
F: Revitalization	4	16.2	22.3
G: System analysis	10	9.2	12.7
H: Emission control	6	2.6	3.6
J: Remote sensing	14	4.5	6.2
Total	100	72.5	

A. FIW Research Strategy

The FIW research strategy (BMWF, 1986) put an emphasis on

concentration of field research to a very few sites,
avoidance of duplicating research projects, and
permanent update of priorities according to new results.

To discuss research needs and results scientists met in three main task groups. Task group I held the central position of research interests and covered research fields A through G (see Table 3–6). Parallel task groups II and III were formed for research fields J and H, respectively. The main research interests of the task groups were defined as follows:

1. Task group I

Task group I was concerned with the problems of diagnosis and causal analysis of the "new" forest damage and the general effects of air pollution on trees and forests. The main questions to be answered were

1. Can the "new" forest damage be distinguished from the old, "classic," and well-studied tree and forest damage caused by high levels of air pollutants (SO_2, HF) in known emission regions? Can they be distinguished from other biotic (fungi, parasites) and abiotic (climate, nutrition) damages?
2. What is the role of air pollution? Can the responsible sources be traced?
3. Are there long-term risks of subchronic levels of air pollutants for trees and/or forest ecosystems?
4. Which short-term and long-term measures can be taken for a stabilization of tree health and the forest ecosystem?

2. Task group II

The goal of task group II was the development of remote sensing methods which may be employed for the determination and interpretation of the extent of forest damage. Although color-infrared (CIR) photographs from airplanes and tree-by-tree CIR interpretation were already practically used for damage evaluation on a routine basis (see Section II D.3), further scientific interest was focused on

1. Development of computer-assisted methods for interpreters of CIR photographs to improve reproducibility, standardization, and speed of the interpretation.
2. Further progress in automatic picture analysis (cluster-analytical methods).
3. Utilization of satellite data (LANDSAT 3-TM, SPOT) for the detection of different damage areas and damage patterns.

3. Task group III

It was the main interest of task group III to develop strategies for emission reductions. However, the development of emission reduction technologies was not included: Clean technology research is by far too extensive for a forest damage research program. In 1986 an "Environmental Technology" stimulation program was defined in which air pollution reduction technologies play a major role. Research priorities of task group (III) were

1. Update of emission inventories,
2. Source-receptor modeling and scenario development, and
3. Cost-benefit assessment for restoration of problem regions.

B. Principal Field Research Locations

Environmental monitoring and research data for specific sites normally represent the conditions of a very small region around each monitoring site only. It has been mentioned that in Austria this problem is extremely important; sometimes data from the very same area but at various elevations on mountain slopes show uncomparable results for air pollution and deposition as well as for climatic and geology preconditions. A great local variability is also noticed for the genetic characteristics of trees. It is not surprising that forest damage symptoms and causes vary widely according to region, climatic environment, and stand conditions. In causal research, only data from the very specific same sampling area may be compared and connected. Thus a countrywide description of depositions or a representative causal analysis of the forest damage situation cannot be expected from any reasonable number of research sites.

FIW research was concentrated on only three sites. It was agreed that with limited research resources the thorough study of few sites is preferable to a less intensive study of a great number of sites. The locations of the principal research areas are indicated in Figure 3–1: (1) Schöneben/ Böhmerwald (Upper Austria); (2) Judenburg (Styria); and (3) Rosalia (Lower Austria).

1. Schöneben/Böhmerwald

The Schöneben research area is part of the large Bohemian Forest and is situated in the very north of Austria (150 km north of the Alps) close to the Czechoslovakian border in the province of Upper Austria. The elevation of the area is 800 to 1000 m. Three study sites have been set up at 860, 880, and 990 m. There is no considerable emission source of SO_2 and NO_x on the Austrian side within 50 km. There is probably no larger source within 50 km on the Czechoslovakian side, but data are scarce.

Climatic conditions can be characterized as typical for a continental summer rain region with an average precipitation of 1100 to 1200 mm per year; the region has an average snow cover for approximately 150 days and about 60 fog days per year. Spruce dominate with 74% of all trees; beech account for 20%. The area has very poor and acidic soils on granite (Bohemian Crystallin) and quite obvious visible forest damage (spruce: yellowing, needle loss).

Air quality data indicate low SO_2 and NO_2 during most of the year except some short moderate/high episodes in winter; 80% and 90% of all half-hour readings are below 20 µg m^{-3} SO_2 and NO_2, respectively (Kolb, Scheifinger, Bogner, and Weihs, 1988). On the other hand ozone poses a certain problem. Two percent of all half-hour readings are above 140 µg m^{-3}, 17% above 100 µg m^{-3} (Kolb and Bogner, 1988). Monthly average ozone levels generally exceed 70 µg m^{-3} from April to September (see Table 3–7). Maximum daily averages in the summer half-year are 120 to 150 µg m^{-3} (OÖLR, 1984–1989).

Table 3–7. Air quality characteristics of the three main forest damage research sites (1984 to 1989).[a]

Site		SO$_2$ (μg m^{-3})		NO$_2$ (μg m^{-3})		O$_3$ (μg m^{-3})	
		maxDM	MM	maxDM	MM	maxDM	MM
Schöneben (1)	Winter	60–120	10–40	60–70	10–20	40–100	40–80
	Summer	20–40	<10	10–20	<10	120–150	70–100
Judenburg (2)	900m Winter	100–150	40–60	n.a.	20–35[b]	n.a.	30–70
	1250m Winter	n.a.	n.a.	n.a.	n.a.	n.a.	n.a.
	900m Summer	30–50	10–40	n.a.	10–20[b]	n.a.	70–110
	1250m Summer	n.a.	n.a.	n.a.	<10	n.a.	100–130
Rosalia' (3)	Winter	50–110	10–30	15–25	15–20	60–90	50–80
	Summer	10–20	<10	10–15	<10	130–150	100–150

[a] Numbers refer to site locations as indicated in Figure 3–1. (maxDM = maximum Daily Mean, MM = Monthly Mean of half-hour readings from automated monitoring stations). Data from (1) OÖLR (1984–1989), (2) and (3) Kolb and Werner (1985), Kolb and Scheifinger (1986), Kolb et al. (1988), Krapfenbauer et al (1989).

[b] Including 20% to 50% NO.

Deposition below the canopy is 18 to 30 kg SO_4–S ha^{-1} yr^{-1} (as S); 14 to 24 kg total N ha^{-1} yr^{-1} (as N); and 5 to 11 kg NO_3–N ha^{-1} yr^{-1} (as N), which is moderate compared to other European sites, but unexpectedly high for a remote site far from emission sources (Glatzel et al., 1988). The main theme of the research efforts in Schöneben is the problem of tree nutrition connected with poor and acidic soils.

2. Judenburg

The Judenburg research area is located on a mountain slope in an industrialized valley in the central Alps. There have been local emission sources for SO_2 and NO_x for many years (pulp and paper mills, iron and steel industry, coal-fired thermal power station). The elevation of the area stretches from 800 to 1250 m above sea level. The European continental climate includes strong warm winds from the south across the Alps (*Föhn*). Annual precipitation is about 800 mm; soils are generally good. Predominant trees are spruce (80%) and larch (11%). Local forest damage connected with high levels of SO_2 in the lower slopes of the valley have been known for at least 20 years. Since the early 1980s, when local SO_2 emissions decreased, a corresponding forest production increase (radial growth increment) and decrease in needle sulfur content was noticed in the lower valley regions (800 to 950 m) but visible forest damage (crown needle thinning) seemed to increase (Eckmüllner and Sterba, 1987). In higher altitudes where SO_2 and NO_x levels have always been low, "new" visible forest damage (needle yellowing) was reported recently. The present air quality situation is influenced by frequent meteorological inversion episodes in winter with fog up to the inversion zone at 950 m. The 900 m site (Table 3–7) is characterized by moderate SO_2, NO, and NO_2 average levels (winter monthly mean: 40 to 60 µg m^{-3} SO_2, 10 to 15 µg m^{-3} NO, and 10 to 20 µg m^{-3} NO_2) and some high SO_2 levels at inversion episodes (maximum daily average: 100 to 150 µg m^{-3} SO_2). Ozone levels are high: Monthly mean concentrations are 30 to 70 µg m^{-3} in winter and 70 to 110 µg m^{-3} in summer. Two percent of all half-hour readings exceed 140 µg m^{-3}; 21% exceed 100 µg m^{-3} (Kolb and Bogner, 1988).

At the high elevation monitoring site (1250 m) where no winter data are determined, NO and NO_2 are very low, but again very high ozone was recorded: Summer monthly mean is 100 to 130 µg m^{-3} (Kolb et al., 1988). Eleven percent of all summer readings were above 140 µg m^{-3}, 50% above 100 µg m^{-3} (Kolb and Bogner, 1988). Interestingly, deposition data do not vary very much with height: Inputs below the canopy are 21(21) kg SO_4–S ha^{-1} yr^{-1} (as S), 15(12) kg total N ha^{-1} yr^{-1} (as N), and 6(6) kg NO_3–N ha^{-1} yr^{-1} for 900 m (and 1250 m), respectively (Glatzel et al., 1988). It should be noted that although there are local emission sources, throughfall depositions are not higher than in the background site Schöneben (see Section III B.1.). The main research themes for the Judenburg region are the impact of acidic air pollutants on tree physiology and the evaluation of historically varying deposition conditions on tree and forest production.

3. Rosalia

The Rosalia research area is typical for forests in low hills east and north of the Alps in eastern Austria. The research region is located about 60 km south of Vienna and covers about 930 hectares of mixed deciduous/coniferous forest (43% spruce, 37% beech, 9% pine, 4% larch) on a slope from 350 to 750 m. With an annual temperature average of 7.9°C and a yearly precipitation of only 700 to 800 mm, the region may be characterized as warm and dry. Vast visible forest damage was not monitored until 1983 but then increased very much. Classified as damaged in 1985 were 77% of spruce, 45% of beech, and 83% of pine. Air quality data (Table 3–7) indicate very low SO_2, NO_2, and NO levels with rare winter inversion episodes (maximum daily mean: 50 to 110 µg m^{-3} SO_2). Eighty percent and 90% of all half-hour readings are below 20 µg m^{-3} SO_2 and NO_2, respectively (Kolb et al., 1988). Yet ozone is certainly a serious problem: 42% of all half-hour readings in 1986–1987 exceeded 100 µg m^{-3}; 12% exceeded even 140 µg m^{-3} (Kolb and Bogner, 1988). Monthly means are in the range of 50 to 80 µg m^{-3} in winter (October to March) and 100 to 150 µg m^{-3} in summer (April to September) (Kolb et al., 1988). Deposition data for the Rosalia area are not yet available but the air pollution monitoring data and the meteorological characteristics suggest a considerably lower amount of deposition than in the Schöneben and the Judenburg sites. The main research interests in the Rosalia area are clearly the problems connected with the high ozone levels: The emphasis of the project is on the uptake of ozone and oxidants by leaves and the physiological effects on trees. For this purpose two 39-m towers were erected in dense forest, where meteorological and deposition parameters are measured above, within, and below the canopy. Because of the various available data this area was also chosen for the system analysis studies for causal analysis of the "new" forest damage (see Section III D).

C. Tree Physiology Research

Understanding tree physiology is crucial for understanding mechanisms of the various symptoms of the "new" forest damage. Unfortunately, although general plant physiology has been extensively covered scientifically, and tremendous progress has been made in the last two to three decades, there are big gaps in our knowledge of the physiology of forest trees. Almost 20% of the financial resources of the FIW is spent for tree physiological and morphological research, mainly on spruce (see Table 3–5, research fields B and C). However, results from tree physiology oriented research may only be interpreted in a very limited way as long as basic facts of the physiological behavior of "healthy," "normal" trees and their natural variability are not known. Comparisons of up to 20 visible "damaged" and "healthy" trees on the same site made clear that the variability within one group was much larger than the expected significant differences on physiological parameters between the two statistical groups (Albert and Hübler, 1987). Those results led to the conclusion that the genetic variability of trees has been underestimated, and a tree physiological approach to the problems of the "new" forest damage is not recommended because too little basic scientific knowledge is available. One of the objectives in the beginning of the research was to find stress-specific physiological parameters in trees as early warning diagnostic tools for various causes of damage. After four years of intense work it had to be accepted that most of the following tested parameters failed to give reliable answers: sulfhydril system compounds (glutathion); pigments (chlorophylls, xanthophylls, carotinoids); ascorbic/dehydroascorbic acid; peroxidase system; organic/inorganic anions; or the Haertel-test. Obviously physiological parameters are influenced by many biotic and abiotic effects, so changes cannot be regarded as air pollution effects. As there are no homogenous stand conditions for trees in Austria, every single tree must be considered as controlled by different environmental conditions Bolhar-Nordenkampf, 1989). A survey of nutrient and micronutrient levels in needles showed very different levels in almost every analyzed tree and did not reflect the influence of air pollution but rather different soil (micro) conditions (Horak and Zvacek, 1987). Despite the uncertainty of physiological results, a promising research field seems to be the diagnosis of damages on a level of needle tissue ultrastructure combined with ion level determination. There are hints of different damage types in the chloroplast environment caused by nutrient deficiency on the one hand and by air pollution effects on the other (Gailhofer and Zellnig, 1987).

In general, tree physiological results proved that the visible conditions of the crowns (needle loss, yellowing) are not reliable linear indicators for evaluating the actual "health" of trees. For instance, needle loss up to 25% may be regarded as a normal variability for spruce, and the observed type of yellowing may be attributed to nutrient deficiency in soils.

Although it is not possible to exclude tree physiology from the research program, future strategy for the FIW will emphasize forest-ecosystem-oriented projects rather than tree physiological projects. As *Waldsterben* seems to be a complex ecosystem disorder, an ecosystem approach seems to be more pertinent. This means a preference for projects that do not rely on tree physiological basic knowledge for interpretation, but rather employ parameters of which a budget input-soil-tree-output may be calculated and fluxes may be followed. On the other hand, tree physiology will become a focal point in future basic research funding even if not directed toward the forest damage problems in the first place.

D. Integrative Ecosystem Research

Because the Rosalia research area has been used as the university experimental forest of the Vienna University of Agriculture since 1972, and much data have been collected, the area was chosen for an integrated approach of forest damage research, a system analytical geographic analysis (Cabela, Grossmann, Kopcsa, and Weber, 1986; Grossmann, 1988).

As mentioned earlier, research results show that caution should be exercised in the interpretation of isolated projects, tree physiological or any other. Instead it must be anticipated that many environmental stresses affect forest ecosystem stability, each posing a risk factor to trees. Only where multiple risk factors exceed a damage limit do visible damages of trees occur (Führer, 1985); trees exposed to the most severe risk levels will be among the first damaged. This concept formed the basis of the Integrative Rosalia Project as a means of an assessment and a causal analysis of the forest damage. Basically an attempt was made to evaluate all possible known risk factors on a detailed geographical basis and calculate patterns of tree damage risks by using risk-factor assessment formulas according to various forest damage hypotheses. Then the calculations were compared to the actual damage patterns and conclusions were drawn on the validity of various hypotheses. To perform the spatial distribution analyses an area has to be large enough and has to have differentiated stand conditions. If the test region is sufficiently heterogeneous with respect to soil conditions, stand age, pollutant distribution, and other meteorological, orographical, and biological factors, the test area may be relatively small (ca. 1000 ha) as it was in the Rosalia project. Most important, enough relevant spatial data must be available and, for geographical calculations (map overlay), a geographical information computer system is essential (see Figure 3–7). For the Rosalia Project thematic maps of the following data were used:

Geology
Stand characteristics:
 incline, basins, and ditches
 soil type
 soil chemistry (pH, nutrients, elements)
Forest data:
 tree type
 tree ages
Meteorology/air pollution:
 SO_2, NO_x (calc. levels)
 Ozone (calc. levels)
 precipitation distribution
 fog frequency

All of the data sets were used to calculate risk maps driven by time-series models adjusted to various hypotheses. From the comparison of calcu-

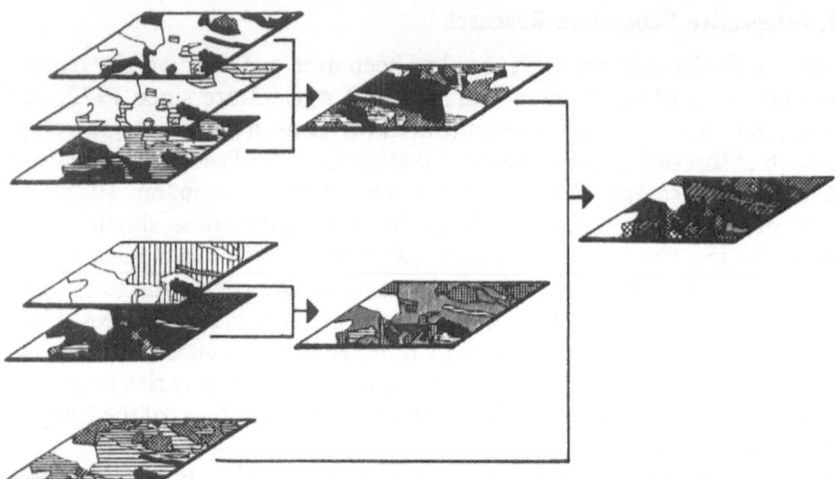

Figure 3–7. Combination and overlay of thematic risk maps with a geographical information system. From Cabela et al. (1986) with kind permission from W.D. Grossmann.

lated distribution of forest damages with the actual tree damages some conclusions could be drawn: Tree age and soil type could not sufficiently explain the actual damage. However, soil conditions, notably pH, had some influence on the distribution of damages. Direct effects of SO_2 and NO_2 could be excluded as primary damage causes because of the very low concentrations for the Rosalia area. The best fitting model was achieved when damage effects by ozone distribution (increase with altitude) and effects of water soluble oxidants in fog areas (increase in valleys) were taken into major consideration with soil acidity and tree age as minor influences (Cabela et al., 1986).

IV. Summary

With most soils and watersheds sufficiently buffered no acute acidification effects like lake dying or fish decline were reported until 1989. However, massive visible forest tree damage in vast areas of the country since 1981 to 1983 were attributed to acidic precipitation and air pollution. As forests make up almost 45% of Austria's surface and are economically and ecologically vital to large parts of the country, research activities have been dominated by the problem of forest damage.

Emissions of SO_2, NO_x and VOCs are moderate on a nationwide basis, but high emission densities may occur in some mountain valleys. SO_2 was cut almost by 60% from 1980 to 1985; measures to reduce NO_x and VOCs have been introduced but have not yet been effective. Air quality is monitored continuously in 153 automated monitoring stations (145 $\times$ SO_2, 95 $\times$ NO/NO_2, 37 $\times$ O_3). Thirty-two stations are located in forested and rural areas. SO_2 and NO_x ambient air concentrations are regarded as problems mainly in urban areas or in mountain valleys close to emission sources. O_3 seems to be very high in most parts of the country. Rural areas show summer monthly means of 70 to 150 µg m^{-1} (35 to 75 ppb) O_3. European synoptic data show Austria and Switzerland to have the highest O_3 levels from all network stations, so the high O_3 may be a serious transboundary continental problem. However, detailed synoptic views of the air quality situation in Austria are not possible because of the countries unhomogenous surface. Wet and bulk deposition is monitored with 69 stations; 19 stations have been set up below the canopy in forests. There is a very high variability of deposition rates. In rural open fields deposition is 7 to 15 kg ha^{-1} yr^{-1} SO_4–S (as S) and 3 to 7 kg ha^{-1} yr^{-1} NO_3–N (as N). There is a trend of lower S– and N– input with altitude. Forest deposition rates are much higher: 15 to 54 kg ha^{-1} yr^{-1} SO_4–S (as S), 5 to 24 kg ha^{-1} yr^{-1} NO_3–N (as N), and 12 to 49 kg ha^{-1} yr^{-1} total N. From the data it may be concluded that acidic depositions and N– inputs pose a very high long-term risk to forest ecosystems and may not be tolerated in the view of stable forest. A deposition-related forest soil acidification within the last 25 years has been confirmed.

When sensitive streams, rivers, and lakes were surveyed for acidification from 1983 to 1986, no acute effects were found. The lowest recorded pH in rivers was 4.8. A considerable risk was reported for the very labile ecosystems in sensitive small high altitude lakes: In 10% of the surveyed lakes the carbonate buffer capacity was exhausted. There are two annual nationwide forest damage monitoring systems: In the 16 $\times$ 16 km grid Bioindicator Grid, needle sulfur, fluorine, and nutrients are determined in tree leaves and needles to identify areas of SO_2 and HF pollution. The 4 $\times$ 4 km grid Forest Condition Inventory (WZI) refers to leaf/needle losses/discoloration, and so on, with the goal to get information on visible tree health regardless of possible causes. Since 1985 when the WZI started, no general deterioration of forest conditions was noticed; from 1986 to 1989 tree appearance rather seemed to improve. Because of its high cost the WZI was partly replaced in 1989 by an annual IRC aerial photography inventory.

Forest damage causal analysis is done within a coordinated research program (FIW). Part of the research strategy of the program from 1984 to 1989 was to concentrate all field experiments in only three different locations. Results show that air pollution is only one of several factors for the destabilization of forest ecosystems but of all air pollutants, O_3 and other photooxidants may be the most important. Tree physiological research did not bring any significant results with respect to damage symptom causality because too many unknown anthropogenous and natural factors influence open ecosystems; thus interpretation of data is difficult. To overcome this dilemma a system analytical multirisk mapping project was tried and gave encouraging results.

Appendix

Principal Researchers (Institutions)

Meteorological Data:	H. Kolb	(3)
Air Quality Monitoring Networks:	K. Kienzl	(2)
Wet/Bulk Deposition Data:	S. Smidt	(1)
Aquatic Ecosystems:	R. Psenner	(6)
Bioindicator Network:	K. Stefan	(1)
Forest Condition Inventory:	J. Pollanschütz	(1)
Forest Soil Data:	G. Glatzel	(5)
FIW	E. Führer	(5)
A: Basic Data	G. Glatzel	(5)
B: Morphology	M. Gailhofer	(7)
C: Physiology	D. Grill	(7)
D: General Forest Ecology	E. Führer	(5)
E: Soil/Rhizosphere	W.E.H. Blum	(5)
F: Revitalization	J. Schmidt	(4)
G: System Analysis/Coordination	R. Orthofer	(4)
H: Emission Control Strategies	E. Cabela	(4)
J: Remote Sensing	K. Zirm	(2)

The numbers refer to affiliations and addresses below:

Institutions

1. Forstliche Bundesversuchsanstalt
   (Federal Forest Research Station)
   Tirolergarten   Tel: +43-(222)-823446
   A-1131 Vienna   Fax: +43-(222)-8045907
2. Umweltbundesamt
   (Federal Environmental Agency)
   Radetzkystrasse 2   Tel: +43-(222)-71158-0
   A-1031 Vienna   Fax: +43-(222)-71158-4258
3. Zentralanstalt für Meteorologie
   und Geodynamik
   (Central Meteorological Service)
   Hohe Warte 38   Tel: +43-(222)-364453
   A-1190 Vienna   Fax: +43-(222)-3691233
4. Österreichisches Forschungszentrum
   (Austrian Research Center)   Tel: +43-(2254)-80-0
   A-2444 Seibersdorf   Fax: +43-(2254)-802118

5. Universität für Bodenkultur
   (University of Agriculture)
   Gregor Mendel-Strasse 33   Tel: +43-(222)-342500-0
   A-1180 Vienna   Fax: +43-(222)-3691659
6. Institute of Limnology
   Gaisberg 116   Tel: +43-(6232)-125
   A-5310 Mondsee   Fax: +43-(6232)-578
7. Universität Graz
   Schubertstrasse 51   Tel: +43-(316)-380-0
   A-8010 Graz   Fax: +43-(316)-382130

Acknowledgments

Thanks to S. Smidt from the Federal Forest Research Station for cooperation. Part of the work was funded with a grant from the Federal Ministry of Science and Research.

References

Albert, R., and K.M. Hübler. 1987. Einige Aspekte zum Ionenhaushalt und Mineralstoffwechsel von Fichten. In E. Führer and F. Neuhuber (eds.), *FIW-Bericht 1987: Ergebnisse aus der Immissionsforschung*, 100–117. Wien: BMWF.

Bermadinger, E., D. Grill, and H. Pfeifhofer. 1987. Epicuticularwachse bei Fichten aus Gebieten mit "klassischen" und "neuartigen" Waldschäden. In E. Führer and F. Neuhuber (eds.), *FIW-Bericht 1987: Ergebnisse aus der Immissionsforschung*, 89–97. Wien: BMWF.

Blum, W.E.H., and N. Rampazzo. 1987. Chemisch-mineralogische Bodenzustandsänderungen im Einflußbereich des Buchenstammablaufes. In E. Führer and F. Neuhuber (eds.), *FIW-Bericht 1987: Ergebnisse aus der Immissionsforschung*, 69–81. Wien: BMWF.

BMFT. 1987. *Statusseminar Ursachenforschung zu Waldschäden KFA Jülich 30.März-3.April 1987*. Tagungsbericht Jül-Spez-413, Jülich 1987. 350 pp.

BMWF. 1986. *Forschungsstrategien gegen das Waldsterben*. Federal Ministry of Science and Research, Wien. 71 pp.

Bolhar-Nordenkampf, H. R. (ed.) 1989. Stress-physiological Ecosystem Research Altitude Profile Zillertal. *Phyton* 28 (3) Special Issue, 302 pp.

Butz, I., and M. Rydlo. 1987. *Fischereibiologische Untersuchungen hinsichtlich Gewässerversauerung in einigen Mühlviertler Bächen*. Paper presented at workshop "Fischethologie and Fischökologie," April 24–26, 1987, Innsbruck.

Cabela, E., W.D. Grossmann, A. Kopcsa, and W. Weber. 1986. *Anwendung der Methodik der Zeitkarte auf Österreich. Pilotprojekt Lehrforst Rosalia*. OEFZS-A—0932, Seibersdorf. 121 pp.

Cowling, E., B. Krahl-Urban, and C. Schimansky. 1986. Wissenschaftliche Hypothesen zur Erklärung der Ursachen. In E. Papke, B. Krahl-Urban, K. Peters, and C. Schimansky (eds.), *Waldschäden. Ursachenforschung in der Bundesrepublik Deutschland und den Vereinigten Staaten von Amerika*, 120–125. KFA Jülich. 137 pp.

Eber, G., and H. Hojesky. 1987. *Luftmeßnetze in Österreich—Übersicht.* UBA-Report, Wien. 39 pp.

Eckmüllner, O., and H. Sterba. 1987. Zuwachsverlagerungen als sensibler Indikator für künftige Zuwachsverluste. In E. Führer, and F. Neuhuber (eds.), *FIW-Bericht 1987: Ergebnisse aus der Immissionsforschung,* 100–117. Wien: BMWF.

Ellenberg, H., R. Mayer, and J. Schauermann. 1986. *Ökosystemforschung. Ergebnisse des Sollingprojekts 1966–1986.* Stuttgart: Ulmer. 507 pp.

EMEP. 1984. *Co-operative programme for monitoring and evaluation of the long-range transmission of air pollutants in europe.* EMEP/MSC-W Report 1/84.

Energiebericht. 1986. *Energiebericht 1986 der österreichischen Bundesregierung.* Federal Ministry of Trade and Industry, Wien.

Führer, E. 1985. Integrierte Waldschadensforschung—Integrierte Waldschadensvorsorge. In E. Führer (ed.), *FIW-Bericht 1985: Ergebnisse aus der Immissionsforschung,* 178–222. Wien: BMWF.

Gailhofer, M., and G. Zellnig. 1987. Die Ultrastruktur der Chloroplasten aus dem Mesophyll schadstoffbelasteter Nadeln von Picea Abies (L.)Karst. In E. Führer, and F. Neuhuber (eds.), *FIW-Bericht 1987: Ergebnisse aus der Immissionsforschung,* 118–123. Wien: BMWF.

Glatzel, G., M. Englisch, and M. Kazda. 1985. Forschungsarbeiten zum Schwefelhaushalt von Waldökosystemen. In E. Führer (ed.), *FIW-Bericht 1985: Ergebnisse aus der Immissionsforschung,* 72–81. Wien: BMWF.

Glatzel, G., K. Katzensteiner, M. Kazda, Kühnert, G. Markart, and D. Stöhr. 1988. Eintrag atmosphärischer Spurenstoffe in österreichische Wälder; Ergebnisse aus vier Jahren Depositionsmessung. In E. Führer, and F. Neuhuber (eds.), *FIW-Symposium 1988,* 60–72. Wien: BMWF.

Glatzel, G., M. Kazda, D. Grill, G. Halbwachs, and K. Katzensteiner. 1987. Ernährungsstörungen bei Fichte als Komplexwirkung von Nadelschäden und erhöhter Stickstoffdeposition—ein Wirkungsmechanismus des Waldsterbens? *Allg.Forst-u.J.-Ztg.* 158(5/6): 91–97.

Glatzel, G., M. Kazda, and G. Markart. 1986. Winter deposition rates of atmospheric trace constituents in forests. Assessment of total input. In H.W. Georgii (ed.), *Atmospheric Pollutants in Forest Areas,* 101–108. Dordrecht: D. Reidel.

Glatzel, G., E. Sonderegger, M. Kazda, and H. Puxbaum. 1983. Bodenveränderungen durch schadstoffangereicherte Stammablaufniederschläge in Buchenbeständen des Wienerwaldes. *Allgemeine Forst Zeitschrift 1983* (26–27), 693–694.

Grennfelt, P., J. Saltbones, and J. Schjoldager. 1988. *Oxidant data Collection OECD-Europe 1985–87 (Oxidate).* NILU 31/88, O-8535, Lillestrom. 175 pp.

Grossmann, W.D. 1988. Products of photo-oxidation as a decisive factor of the new forest decline? Results and considerations. *Ecological Modelling* 41: 281–305.

Honsig-Erlenburg, W., and R. Psenner. 1986. Zur Frage der Versauerung von Hochgebirgsseen in Kärnten. *Carinthia II,* 176(96): 443–461.

Horak, O., and L. Zvacek. 1987. *Mikronährstoffe, toxische Schwermetalle und Aluminium in Waldökosystemen.* OEFZS-A-0953, Seibersdorf. 87 pp.

Janauer, G.A. 1986. *Untersuchungen zur Versauerung österreichischer Fließ—gewässer.* BMLF-Forschungsarbeiten Wasserwirtschaft Wasservorsorge, Wien. 50 pp.

Kazda, M. 1990. Sequential stemflow sampling for estimation of dry deposition and crown leaching in beech stands. In A.F. Harrison, P. Ineson, and O.W. Heal (eds.), *Nutrient Cycling in Terrestrial Ecosystems.* London: Elsevier.

Kazda, M., and G. Glatzel. 1986. Schadstoffbelasteter Nebel fördert die Infektion von Fichtennadeln durch pathogene Pilze. *Allgemeine Forst Zeitschrift 1986* (18): 436–438.

Kolb, H., and M. Bogner. 1988. Ozonbelastung an den FIW-Stützpunkten—eine Betrachtung im synoptischen Scale. In E. Führer and F. Neuhuber (eds.), *FIW-Symposium 1988,* 73–86. Wien: BMWF.

Kolb, H., and H. Scheifinger. 1986. *Immissionsklimatologische Studie zum Lehrforst Rosalia.* BMWF-Forschungsbericht 36.021, Wien. 70 pp.

Kolb, H., H. Scheifinger, M. Bogner, and P. Weihs. 1988. *Immissionsklimatologische Untersuchungen der Meßstellen der FIW.* BMWF-Forschungsbericht 36.018/2, Wien. 285 pp.

Kolb, H., and R. Werner. 1985. Ergebnisse immissionsklimatologischer Untersuchungen in den F.I.W. Untersuchungsgebieten. In E. Führer (ed.) *FIW-Bericht 1985: Ergebnisse aus der Immissionsforschung,* 101–110. Wien: BMWF.

Krapfenbauer, A., W. Grossmann, J. Gasch, G. Gutschik, H. Wagner, and E. Ulrich. 1989. Ökosystemorientierte kausalanalytische Studien zu Fragen des Waldsterbephänomens. BMWF-Forschungsbericht 36.052, Wien. 206 pp.

McLaughlin, S.B. 1985. Effects of air pollution on forests. A critical review. *JAPCA* 35(5), 512–534.

NAPAP. 1987. *NAPAP interim assessment. Vol. IV: Effects of acidic deposition,* Washington, DC.

Nielsson, S.I., H.G. Miller, and J.D. Miller. 1982. Forest growth as a possible cause of soil and water acidification: An examination of the concepts. *Oikos* 39: 40–49.

OÖLR. 1984–1988. *Monatsberichte des automatischen Luftüberwachungsnetz Oberösterreich.* Monthly Reports. Amt der oberösterreichischen Landesregierung, Linz.

Orthofer, R., and G. Urban. 1989. *Abschätzung der Emissionen von flüchtigen organischen Verbindungen in Österreich.* OEFZS—4492, Seibersdorf. 163 pp.

Pollanschütz, J. 1987. Periodische Luftbildinventur—ein Teilprojekt des österreichischen Waldschadens-Beobachtungssystem. *Österr Forstzeitung* 8: 74–76.

Pollanschütz, J., and M. Neumann. 1987. *Waldzustandsinventur 1985 und 1986. Gegenüberstellung der Ergebnisse.* FBVA-Berichte 23/1987, Wien. 98 pp.

Psenner, R. 1987. *Depositioni acide in Austria ed effetti sulle acque dolci e sulla vegetazione.* Simposio "Deposizioni acide, un problema per acque e foreste." Pallanza (Italy), April 9–10, 1987.

Psenner, R., K. Arzet, A. Brugger, J. Franzoi, F. Hiesberger, W. Honsig-Erlenburg, F. Horner, W. Müller, U. Nickus, P. Pfister, P. Schaber, and F. Zapf. 1988. *Versauerung von Hochgebirgsseen in kristallinen Einzugsgebieten Tirols und Kärntens.* BMLF-Forschungsarbeiten Wasserwirtschaft Wasservorsorge, Wien. 335 pp.

Puxbaum, H., and E. Ober. 1987. *Backgroundstation Exelberg.* UBA-Report, Wien. 141 pp.

Puxbaum, H., C. Rosenberg, M. Gregori, C. Lanzersdorfer, E. Ober, and W. Winiwarter. 1988. Atmospheric concentrations of formic and acetic acid and related compounds in eastern and northern Austria. *Atmospheric Environment* 22(12): 2841–2850.

Puxbaum, H., M. Weber, and G. Pech. 1985. *Occurrence of Inorganic and Organic Components at Four Source-Dominated Sites in Europe.* Proceedings Symposium on Heterogenous Processes in Source Dominated Atmospheres. LBL 20261, CONF-85, 1077, UC 11. New York.

Rosenberg, C., W. Winiwarter, M. Gregori, G. Pech, V. Casensky, and H. Puxbaum. 1988. Determination of inorganic and organic volatile acids, NH_3, particulate SO_4^{2-}, NO_3^- and Cl^- in ambient air with an annular diffusion denuder system. *Fresenius Z Anal Chem* 331: 1–7.

Schöpfer, W., and J. Hradetzky. 1984. Der Indizienbeweis: Luftverschmutzung ist die maßgebliche Ursache der Walderkrankungen. *Forstw Centralblatt* 103(4–5): 231–248.

Smidt, S. 1983. *Untersuchungen über das Auftreten von Sauren Niederschlägen in Österreich.* Mitteilungen Forstliche Bundesversuchsanstalt 150, Wien. 88 pp.

Smidt, S. 1986. *Bulkmessungen in Waldgebieten Österreichs.* FBVA-Berichte 13/1986. 32 pp.

Smidt, S. 1988a. Luftschadstoffmonitoring in österreichischen Waldgebieten. In E. Führer, and F. Neuhuber (eds.), *FIW-Symposium 1988*, 39–59. Wien: BMWF.

Smidt, S. 1988b. *Messungen der nassen Deposition in Österreich. Meßstellen, Jahresmeßergebnisse, Literatur.* FBVA-Berichte 27/1988.

Stefan, K. 1987. *Ergebnisse der Schwefel- und Nährstoffbestimmungen in Pflanzenproben des österreichischen Bioindikatornetzes.* VDI-Berichte 609, 555–580.

Stefan, K. 1989. Grundnetz des österreichischen Bioindikatornetzes—Ergebnisse der Schwefelanalysen der Probenahme 1988 und Vergleich der Resultate der von 1983 bis 1988 bearbeiteten Grundnetzpunkte. FBVA-Bericht BIN-S 42/1989, Federal Forest Research Station, Vienna 1989.

Stefan, K. 1990. Grundnetz des österreichischen Bioindikatornetzes-Ergebnisse der Schwefelanalysen der Probennahme 1989 und Vergleich der Resultate der von 1983 bis 1989 bearbeiteten Grundnetzpunkte. FBVA-Berichte BIN-S 52/1990, Vienna.

Stöhr, D. 1984. *Waldbodenversauerung in Österreich. Veränderung der pH-Werte von Waldböden während der letzten Dezennien.* BMGU Forschungsbericht, Wien. 165 pp.

Ulrich, B., and E. Matzner. 1983. *Abiotische Folgewirkungen der weiträumigen Ausbreitung von Luftverunreinigungen.* UBA-Forschungsbericht 104:02:65, Berlin. 221 pp.

Zirm, K., F. Fibich, J. Hackl, H. Malin, H. Mauser, and M. Weinwurm. 1985. *Erhebung der Vitalität des Waldes in Vorarlberg.* ÖBIG-Report, Wien. 83 pp.

Acidification Research in the Federal Republic of Germany

H.D. Gregor*

Abstract

Acidification research in the Federal Republic of Germany is integrated into the overall air pollution monitoring and research program and deals with all subjects from emission, transmission, and deposition to environmental effects and abatement strategies as well as control technologies. The Federal Environmental Agency (*Umweltbundesamt,* UBA) plays an important role in the realization of the Environmental Research Plan (UFOPLAN).

After the onset of serious forest damage a massive research program was initiated around 1983. It is coordinated by the Interministerial Working Group "Forest Damage/Air Pollution" (IMA), combining federal as well as Laender ministries. The program is part of the national "Action Program to Save the Forest" (1985) and covers approximately 660 projects (1982 to 1988) in 15 research areas including laboratory work and field experiments at more than 140 forest sites. The state of science is reviewed by the Scientific Advisory Council (FBW). As in the overall environmental research in the Federal Republic, international cooperation is also an important part of the forest damage research program. In addition to research on health effects, environmental damage to materials, and effects on soils and ecosystems, dispersion and deposition models as well as monitoring and control technologies are investigated. Gene ecological research is expected to supply further guidance for the protection of genetic resources.

*Umweltbundesamt (Federal Environmental Agency), Bismarckplatz 1, D-1000 Berlin 33, Federal Republic of Germany.

139

I. Introduction

Acidification research has to be seen in the context of research done in the field of air pollution effects research as a whole. It is the basis for legislative measures for the reduction of environmental pollution.

Because air pollution is also a transboundary problem, the Federal Republic of Germany is actively involved in research cooperation bilaterally or, for example, in the framework of UN-ECE International Cooperative Programs (ICPs). The Federal Republic leads or participates actively in the ICPs on forests, crops, materials, freshwaters, and integrated monitoring. In its international cooperation, the Federal Republic also supports the critical levels/loads concept (de Vries, 1988; Federal Environmental Agency, Umweltbundesamt, 1988a; Gregor, 1988; Nilsson and Grennfelt, 1988) as a scientific basis for environmental policy supplementing the best available technology approach (Jost, von Moltke, and Persson, 1988). It funds research to fill in gaps of knowledge in this field (e.g., long-term low concentration fumigation experiments and the influence of natural stressors). There are still some problems with the critical levels/loads concept, as critical levels for SO_2 and NO_x are generally exceeded in industrial areas and for O_3 in several regions (e.g., at higher altitudes), and critical loads are exceeded over large areas. The present policy of emission control in the Federal Republic of Germany— although very effective—is hardly capable of responding to small-scale differentiations especially regarding critical loads, as measures are more oriented to dealing with larger regions, or with single sources and source categories.

It is, for instance, true that deposition in forest areas is highly differentiated on a small scale, but the input from emission sources is relatively homogeneous, with differentiation occurring as a result of varying receptor characteristics such as at forest margins, and so on.

Because present means for the small-scale control of deposition criteria—determined by specific receptor characteristics—are lacking, critical levels/loads are at the moment principally regarded as just guidelines for establishing "more realistic" target loads.

II. Research on the Effects of Air Pollution

A. Effects on Human Health

Several issues are important in the area of air pollution's effect on human health, although they are not elaborated on in this chapter. In the area of environmental carcinogens and environmental mutagens, substances under investigation (and consideration for further regulation) include asbestos, PAHs, benzene, cadmium, arsenic, nickel, chromium and their compounds, among others.

Recently more effort has been aimed toward indoor pollution (SRU, 1987a); and effects of smog episodes are still being explored by a large number of investigations (BMU, 1988; UBA, 1989a).

B. Environmental Effects

1. Air Pollution Effects on Forests

Six consecutive annual full-scale forest damage surveys (BML, 1988; Federal Ministry for Food, Agriculture and Forestry [BML], 1987) in the Federal Republic of Germany from 1983 to 1988 resulted in the following picture.

The extent of total damage increased drastically until 1984, when 50% (in 1983: 34%) of the total forest area showed signs of damage (damage classes 1 to 4; see Table 4–1). In the following years only minor changes in forest vitality were recorded (1985, +2%; 1986, +2%; 1987, −2%). In 1988 no changes in total damage (all tree species, all damage classes) were found. The affected area now covers 52% of the total forest area (i.e., 3.9 out of 7.5 million ha) (see Figure 4–1). If damage stage 1 is treated as a warning stage only and is not considered, 15% (1.1 million ha) of the total forest area shows more severe signs of damage.

From the early 1980s damage trends appeared differently in different tree species. First coniferous trees (silver fir, Norway spruce, Scots pine) were more intensely affected than broad-leaved species such as beech and oak. But since 1986 beech and oak have shown higher damage figures than the coniferous species, which in turn showed some recovery after 1985–1986. Like 1984, 1985, and to some extent 1986, 1987 and 1988 were also years with rather favorable climatic conditions for forest growth. Nevertheless signs of recovery are limited. Therefore the situation of forests in the Federal Republic of Germany remains a matter of

Table 4–1. Key for the determination of damage classes (BML, 1988).

Needle/ leaf loss	Discoloration			
	<10%	11%–25%	26%–60%	>60%
	Damage class			
<10%	0	0	1	2
11%–25%	1	1	2	2
26%–60%	2	2	3	3
61%–99%	3	3	3	3
Dead	4	4	4	4

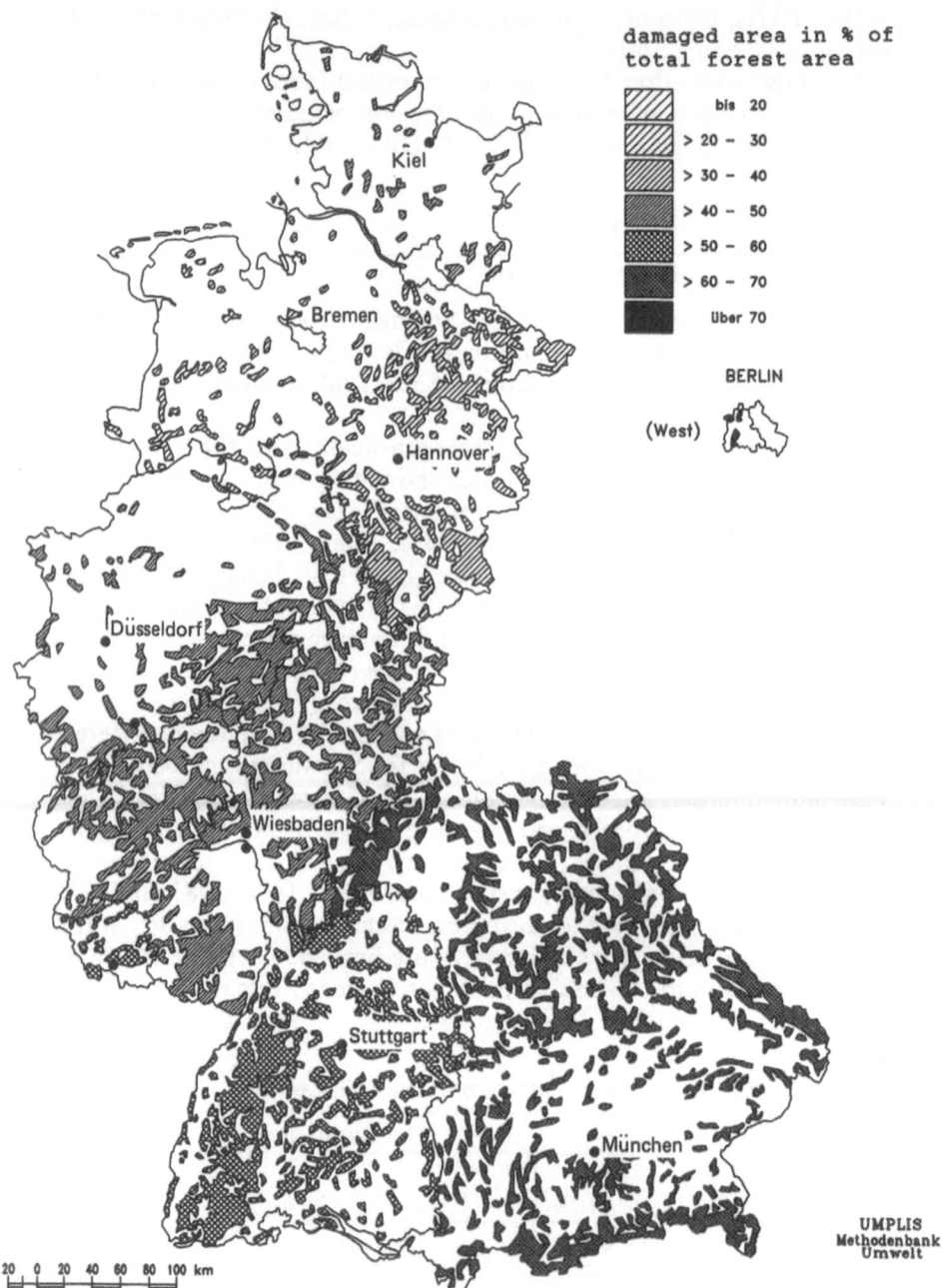

Figure 4–1. 1987 Forest damage survey in the Federal Republic of Germany—all tree species, all damage classes.

great concern. This may also be true for a number of ECE countries which have participated in the ECE-wide forest damage assessment program (BML, 1988; Global Environment Monitoring System, 1987) and now find themselves in the group with high damage figures (see Figure 4-2).

It has to be realized at this point that the assessment of damage to forest ecosystems until today has only considered a few visible symptoms (e.g., discoloration, needle/leaf loss), whereas other, especially invisible effects, have not been assessed. Furthermore, other components of the forest ecosystems (Gregor, 1984) were not at all included. They are often disregarded even in forest damage research.

When forest damage increased drastically in the early 1980s, research efforts were intensified, and considerable funds were directed toward investigation of the causes of forest decline as well as strategies for their

Figure 4-2. 1987 Forest damage survey in Europe.

protection from further deterioration. At the same time the "Intermin-isterial Working Group" (IMA) was established by the cabinet ministers as an effective coordinating body. This coordination has proven to be very successful and unparalleled in any other field of cooperation. To-gether with its Scientific Advisory Council (*Forschungsbeirat "Waldschäden/Luftverunreinigungen,"* FBW) and its Secretariat (at the Federal Environmental Agency) IMA directed funds of approximately DM 250 million to more than 660 individual projects in the 15 major disciplines of forest damage research (see Figures 4–3 and 4–4) between 1982 and 1988.

Main areas of research are

improvement of damage assessment methods to include ecosystem
    parameters,
clarification of damage mechanisms in forest tree species,
deposition monitoring in forests,
atmospheric chemistry.

An overview about all current forest damage research activities was pub-lished by the IMA secretariat (UBA, 1988b).

In future projects, emphasis will be put on the improvement of diag-nostic tools and remote sensing techniques for application in forest dam-age assessment and the harmonization of sampling, measuring, and monitoring techniques, aiming at the same time at a better understand-ing of biochemical/biophysical parameters and of combination effects between air pollution and other stress factors such as climate, pests, and so on.

The overall research program has efficiently combined scientists from fields that had never before cooperated on projects. Cooperation was en-hanced by selecting common research sites (Bundesminister für Forschung und Technologie [BMFT], 1985a; Umweltbundesamt, 1987a) and by holding interdisciplinary as well as specific seminars for the bene-fit of rapid information exchange among the scientists involved. Also in-ternational cooperation and comparison of research strategies was steered and/or supported by IMA or its members (Bundesminister für Forschung und Technologie [BMFT], 1985b) on the basis of numerous international agreements between the Federal Republic and the United States (Papke, Krahl-Urban, Peters, and Schimansky, 1987), Canada, the Netherlands, and so on. Today forest damage and forest damage research is included as a topic in many bilateral agreements, for example, with France (Silva), Czechoslovakia, or the United States, and there are many good examples of cooperative research projects. Supranational efforts in the framework of the EEC should also be mentioned here (e.g., the EEC fumigation program using open-top chambers).

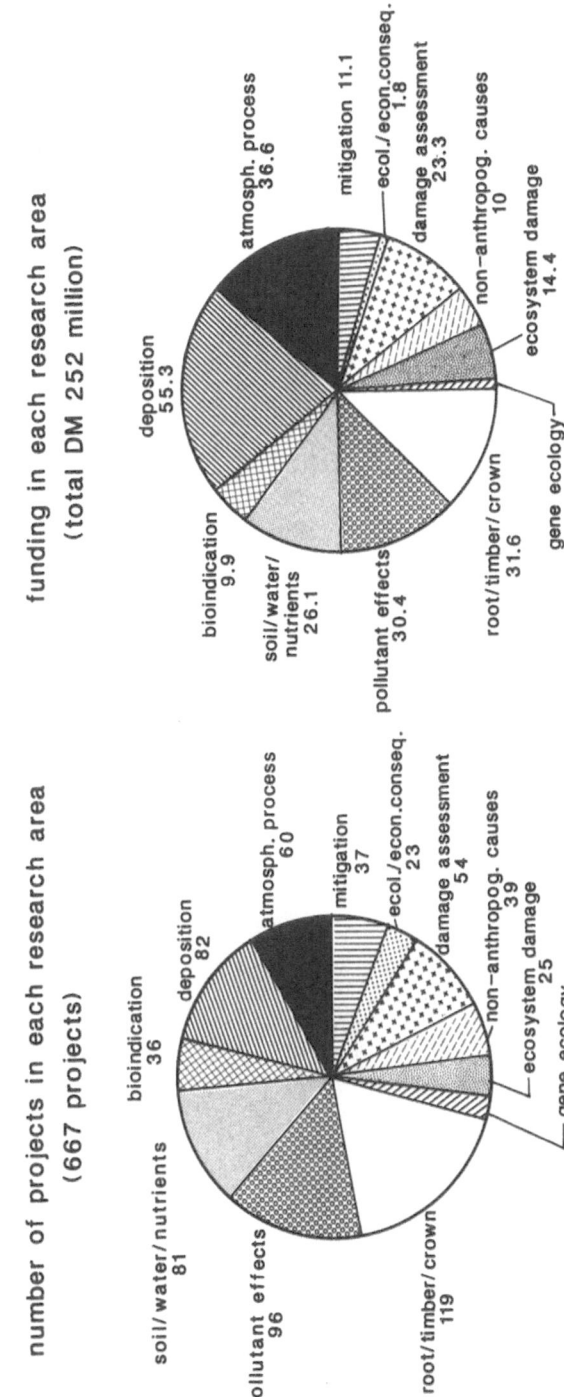

Figure 4–3. IMP-coordinated forest damage research: research areas, number of projects, and funding. (UBA, 1989a.)

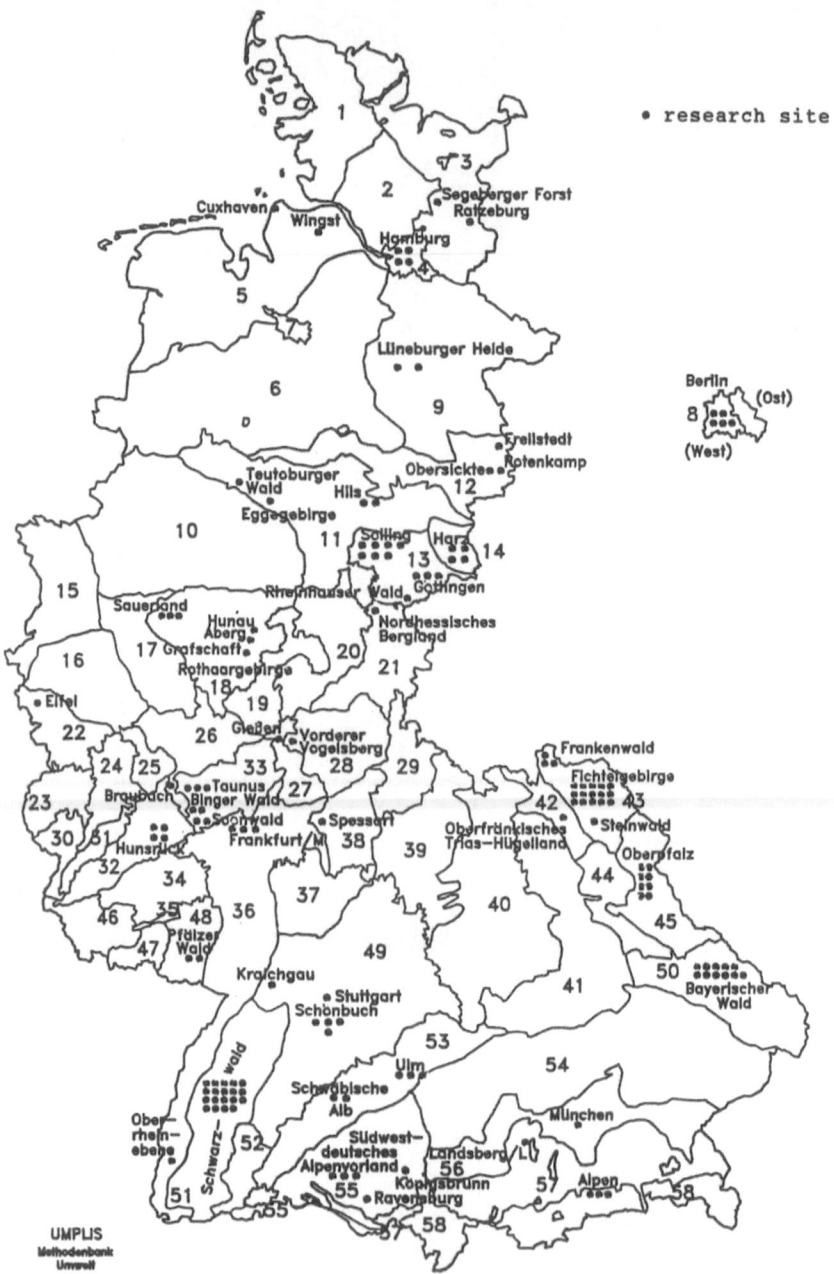

Figure 4–4. Location of forest damage research experimental sites (as of 1986).

The state of knowledge concerning the causes for forest decline in the Federal Republic of Germany has been assessed almost annually by several expert groups. From 1982 to 1984 several reports to the Conference of Environmental Ministers (BML, 1982) and the federal government (FBW, 1984; SRU, 1983) concluded, "that—although scientifically not totally proven—there was sufficient evidence for a major role of air pollution (especially SO_2, NO_x and their atmospheric conversion products)." At the scientific symposium "New Hypotheses for the Causes of Forest Damage" (Federal Environmental Agency (UBA), 1986b), a number of further hypotheses were discussed including

organic compounds (aromatic, chlorinated, etc.),
microwaves and radiation,
climate and site effects,
pests, and
ammonium.

The results of this meeting strengthened the established view of forest researchers and the Scientific Advisory Council (FBW, 1986) also stated that "hypotheses satisfactorily explaining forest decline with factors excluding air pollution had not been presented so far."

The Report of the Council of Environmental Advisors (SRU, 1987b) added special reference to the role of ammonia in forest damage. Most recently the Scientific Advisory Council edited a diagnostic handbook (Hartmann, Nienhaus, and Butin, 1988) of typical symptoms of the new type of forest damage together with symptoms caused by factors such as nutrient deficiencies, diseases, climatic stress, and so on. The Federal Action Program "Save the Forest," Phase III (BMU, 1989) not only describes the present situation in the forests but also covers all measures planned on the basis of forest damage research since 1983. Main subjects are

the state of the forests in the Federal Republic of Germany,
main areas of research,
effectiveness of pollution control aimed at the protection of forests, and
planned activities (national/international).

Topics receiving particular attention include

the situation of high elevation forests,
forests close to industrialized areas,
genetic and selective effects in tree species.
Special reference is made to the Federal/Laender program for the protection of gene resources.

2. Genetic Ecology

A new major research area has developed in the genetic field since the Federal Environmental Agency started a number of projects in 1983 dealing with changes of the genetic structure of forest tree populations (e.g., beech, spruce) under environmental stress. Since then about DM 2 million have been spent on investigations of gene ecological effects of air pollution on forest stands, on genetic effects of simultaneous impact of air pollution on root and shoot, on effects on reproduction processes, and on adaptation processes.

As it was found to be true that air pollution causes selection processes and thus leads to losses in genetic variability, precautions for the protection of gene resources have to be taken in order to maintain a potential for adaptation of tree populations to future environmental conditions. Therefore research will be intensified in the field of causes and mechanisms of gene ecological effects, as well as policy-oriented investigations of measures for the protection of genetic variability. This research involving federal research centers as well as university institutes will have to be highly interdisciplinary and cooperative. The Federal Action program "Save the Forest" (Bundesminister des Innern [BMI], 1985b) in its third phase, after approval by the cabinet ministers (Bundesminister für Umwelt, Naturschutz und Reaktorsicherheit [BMU], 1989) lists such research as an activity with high priority.

3. Effects on Ecosystems

Effects research is aimed at establishing a scientific basis for the definition of air quality criteria. Until recently, research was primarily concerned with the effects of single pollutants on individual plants/species; now attention is being directed more toward the effects of combinations of pollutants on plant communities and ecosystems, better reflecting actual field conditions. Many such projects were also discussed at the critical levels/loads workshops (Nilsson and Grennfelt, 1988; UBA, 1988a), and their results substantially supported the decisions reached there.

Ecological monitoring, using standardized biological test procedures (e.g., standard grass culture) or monitoring selected naturally occurring species is to be regarded as complementary to purely analytical procedures. It is used to record pollutant loads as well as to evaluate the effectiveness of control measures. Research deals with changes in ecosystems such as adverse effects on individual species or changes in species abundance and distribution. Further concerns include projects on integrated monitoring and ecosystems research. Two pilot studies, "The Influence of Man on High Altitude Ecosystems" (Haber, Schaller, and Spandau, 1986) and "Integrated Environmental Monitoring" (Fränzle, 1987), are aimed at the selection of indicators capable of sufficiently characterizing specific ecosystems.

The Federal Environmental Agency contributes to the ECE-Integrated Monitoring activity by extending its activities at permanent air quality measuring stations (background stations) to include soil, water, and bioindication.

4. Effects on Soil

Policy and research on soil effects are centered on the realization of the federal government's "Soil Protection Concept" (BMI, 1985a).

Activities are also directed to nationwide soils analysis programs, gathering data for agricultural soils and sampling forest soils in a dense grid. Research on procedure standardization is still in progress; in some fields data are already available.

The Federal Republic of Germany will intensify soil research by promoting R&D in the following areas:

deposition measurements with regard to nutrient supply and buffering capacity,
element cycling,
transfer processes between plants/soil/groundwater,
combination effects,
deposition pathways, and
establishing a basis for determining ecological and toxicological quality criteria for water, air, soils, plants, and food and fodder materials.

5. Effects on Materials

Research is concentrated on the establishment of damage mechanisms and dose-response relationships for various materials. And as there is no no-effect level for pollutant effects on materials, passive protective measures are being investigated for a range of cultural and industrial materials.

Major successes in protecting medieval stained glass have been achieved, and research in the field of conservation of historic monuments also continues to be actively promoted. In this context the Federal Republic of Germany contributes to the ECE-ICP on materials by running 6 of the 38 measuring stations and supplying copper and bronze reference samples.

Further activities include projects on the investigation of effects on mortar, brick structures, and other materials. The Federal Republic of Germany pilots the NATO-CCMS Study "Conservation of Historic Brick Structures" and funds research on the effects of air pollution on polymers, on concrete, and on metal objects of artistic and historic importance.

A major tool for better coordination of efforts for the conservation of cultural monuments in the Federal Republic of Germany, which could easily serve as an example for activities in other countries or internationally, is the "Information and Documentation Center on Environmental Effects on Cultural Monuments" (KUD), presently run as a pilot project at two institutions in the Federal Republic of Germany, one of which is the Federal Environmental Agency. Here, data on effects, restorative measures, ambient air quality, and the objects' history are documented and made available using specially developed electronic data processing (Fitz, 1989).

III. Air Quality Monitoring Oriented Research

Air pollutants cause various adverse effects (BMI, 1984). Their control requires information on the source, distribution, conversion, deposition, and type of pollutant (BMU, 1988, UBA, 1989a). To this end, the Federal Republic of Germany promotes laboratory, field, and pilot experiments to supplement monitoring and measuring activities on a larger scale (see Table 4-2).

The ECE concept of critical levels/loads demands a sophisticated tool to evaluate emissions and control strategies on a European scale. In order to determine the effectivity of abatement measures and to support envisaged international control strategy development, long-range dispersion models have to be developed and applied.

The Dutch-German research project PHOXA (Photochemical Oxidants and Acid Deposition Model Application within the Framework of Control Strategy Development), which started in 1984 (Pankrath, 1987), is especially designed for this purpose and consists of three branches: photochemical oxidant modeling, acidic deposition modeling, and databases.

Within the concept of critical levels/loads, regional differentiation and the required small-scale data on atmospheric input require the refine-

Table 4-2. Emission-control (air)-oriented research in the Federal Republic of Germany, 1983 to 1986. (BMU, 1988.)

Topic	Number of projects	Total funding (DM)
Emissions, emission control	289	660 Mio
Air quality monitoring	278	113 Mio
Effects research[a] of which	128	36 Mio
General effects	12	3 Mio
Health effects	13	5 Mio
Effects on ecosystems	73	22 Mio
Climate effects	9	2 Mio
Effects on materials	21	4 Mio

[a] Research on acidification of ground and surface waters is not elaborated on in this chapter.

ment of larger-scale dispersion calculation to evaluate source-oriented measures. Two methods of achieving this regional refinement are possible, either postprocessing or nesting. First experience of applying postprocessors has been gained within the application of the complex German-Canadian dispersion model TADAP (Transport and Deposition of Acidifying Pollutants) (Pankrath, Stern, and Builtjes, 1988). By use of PHOXA's modeling tools, including the results of IIASA's RAINS project (Alcamo et al., 1987), it is the intention to apply and extend these results to the ECE assessment criteria for critical levels/loads.

IV. Air Quality Monitoring in the Federal Republic of Germany

Monitoring of air pollution in the Federal Republic of Germany is mainly carried out by two groups: the individual states (Laender) within their own territories, and the Federal Environmental Agency on behalf of the federal government with its national monitoring network (see Table 4–3 and Figures 4–5 to 4–8).

Laender

activities:
determination of air pollution in areas with high population density or in
    highly industrialized areas; recently additional stations in rural areas
    ("forest stations") as a result of forest damage.

coverage:
monitoring station density regionally adequate

Federal Environmental Agency

activities:
national network initially installed for rural ambient air pollution moni-
    toring ("background data"), recently with more concentration on
    large-scale transboundary pollutant transport.

coverage:
monitoring station density roughly 60 × 60 km

Early Warning System

A new major effort in this field concerns the development of an "Early Warning System" for episodes of transboundary transport of pollutant concentration peaks, and of a dynamic information system on pollutant loads and special events, providing a basis for decisions to be made by the federal or Laender governments (UBA, 1987b).

Figure 4–5. Air quality and deposition monitoring network in the Federal Republic of Germany.

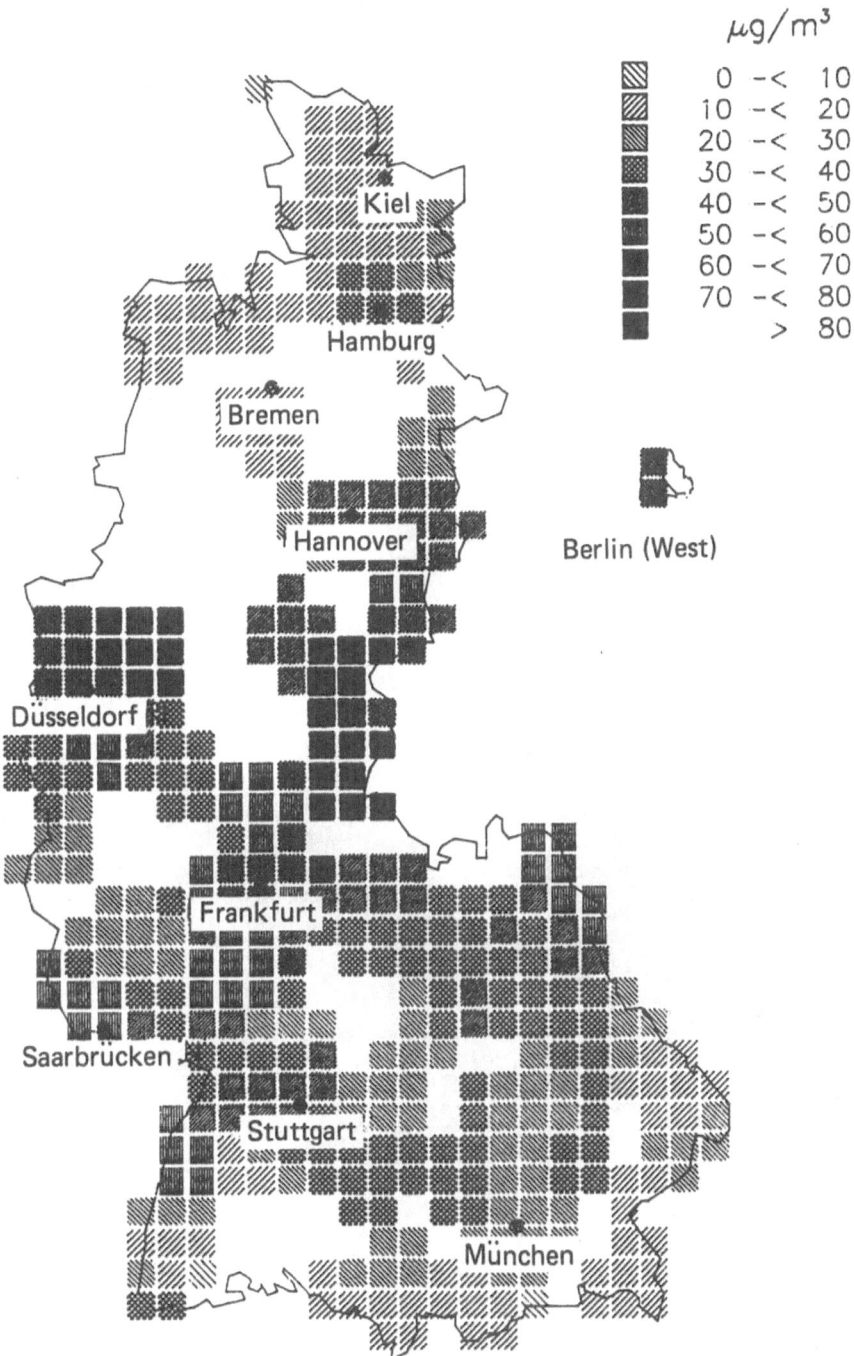

Figure 4–6. SO₂: annual mean ambient concentrations in the Federal Republic of Germany, 1985. (Adapted from UBA, 1986a.)

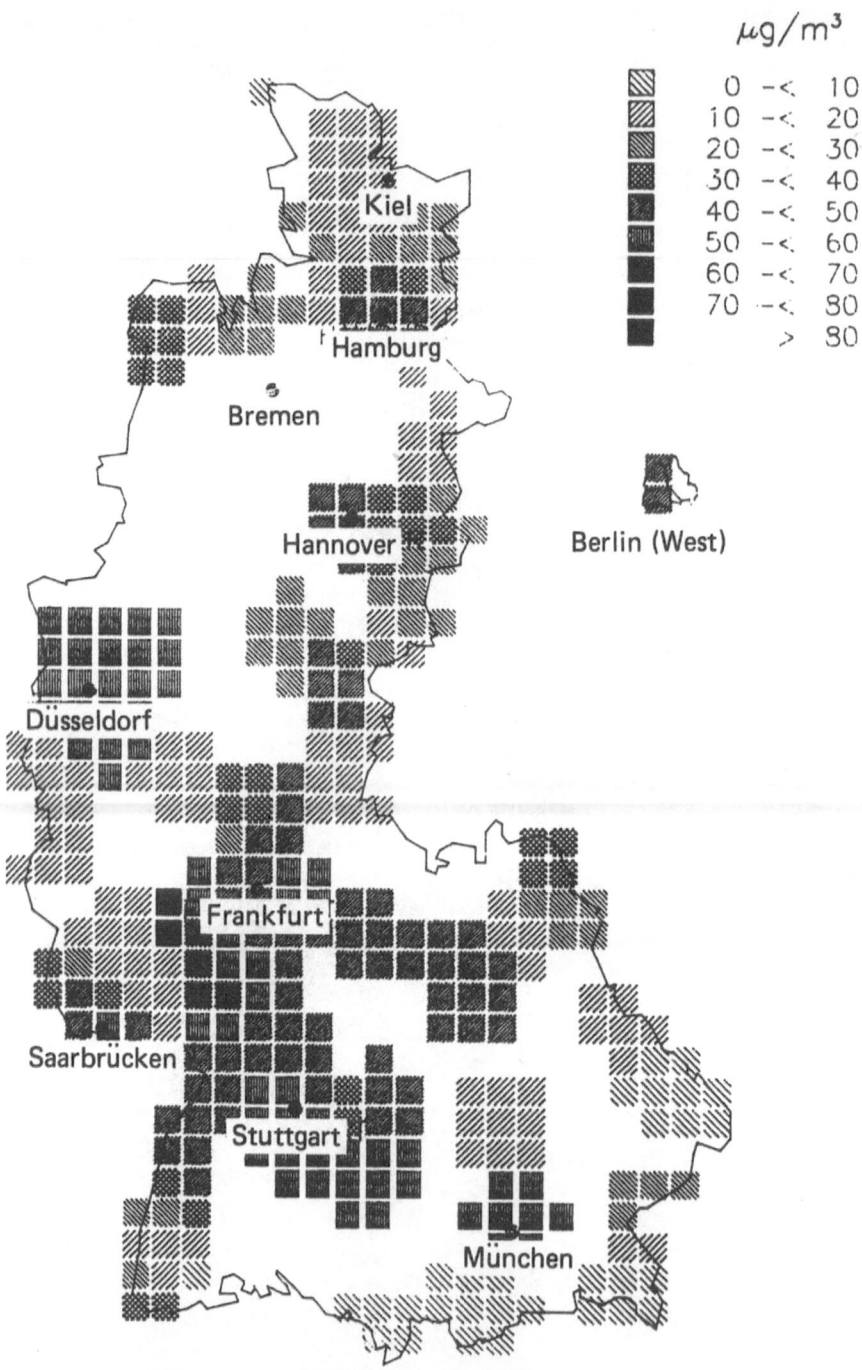

Figure 4–7. NO$_x$: annual mean ambient concentrations in the Federal Republic of Germany, 1985. (Adapted from UBA, 1986a.)

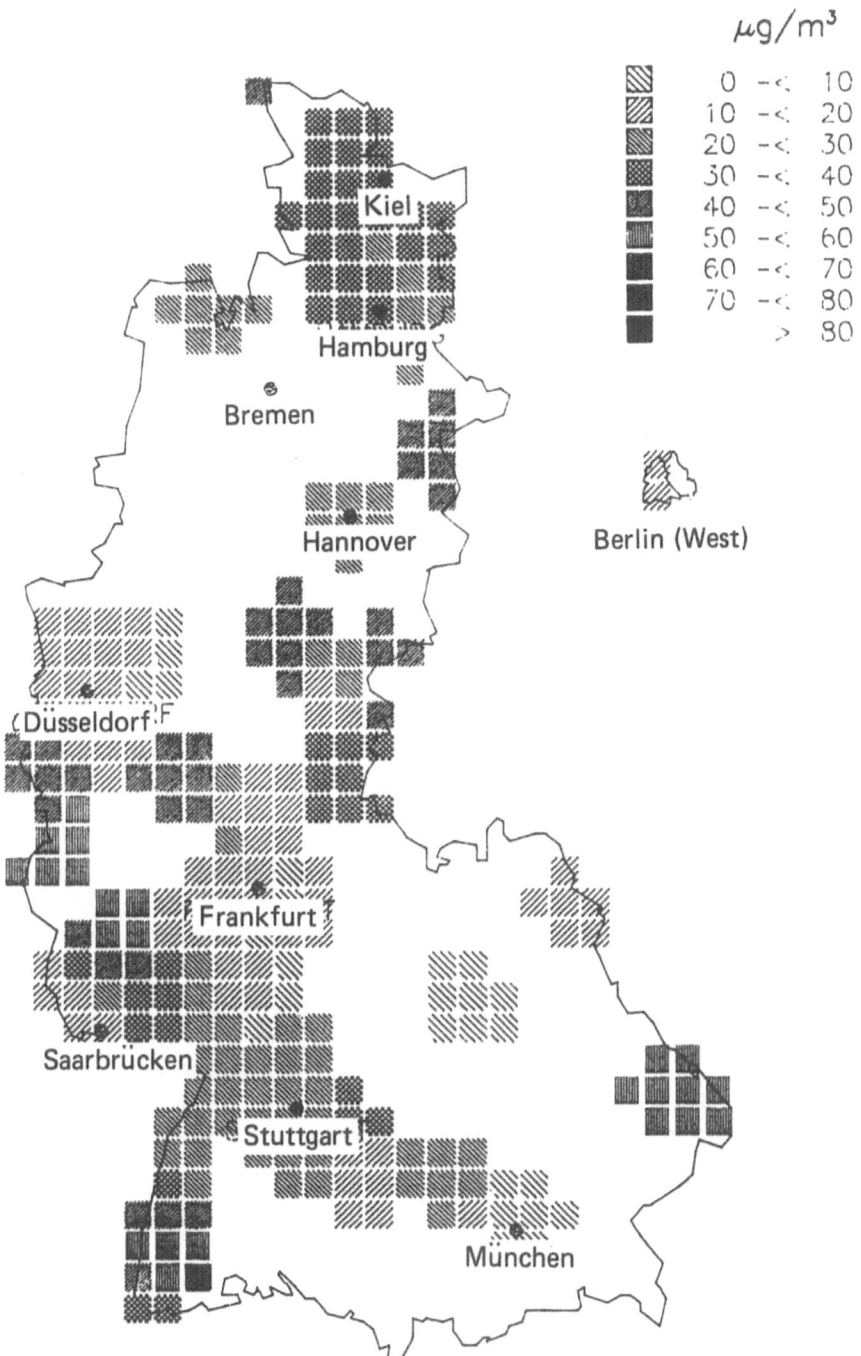

Figure 4–8. Ozone: annual mean ambient concentrations in the Federal Republic of Germany, 1985. (Adapted from UBA, 1986a.)

Table 4–3. Air quality monitoring stations in the Federal Republic of Germany: number, location, measured components. (UBA, 1987b.)

SO$_2$	CO	NO$_2$	NO	CnHm	O$_3$	CH$_4$	CO$_2$	TSP	MET	Laender & Cities*
										Components
				Number of monitoring stations						
36	33	36	36	9	36	—	16	33	26	Baden-Wuerttemberg
73	36	31	31	26	16	—	—	42	36	Bavaria
2	2	2	2	—	2	—	—	2	1	Bremen*
23	8	23	23	13	3	—	—	23	—	Hamburg*
33	25	33	33	7	23	—	—	22	23	Hesse
37	21	37	37	27	13	—	—	36	37	Lower Saxony
66	63	66	66	—	21	—	—	66	28	Northrhine-Westph.
7	—	7	7	7	7	7	—	7	7	Schleswig-Holstein
15	7	8	7	3	3	3	—	8	6	Saarland
16	14	18	18	8	12	1	—	14	10	Rhineland-Palatin.
43	21	13	13	1	7	—	1	18	2	Berlin*
17	—	13	13	—	13	—	5	17	13	Federal Environmental Agency

Legend:
CnHm: organic compounds (except methane)
TSP: total suspended particles
MET: meteorological parameters

References

Alcamo, J., Amann, M., Hettelingh, J. P., Holmberg, M., Hordij, K. L. Kälmari, J., Kauppi, L., Kauppi, P., Kornai, G., and A. Mäkelä. 1987. Acidification in Europe: A simulation model for evaluating control strategies. *Ambio* 16: 232–245.

Bundesminister des Inneren (BMI). 1984. *Report of the causes and prevention of damage to forests, waters and buildings by air pollution in the Federal Republic of Germany.* München. 140 pp.

Bundesminister des Inneren (BMI) (ed.). 1985a. *Bodenschutzkonzeption der Bundesregierung* Kohlhammerverlag Stuttgart, Berlin. 232 pp.

Bundesminister des Inneren (BMI) (ed.). 1985b. Aktionsprogramm Rettet den Wald. *Umweltbrief* Nr. 32 Bonn. 40 pp.

Bundesminitser für Ernährung Landwirtschaft und Forsten (BML). 1982. Schriftenreihe A: *Angewandte Wissenschaft* 273 Waldschäden durch Luftverunreinigungen Landwirtschaftsverlag Münster. 65 pp.

Bundesminister für Ernährung Landwirtschaft und Forsten (BML). 1988. *Waldschadenserhebung 1988* Bonn. 83 pp.

Bundesminister für Forschung und Technologie (BMFT). 1985a. *Umweltforschung zu Waldschäden* 2. Bericht Bonn. 80 pp.

Bundesminister für Forschung und Technologie (BMFT). 1985b. *Umweltforschung zu Waldschäden* 3. Bericht Bonn. 136 pp.

Bundesminister für Umwelt, Naturschutz und Reaktorsicherheit (BMU). 1986. *Daten zur Umwelt 1986/87* Erich Schmidt Verlag Berlin. 550 pp.

Bundesminister für Umwelt, Naturschutz und Reaktorsicherheit (BMU). 1988. *Vierter Immissionsschutzbericht der Bundesregierung* Bonn. 116 pp.

Bundesminister für Umwelt, Naturschutz und Reaktorsicherheit (BMU). 1989. *Aktionsprogramm Rettet den Wald,* III. Fortschreibung. Bonn. In press.

de Vries, W. 1988. Critical loads for sulphur and nitrogen. In *Air Pollution in Europe: Environmental Effects, Control Strategies and Policy Options,* Conference Discussion Document Vol. I, Norrtälje and Stockholm 1988. 30 pp.

Federal Ministry for Food, Agriculture and Forestry (BML). 1987. *1987 Forest Damage Survey* Bonn. 81 pp.

Fitz, S. 1989. MONUFAKT—The Federal Environmental Agency's database for the protection of historic monuments and cultural heritage. In *Proceedings Science, Technology and European Cultural Heritage,* Bologna June 1989. In press.

FBW, Forschungsbeirat Waldschäden/Luftverunreinigungen. 1984. In Bundesministerium für Forschung und Technologie (BMFT) (1985) *Umweltforschung zu Waldschäden* 3. Bericht Bonn. 136 pp.

FBW, Forschungsbeirat Waldschäden/Luftverunreinigungen. 1986. 2.Bericht Kernforschungszentrum Karlsruhe, Karlsruhe. 230 pp.

Fränzle, O. 1987. Erarbeitung und Erprobung einer Konzeption für die integrierte regionalisierende Umweltbeobachtung am Beispiel Schleswis-Holstein. In Umweltbundesamt (UBA) (ed.), *Instrumentarium zur ökologischen Planung.* UBA-Texte 14/87.

Global Environment Monitoring System (GEMS). 1987. *Forest Damage and Air Pollution* Report of 1986 Forest Damage Survey in Europe ECE/UNEP (ed.) Geneva. 48 pp.

Gregor, H.D. 1984. Auswirkungen der atmosphärischen Belastung auf Waldökosysteme. *Allgemeine Forstzeitschrift* 51/52:1279–1282.

Gregor, H.D. 1988. Critical levels for the effects of air pollutants on plants, plant communities and ecosystems. In *Air Pollution in Europe: Environmental Effects, Control Strategies and Policy Options,* Conference Discussion Document Vol I, Norrtälje and Stockholm 1988. 38 p.

Haber, W., J. Schaller, and L. Spandau. 1986. *Mögliche Auswirkungen der geplanten Olympischen Winterspiele 1992 auf das Regionale System Berchtesgaden.* Deutsches Nationalkommitee MaB. MaB-Mitteilung 22 220 p.

Hartmann, G., F. Nienhaus, and H. Butin. 1988. *Farbatlas Waldschäden, Diagnose von Baumkrankheiten* Ulmer Verlag Stuttgart. 256 pp.

Jost, D., K. von Moltke, and G. Persson. 1988. Management Strategies and Policy Options, In: *Air Pollution in Europe: Environmental Effects, Control Strategies and Policy Options,* Conference Discussion Document Vol II, Norrtälje and Stockholm 1988. 20 pp.

Nilsson, J., and P. Grennfelt. (eds.). 1988. *Critical Loads for Sulphur and Nitrogen,* Report from Skokloster workshop, Skokloster, Sweden Miljorapport 1988:15.

Pankrath, J. 1987. Modellinstrumentarium zur Untersuchung des großflächigen Stoffeintrags in Waldgebiete. Symposium *"Klima und Witterung im Zusammenhang mit den neuartigen Waldschäden"* GSF-Bericht 10/87 München. 19–50.

Pankrath, J., R. Stern, and P. Builtjes. 1988. Application of long-range transport models in the framework of control strategies: Example of photochemical air pollution. In K. Gefen and J. Löbel (eds.), *Environmental Meteorology*, 588–611. Amsterdam: Kluewer Academic Publishers.

Papke, H.E., B. Krahl-Urban, K. Peters, and C. Schimansky. 1987. *Waldschäden, eine Dokumentation* Projektträgerschaft für Biologie, Ökologie und Energie der Kernforschungsanlage Jülich (KFA) (ed.) im Auftrag des Bundesministeriums für Forschung und Technologie (BMFT) und der U.S. Environmental Protection Agency (EPA) Jülich. 137 pp.

Rat von Sachverständigen für Umweltfragen (SRU). 1983. *Sondergutachten Waldschäden und Luftverunreinigungen* Verlag W. Kohlhammer Stuttgart. 172 pp.

Rat von Sachverständigen für Umweltfragen (SRU). 1987a. Luftverunreinigungen in Innenräumen (Kurzfassung des Sondergutachtens) Bundesminister für Umwelt, Naturschutz und Reaktorsicherheit (BMU) (ed.) *Umweltbrief* Nr. 35 Bonn. 26 pp.

Rat von Sachverständigen für Umweltfragen (SRU). 1987b. *Umweltgutachten 1987* Bundesminister für Umwelt, Naturschutz und Reaktorsicherheit (BMU) (ed.) Bonn. 1362 pp.

Umweltbundesamt (UBA) (ed.). 1986a. *Daten zur Umwelt 1986/87* Erich Schmidt Verlag Berlin. 550 p.

Umweltbundesamt (UBA) (ed.). 1986b. *Wiss. Symposium Neue Ursachenhypothesen* UBA-Texte 19/1986 Berlin. 431 pp.

Umweltbundesamt (UBA) (ed.). 1987a. *Ballungsraumnahe Ökosysteme*; Erster Forschungsbericht Der Senator für Stadtentwicklung und Umweltschutz (ed.) Berlin. 61 pp.

Umweltbundesamt (UBA) (ed.). 1987b. *Monatsberichte aus dem Meßnetz*, May 1987, Berlin. 46 pp.

Umweltbundesamt (UBA) 1988a. *ECE Critical Levels Workshop*, Bad Harzburg March 1988, Final Draft Report 146 pp.

Umweltbundesamt (UBA) (ed.). 1988b. *Gesamtdarstellung der öffentlich geförderten Waldschädenforschung* UBA Texte 27/88.

Umweltbundesamt (UBA) 1989a. *Luftreinhaltung '88.* Berlin: Erich Schmidt Verlag.

Umweltbundesamt (UBA). 1989b. *Jahresbericht 1988* Berlin. 158 pp.

Acidic Precipitation Research in The Netherlands

A.H.M. Bresser[*]

Abstract

Acidic precipitation in The Netherlands is measured with a dense monitoring system. Total deposition has dropped from 6000 mol ha^{-1} a^{-1} to 5000 (in equivalents H$^+$). The distribution over substances, origin of the deposition, and wet and dry deposition is presented.

The Dutch Priority Program on Acidification allocates 80% for effect research, 10% for research on ammonia, and 10% for integrated assessment. It is financed by four ministries, oil refineries, and electricity producing companies. The program is coordinated by the National Institute of Public Health and Environmental Protection (RIVM). Effect research is centered at two forest sites (Douglas) and one heathland site where air, soil, and biota are continuously monitored. The research is supported by laboratory and field experiments. Ammonia research aims at better estimates of emissions from stables and fields. Integrated assessment is the focal point for the program and serves to formulate policy options.

The main concern is for forests. From 1983 to 1988 the percentage of hardly vital plus not vital forest increased from 9.5 to 21. Inputs from the atmosphere are considered important. Both coniferous and deciduous forests are damaged.

Hardly any heathland is left in The Netherlands without substantial growth of grasses, probably due to nitrogen. Small surface waters in sandy regions are almost all acidified. Damage to agricultural crops is estimated at 300 million guilders a year, monetary damage to materials at 200 million. Conclusions are related to critical loads and levels. Critical loads for total acid and nitrogen on forests and heathlands on poor sandy soils are surpassed 5 to 6 times. Heavily exposed objects (forest edges) receive 7 to 8 times the critical loads. Ozone concentrations are above critical levels but below levels where serious damage is expected.

[*]National Institute of Public Health and Environmental Protection (RIVM), P.O. Box 1, 3720 BA Bilthoven, The Netherlands.

159

I. Introduction

Acidic precipitation research in The Netherlands can be divided into three parts. The first part consists of research on the direct effects of atmospheric pollution on human health. It was already recognized in the 1970s that long-range transport contributed largely to concentrations of air pollutants in The Netherlands. In these years emphasis was on sulphur dioxide (SO_2). This led to monitoring of the situation, emission research, and modeling of the transport and transformation processes.

Nowadays the National Air Quality Monitoring System (Anon., 1986a) consists of 17 macro stations where SO_2, NO_x, O_3, CO, particulate matter with heavy metals and organic compounds, concentrations in rainwater, and accumulation of deposited pollutants on the soil are continuously monitored together with effects on a number of biological indicators (tobacco, tulips, and lichens) and groundwater quality. At another 60 stations in rural areas, SO_2 is measured; at several stations ozone and NO_x are also considered. In 15 urban situations the monitoring focuses on emissions from traffic and includes CO and NO_x. Results of all measurements are transmitted to a computer system at the National Institute of Public Health and Environmental Protection (RIVM) where data processing takes place with a computer model that calculates the air pollution situation throughout the country. This situation is broadcasted permanently and updated every three hours in several teletext pages.

Emission research led to an Emission Registration System for large sources in The Netherlands and to emission factors for estimations of annual and peak emissions from smaller sources. The inventory of emissions for this system is updated about every four years. Modeling led to a set of tools for calculations on local, national, and regional air pollution situations for SO_2, NO_x, O_3, NH_x, and other substances. Historical, actual, and future air pollution situations can be simulated with a variable time step and a variable spatial resolution. With these models episodes of high air pollution concentrations can be simulated, but it was also possible to simulate the transport and deposition of radioactivity from the Chernobyl accident over Europe.

This first part of the research sets a firm base for acidic precipitation research because information on both high concentrations and annual deposition of acidifying substances is available for effect research, validated by monitoring. In Section II the air pollution climate in The Netherlands will be described with results from monitoring, emission inventory, and model calculations.

The second part of the acidic precipitation research in The Netherlands originates from ecological research. In various valuable and vulnerable ecosystems some changes had already been detected shortly after World War II. Many of these systems—aquatic and terrestrial—have been studied since then, so a rather large data set is available for further studies on acidification. This field research is supported by laboratory research on specific effects and pollutants. Some of the results from this research will be referred to, for example, in Sections IV and V.

The third part of the research is concentrated in the Dutch Priority Program on Acidification. In the early 1980s it was recognized that acidic precipitation was a great threat. In an inventory this threat was assessed and gaps in knowledge were revealed. In order to fill these gaps a special research program was started. This will be dealt with in Sections III and IV.

In Section VI reference will be made to other ongoing research programs in The Netherlands closely related to acidification research. In Section VII some conclusions will be drawn both regarding policy and further research.

II. Air Pollution Climate in The Netherlands

The air pollution climate in The Netherlands can be described with figures and tables derived from the National Monitoring Network for Air Pollution (Anon., 1986, 1988). The total depositions for a more or less average year (1980) are given in Table 5–1.

The distribution of the deposition of SO_x, NO_x, and NH_x over the country is shown in Figures 5–1 to 5–3. In Figure 5–4 the total acidic deposition (potentially) is given and in Figure 5–5 the total deposition of nitrogen compounds. The trend in deposition over the last seven years is shown in Figures 5–6 and 5–7.

Of the total deposition of sulphur compounds, 25% originates in The Netherlands. The most important sources abroad are in the Federal

Table 5–1. Total depositions in The Netherlands in 1980 (rounded data) in mol ha^{-1} a^{-1}.

Component	Wet	Dry	Total
SO_x	370	1050	1420
NO_x	400	1250	1650
NH_x	630	720	1350
Total potentially Acidic deposition[a]	1770	4070	5840
Total N	1030	1970	3000
			= 42 kg ha^{-1} a^{-1}

[a]Under the following assumptions: All NH_x is nitrificated to NO_3^-, no denitrification and equilibrium in the vegetation-soil systems, and total deposition = $2 \times SO_x + NO_x + NH_x$ (in mol H$^+$ ha^{-1} a^{-1}).

TOTAL SO$_X$ DEPOSITION IN THE NETHERLANDS IN 1980 (MOL . HA^{-1})

200 - 600   600 - 1000   1000 - 1400   1400 - 1800   1800 - 2200   2200 - 2600

Figure 5–1. Total SO$_x$ deposition in The Netherlands in 1980.

TOTAL NO$_X$ DEPOSITION IN THE NETHERLANDS IN 1980 (MOL . HA^{-1})

400 - 800   800 - 1200   1200 - 1600   1600 - 2000   2000 - 2400   2400 - 2800

Figure 5–2. Total NO$_x$ deposition in The Netherlands in 1980.

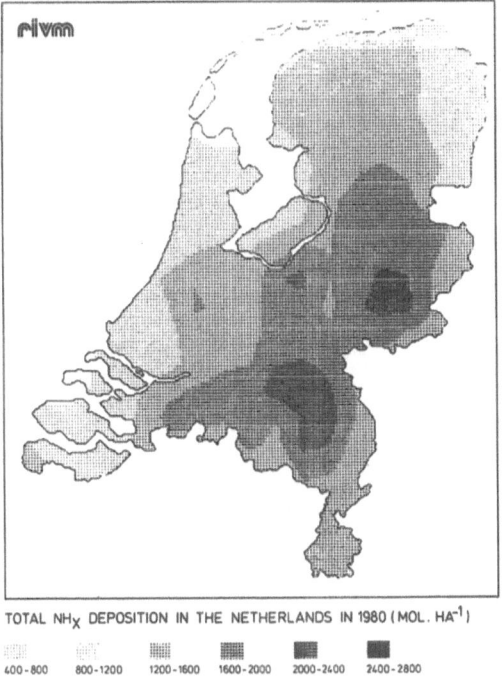

TOTAL NH_X DEPOSITION IN THE NETHERLANDS IN 1980 (MOL . HA⁻¹)

400 - 800   800-1200   1200-1600   1600-2000   2000-2400   2400-2800

Figure 5–3. Total NH$_x$ deposition in The Netherlands in 1980.

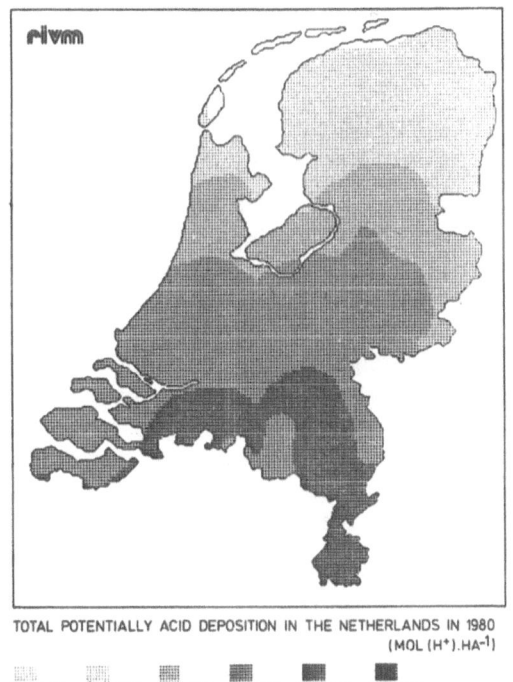

TOTAL POTENTIALLY ACID DEPOSITION IN THE NETHERLANDS IN 1980
(MOL (H⁺).HA⁻¹)

3600-4400   4400-5200   5200-6000   6000-6800   6800-7600   7600-8400

Figure 5–4. Total potentially acidic deposition in The Netherlands in 1980.

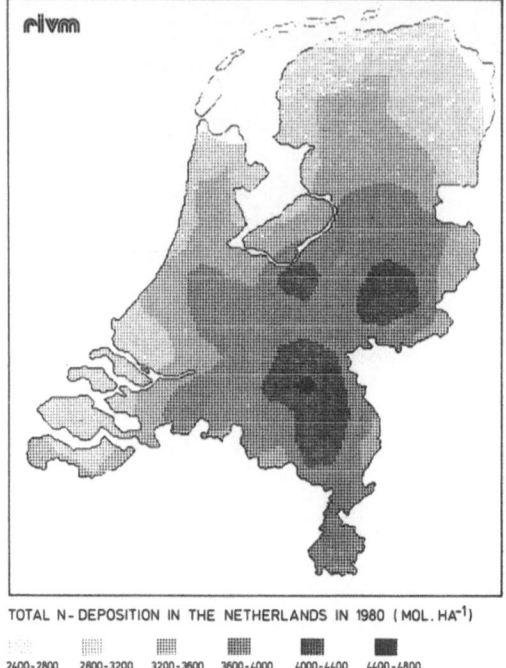

TOTAL N- DEPOSITION IN THE NETHERLANDS IN 1980 (MOL. HA⁻¹)

| 2400-2800 | 2800-3200 | 3200-3600 | 3600-4000 | 4000-4400 | 4400-4800 |

Figure 5–5. Total N deposition in The Netherlands in 1980.

Republic of Germany, the German Democratic Republic, Poland, and the United Kingdom. In 1980 the total emission of sulphur dioxide in The Netherlands was about 465×10^6 kg. Of this amount about 15% to 20% was deposited in The Netherlands; the remainder was exported. Since 1980 emissions in the country have dropped to about 50%.

Of the total deposition of nitrogen oxides, 30% to 35% originates in The Netherlands. The western part of the Federal Republic of Germany is the most important foreign source. Emissions of NO_x in The Netherlands were about 540×10^6 kg in 1980 of which 80% to 85% was exported. There was a slight drop in NO_x emissions in more recent years.

To a large extent, ammonia and ammonium depositions in The Netherlands originate from cattle, pigs, and poultry breeding in the country itself: about 65% to 70%. Estimates of ammonia emissions in The Netherlands and for all of Europe have been made by Buysman (1986). His estimation of 140×10^6 kg for the year 1980 has been proven to be about 40% to 50% too low. More recent estimates arrive at 230×10^6 kg a^{-1} for 1985. When put into model calculations (Asman and Janssen, 1986a, 1986b) and compared with the available measurements of concentrations of NH_3 at several stations over the country this last estimate

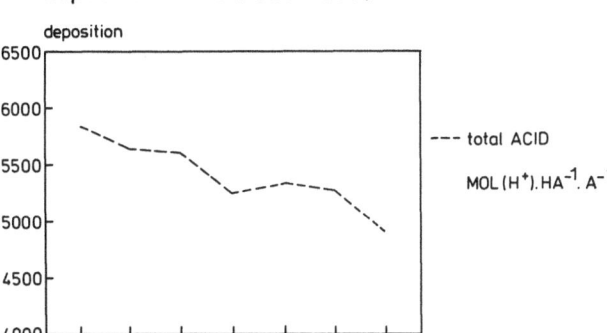

Figure 5–6. Trend in deposition of total potentially acid.

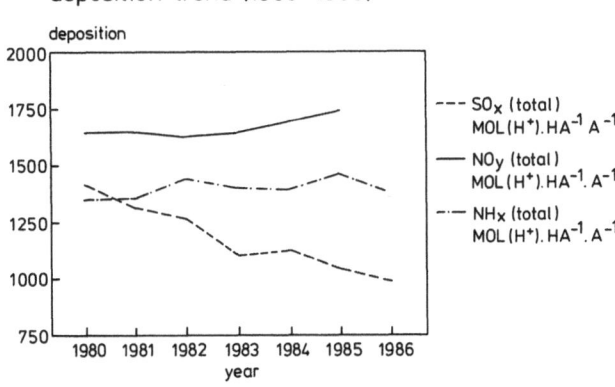

Figure 5–7. Trend in deposition of separate components.

seems quite accurate. About 80% to 85% of the Dutch emissions of NH_3 deposit as NH_3 or NH_4^+ within the borders.

The situation for ozone as a yearly average concentration is shown in Figure 5–8. Figure 5–9 shows the distribution of peak concentrations over the country. The origin of high average and peak concentrations is much more difficult to discern. Episodes clearly originate from high emissions of volatile organic hydrocarbons and nitrogen oxides under stable weather conditions. Long-term average concentrations are determined for a large part also by mixing from higher atmospheric layers and thus by much more widespread regional or global pollution.

Formerly, episodes of high SO_2 concentrations occurred rather frequently. In more recent years episodes with high concentrations of air

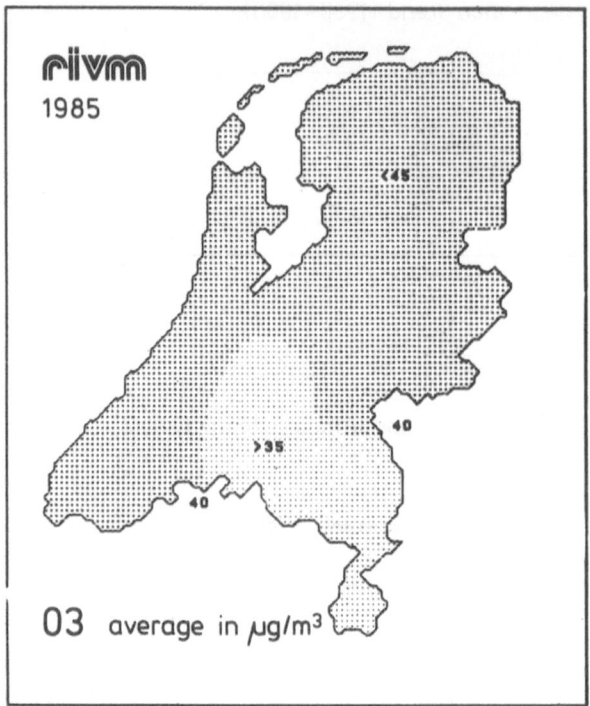

Figure 5-8. Average ozone concentrations over The Netherlands in 1985.

pollutants still occur especially in winter, but now it is a mixture, with NO_x dominating most of the time.

III. Organization of Acidic Precipitation Research

In the Dutch national administration the Minister of Housing, Physical Planning, and Environment is responsible for the quality of the environment. The Minister of Agriculture and Fisheries is responsible for forests and nature preservation. Both ministers have institutes to carry out research on acidification. In 1984 it was recognized that integration of knowledge was necessary in order to arrive at a better understanding of the processes connected with the acidification phenomena. Together with the Ministry of Economic Affairs, responsible for power generation and industry, the refineries, and the Central Electricity Board (SEP) the ministries formed a steering body and provided a budget of about 15 million Dutch guilders for a three-year research program. A programming committee and a project group were formed, which together make up the Dutch Priority Program on Acidification (Schneider and Bresser, 1986a).

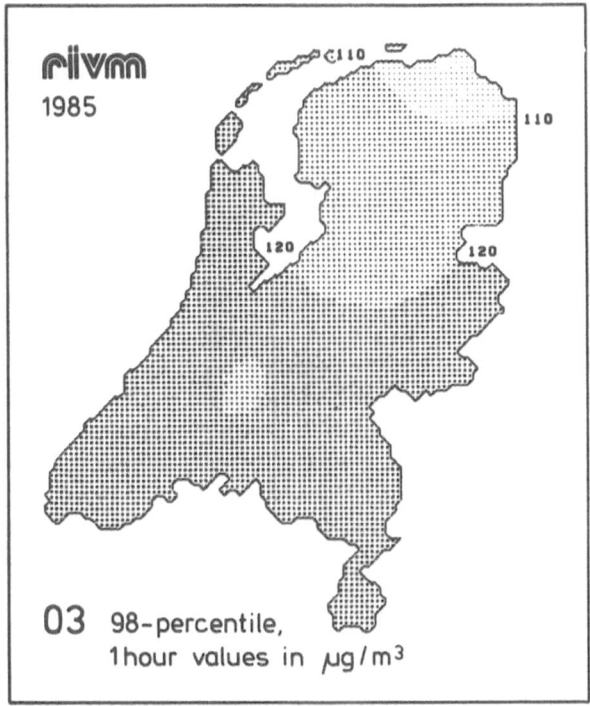

Figure 5–9. Ozone peak concentrations over The Netherlands in 1985.

Several ongoing research projects joined the program. At the end of 1986 about 60 projects were combined into an integrated research effort. Because in most projects only about half the costs are subsidized and sometimes even no extra money from the initial budget was needed, the total amount of money in the program is about 60 million guilders (about U.S. $30 million).

The RIVM is charged with the overall coordination. Program leader is Dr. Ir. T. Schneider, secretary Ir. A.H.M. Bresser. About 40 research groups are participating. There are about 150 scientists with an additional 150 as technical/analytical assistants.

About 80% of the research efforts is put into effect research both in laboratories and in the field. About 10% is put into research on the emission of ammonia. The air pollution climate in The Netherlands, as pointed out in Section II, makes it clear why this attention to ammonia is needed. The remaining 10% of the budget is put into an integrated study on the effectiveness of measures to reduce emissions in view of the effects. These three items will be discussed in Section IV.

Within the program communication is rather intense. Besides the monthly meetings of the project group, several working committees meet regularly on specific subjects such as the integrated measurements at forest sites or the research on the nitrogen cycle in the soil. Once or twice a year the program leader and secretary visit each of the research groups. At the end of each year all project leaders report verbally in a review meeting with discussions on methods and results with all researchers involved in the program. Reports of these meetings are made available to the public (Schneider and Bresser, 1986b, 1987a). Reports from individual projects that are of interest to a larger public are made available either by the research institute or by the program secretary.

Each year the program leader evaluates the progress and results and reports to the steering body. In 1987 a full interim evaluation was asked for. This included an evaluation of research in all countries, the present situation of acidification in The Netherlands, and a prognosis for the future (Schneider and Bresser, 1987b). The first program was finished in mid-1988. The final report of this phase forms the basis of policy actions which will be presented mid-1989 (Schneider and Bresser, 1988).

A second phase has been in progress since early 1988. An overview of the program was presented by Schneider and Bresser (1989). Central items in this second phase are the integrated measuring at two forest stands and a heathland stand (continuation and slight modification), the nitrogen cycle in forest and heathland systems, and integrated assessment.

IV. Dutch Priority Program on Acidification

The three main items within the program have already been mentioned: effect research, ammonia emission research, and integrated assessment.

A. Effect Research in Forests

Although a rather substantial budget had been allocated to research into the processes, it still is impossible to cover all the possible hypotheses and test them in the laboratory and in the field with all the species that seem to be affected by acidification. The main concern was with forest damage. The answers to the questions of whether, how, and to what extent deposition of acidifying substances, oxidants, and nutrients from the atmosphere contributed to the observed decline in forest health were most important. The Douglas (*Pseudotsuga menziesii*) was chosen as the species to be studied in depth both in laboratories and at field sites. The reasons for this choice were mainly the following:

1. The Douglas was considered to be the most important conifer in future forest plantations in The Netherlands;
2. Conifers in general showed more reaction to air pollution than deciduous trees; and

3. The Douglas showed more clear symptoms than the more widespread Scots Pine.

At two Douglas stands in the center of the country, integrated field measurements have been started. Figure 5–10 shows the general layout of these stations. At several heights above, within, and under the canopy, weather and air pollution conditions are measured continuously (SO_2, NO_x, NH_3, O_3). Wet deposition is measured above and beneath the canopy at several places. Occult deposition in mist and dew is measured separately. Dry deposition fluxes are measured periodically and calculated permanently from concentrations. In the soil the water content and composition is measured continuously with suction through porous plates.

Leaf fall and decomposition of litter are measured periodically with baskets and litter bags. Growth of roots and mycorrhizal fungi are observed in excavations with perforated plates and an endoscope. Quantification of root growth is obtained by combination of observations and in-growth cores. Physiological measurements of trees include respiration in the root zone and at the needles, photosynthetic activity, hormonal conditions, needle constitution, and so on. The last measurements are performed in the laboratory with branches and needles which are periodically taken from the trees.

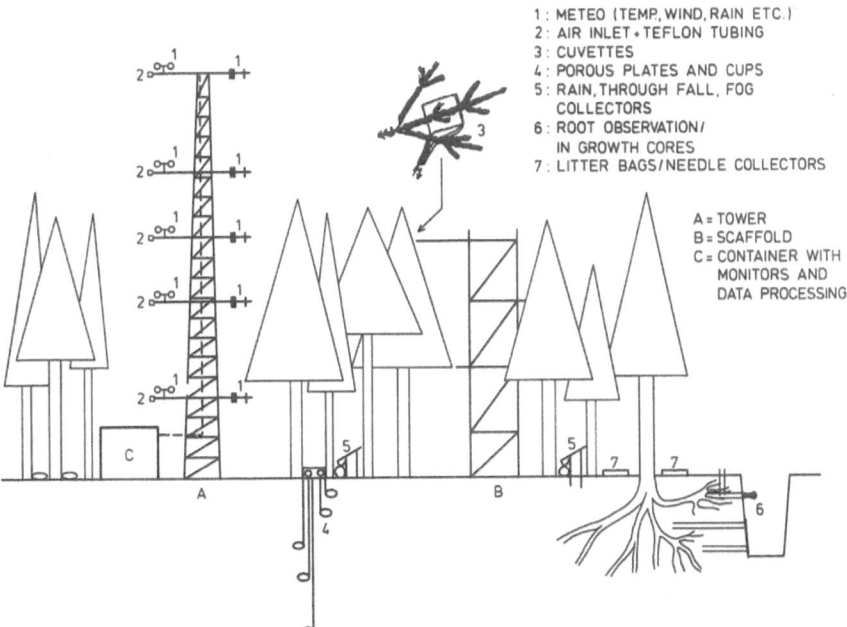

1 : METEO (TEMP, WIND, RAIN ETC.)
2 : AIR INLET + TEFLON TUBING
3 : CUVETTES
4 : POROUS PLATES AND CUPS
5 : RAIN, THROUGH FALL, FOG COLLECTORS
6 : ROOT OBSERVATION / IN GROWTH CORES
7 : LITTER BAGS / NEEDLE COLLECTORS

A = TOWER
B = SCAFFOLD
C = CONTAINER WITH MONITORS AND DATA PROCESSING

Figure 5–10. Layout of integrated monitoring in a forest stand (Douglas).

During air pollution episodes the program is intensified with measurements of other oxidants, organic hydrocarbons, and various nitrogen compounds (HNO_3, PAN).

At the forest sites some experimental setups are installed also. In several trees at various heights branch cuvettes have been placed to measure photosynthesis at ambient conditions (control), at conditions with filtered air, and at conditions with extra ozone supply. Part of one of the plots will be a reference site where trees will be supplied with the water and nutrients needed for optimal growth. The optimal nutrition will be studied using methods similar to the Swedish approach (e.g., Tamm, 1985).

The data from these plots will be used to validate a tree growth model in combination with hydrological and soil models. Several parameters for these models are obtained from supporting experiments in laboratories. Direct damage to needles is observed in fumigation experiments. Root growth is studied with and without mycorrhizal fungi. The fungi are studied separately under varying conditions of soil water. Emphasis in these studies is on the hydrological and nutritional conditions of the trees both under and above ground. Extrapolation to other stands of Douglas will be made possible by an inventory of several stands over the country where soil conditions and tree health are measured a couple of times a year, in combination with the nationwide available data on air pollution concentrations.

Although emphasis is on the Douglas, in several of the projects that joined the program other species were a subject of research. So some data are available for extrapolation to other forests but probably not enough to obtain accurate assessment of future damage for all forests with all soil types.

B. Other Effect Research

Besides forests the heathlands in The Netherlands are also deteriorating quite rapidly. In one of the few remaining healthy heathlands, an integrated measuring plot has been installed more or less similar to the forest sites. Here the causes of abundant grass growth and decline of heather are studied.

Heather is also subjected to high concentrations of gaseous pollutants and high depositions (wet) of nitrogen and sulphur compounds in laboratories in order to study effect mechanisms.

From fumigation experiments it was observed that agricultural crops are sensitive to ozone at concentrations which occur quite regularly in The Netherlands (Van der Eerden, Tonneijck, and Wijnands, 1987). The effect mechanism is studied more in depth, particularly using broad beans (*Vicea faba*). Effects of SO_2 and NH_3 are also studied with beans. In some experiments other species of agricultural crops and seedlings of trees are studied as well.

C. Ammonia Research

In view of the Dutch situation with regard to ammonia emission (see Section II) special attention is paid to sources of atmospheric ammonia and ammonium. Stables for poultry, cattle, and pigs are measured with respect to NH_3 output by the atmosphere. Centrally ventilated stables where the flux can be measured are rather rare for cattle, so at the lee side of such a stable the plume is measured in order to estimate total emission as a factor of stable climate and other varying factors. Also volatilization of NH_3 from storage is measured at varying degrees of coverage. Emission of NH_3 during spreading is measured with an equipped truck that follows right behind a manure spreader. Emissions of NH_3 from meadows with cattle are estimated with tunnels, and emission from arable land (corn) with application of manure and other fertilizers are measured with micrometeorological methods according to Denmead (1983) and Ryden and McNeil (1984).

Several methods for reduction of NH_3 emissions in ventilated air are under research (biofiltration and bioscrubbers). In an overview study estimates have been made of NH_3 emissions in Europe (Buysman, 1986), and a surface depletion NH_x model has been developed for a description of NH_x transport and transformation over Europe (Asman and Janssen, 1986a, 1986b).

D. Integrated Assessment

Within the program a major effort has been adopted to integrate all quantitative research results into an overall model (Bresser, 1986). This part of the program serves two purposes:

1. It forms a focal point for all separate research projects because results are more meaningful in terms of impact on a policy level if they fit into a framework, and
2. The framework assists decision making.

The schematic of the model is given in Figure 5–11. The first parts of the model are operational. Scenarios for emissions in Europe are transformed into depositions and concentrations in 20 regions in The Netherlands (see Figure 5–12) with various receptors (forest, heathland, small surface waters). Effect modules for these receptors including indirect pathways of pollutants via the soil are in development.

This model study contributed to the policy assessment in 1988 with an assessment of the effectiveness of measures to prevent and combat depositions contributing to acidification (Schneider and Bresser, 1988).

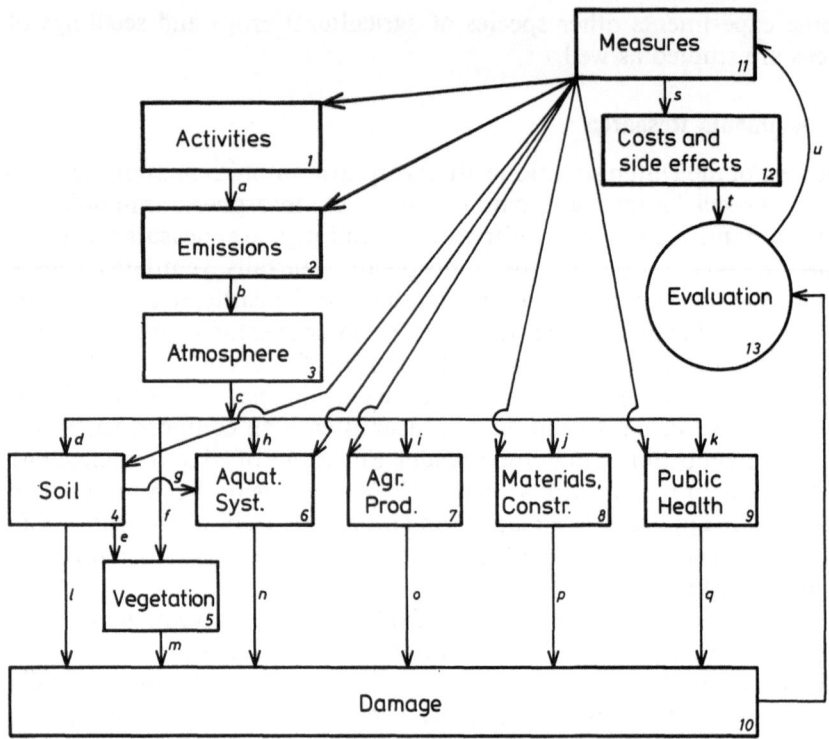

Figure 5–11. Schematic of the Dutch acidification simulation (DAS) model.

V. Related Programs and Projects

Traditionally environmental policy and research is compartmentally oriented: atmosphere, soil, water. Some years ago central themes entered policy and research. Acidification was the first theme to be dealt with in a more integrated way. Other central themes are eutrofication (discharge of fertilizers into the environment), diffusion of dangerous substances through the environment as a whole, removal of wastes, and disturbance. From these other themes eutrofication has already led to a research program which is mostly aimed at the overall effects of intensive agriculture. An overlap with the NH_3 research within the Dutch Priority Program on Acidification was anticipated. Most of the research into the NH_3 problem will be incorporated in the eutrofication theme. Within this theme integrated assessment also plays a central role. In an earlier stage a technological research program had been set up to search for new and better ways to handle excess manure. In several separated projects more local situations with heavy manure loads were studied. All together a quite substantial

13 EMISSION AREAS IN EUROPE

20 EMISSION AND RECEPTOR AREAS IN
THE NETHERLANDS

Figure 5–12. Regionalization in the DAS model.

research effort has been directed toward the problems of intensified agriculture practice in The Netherlands.

For several years a very large research program has been in place to study the possibilities and effects of the reintroduction of coal in energy production both on a large and small scale. The effects on air quality have been studied with measurements at a coal-fired power plant, with laboratory research, and with model calculations (Van Egmond, 1986).

Recently a priority research program on soil research has been set up which includes effects of acidification on soil ecosystems. This program aims at creating centers of excellence for several research topics in order to establish a firm base for scientific soil research. Results from this program are not expected within the next few years.

Internationally the Dutch research teams participate in the EEC-COST 611 and EEC-COST 612 concerted actions (air pollution and effects of air pollution). In this framework internationally coordinated research projects have also been set up with subsidies from the EEC's fourth Environmental Research Program. Cooperation also exists within the UN-ECE framework. The Netherlands actively participate in the Environmental Monitoring and Evaluation Program (ECE-EMEP). An international counterpart for integrated assessment was found at the International Institute for Applied Systems Analysis (IIASA) where the Regional Acidification Information and Simulation (RAINS) model has been developed to calculate the effects of various SO_2 scenarios (Alcamo, Hordijk, Kämäri, Kauppi, Posch and Runca, 1985). Several participating institutes from the Dutch program participated at IIASA especially for studies on the effects of air pollution on soil, and the IIASA approach serves as an important example for the Dutch study.

In order to facilitate information exchange and international coordination of research the first Meeting of Acidification Research Coordinators was organized in The Netherlands (MARC 1) in 1986. This informal meeting had a follow-up in the United Kingdom where MARC II was held in September 1987, and in Canada in 1988 (MARC III).

VI. Forest Damage Inventory

Since 1984 the State Forestry Service has been reporting at the end of each year on the state of the Dutch forests. From the inventory of 1987 (Anon., 1987) the main conclusions will be presented here.

The vitality of the Dutch forests, measured as the percentage of needle or foliage loss and the degree of yellowing, has remained stable compared to 1986. In 1986 and 1985 the situation already had worsened compared to 1984. The percentage of vital forest has decreased from 47 to 43. Hardly vital forests increased from 16% to 17%, and not vital forests decreased from 5.1% to 4.7%. From 1985 to 1987 the sum of hardly vital and not vital forests increased from 9.5% to 21%. In the northern and western parts of the country forests are generally healthy. In Figure 5–13 some results are visualized in diagrams. The more densely forested regions in the central, eastern, and southern parts of the country, however, are more damaged. The sandy regions in Brabant (see Figure 5–14) are especially heavily damaged. This region has a very high density of intensive cattle and pig breeding farms and also has a rather strong deposition of substances from the Ruhr area and central Europe.

It had already been established in 1983 and 1984 that air pollution had to be added to the factors that normally determine the vitality of forests. Vital forests are considered capable of recovery from damage due to weather, insects, fungi, and so on. Air pollution is synergistic to these natural damage factors.

As for the species, the main decrease is in Inland Oak, Douglas, and Corsican Pine; Scots Pine show an encouraging recovery from extended earlier damage.

Damage from insects and fungi is considered to be extremely high: 7% of Scots Pine is damaged by insects. Health decrease of Inland Oak mainly seems to be due to several years of heavy damage by insects. Fungal damage of Corsican Pine is very widespread. *Sphaeropsis sapinea* (a fungus causing the death of branches) and *Brunchorstia* are found in a large part of the forest stands with this species. The death of branches in Scots Pine stands seems also to be due to *Sphaeropsis sapinea*. This occurs to a slight degree over almost all The Netherlands. Yellowing of needles and leaves, characteristic of a relative deficiency of certain nutrients, is found in all pine species, Douglas, and oak. Due to favorable weather conditions for forest growth in the spring and summer of 1987 (much rain) a number of species show some recovery.

VII. Other Observed Effects

Almost all heathlands, small surface waters, and other more or less natural ecosystems show a decrease in number of species and in vitality. Only with lichens can some restoration be seen, probably due to the neutralizing effect of NH_x on the bark of trees before nitrification can take place.

The growth of grasses in heathland is progressing fast. Hardly any heathland without grass can be found in The Netherlands. Acidification and eutrofication of fens and peat bogs have already led to an almost complete loss of the original vegetations in these systems.

Acidification of natural areas also progresses toward the groundwater underneath. At several locations concentrations of nitrate are high and increasing even at depths of 10 and 20 m. This indicates nitrogen

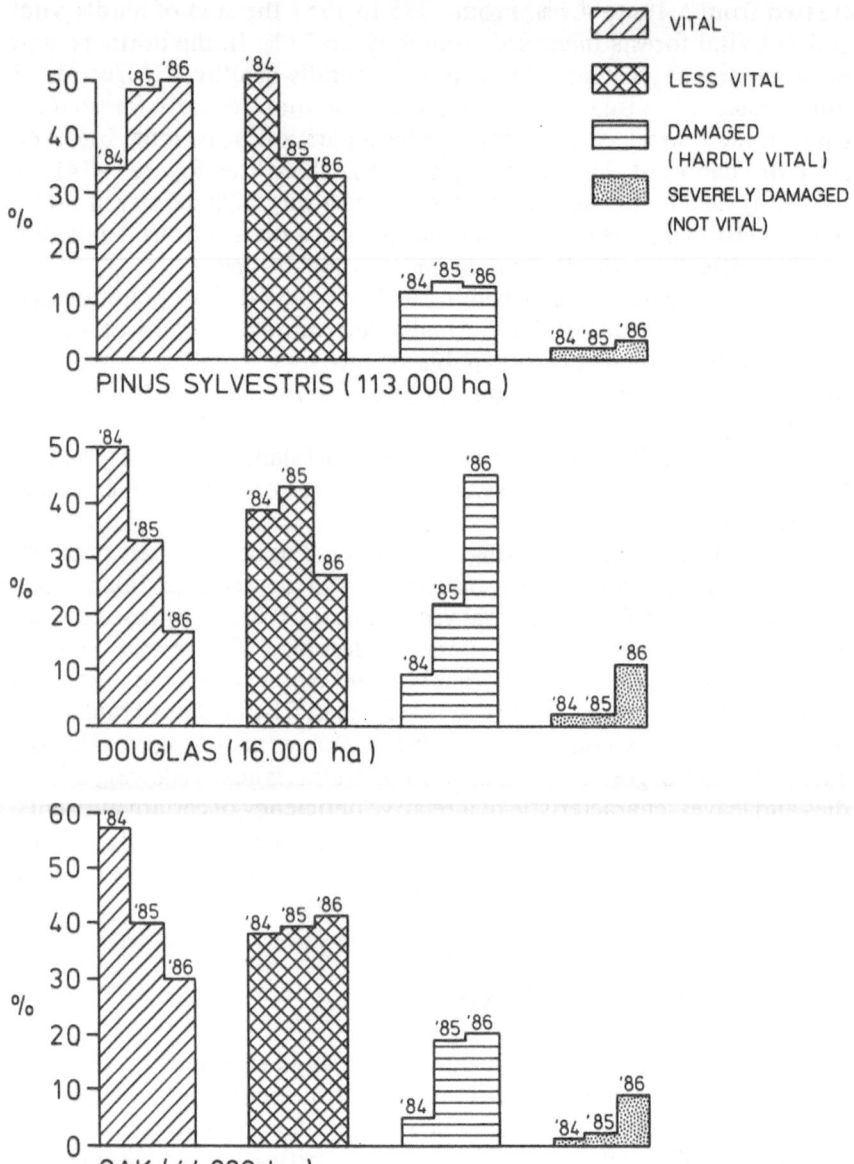

Figure 5-13. Some results of forest vitality surveys.

saturation of the forest soil above. Also decreasing pH and increasing concentrations of sulphate and aluminum point to a rather rapid acidification of the sandy soils in The Netherlands.

The estimate of damage to agricultural crops was about 4% to 5% of the yield in 1983. Due to favorable weather conditions in summer and winter of the past years (1986–1988), both SO_2 and ozone concentrations have been substantially lower than in former years (early 1980s). Total damage to agricultural crops now is estimated at about 300 million Dfl, to be paid by the consumers. Approximately three-fourths is caused by ozone and one-fourth by SO_2. Monetary value of the damage to the cultural heritage is 30 to 60 million Dfl per year. Damage to goods by SO_2 alone is about 170 million Dfl (Dutch guilders) a year. Addition of antiozonants to rubber products (tires) to prevent ozone damage costs the consumers 50 to 100 million Dfl each year.

VIII. Some Conclusions

Although a large part of the program only started quite recently, some overall conclusions already can be drawn. The conclusions are of a varied nature.

With regard to the deposition of acidic substances it can be concluded that the level above which coniferous forests and heathland systems on nutrient-poor sandy soils in The Netherlands are expected to have a loss in vitality is about 1400 mol(H^+) ha^{-1} a^{-1} (total potential acid). For deciduous forests on richer soils the critical load for acidic deposition is about 2400 mol(H^+) ha^{-1} a^{-1}. At these levels not all the changes in ecosystems can be excluded. If even the most sensitive systems should be protected the level of deposition can be only about 400 to 700 mol(H^+) ha^{-1} a^{-1}. The present deposition level as an average over the country is about 4 times the critical level. In the regions with sensitive soils (see Figure 5–14) and especially at the edges of forests and heathlands the deposition can be as high as 10,000 mol(H^+) ha^{-1} a^{-1}. From all field research and experiments in laboratories nitrogen shows up as one of the main components for the explanation of forest decline in The Netherlands. Critical levels of nitrogen deposition (NO_x + NH_x) with regard to changes in vegetation from natural systems to systems dominated by nitrofilic species are only slightly above background levels (400 mol ha^{-1} a^{-1}). If the ratios between ammonium and some vital nutrients have to be kept at a level where uptake of those nutrients is still possible, then the NH_x deposition should be below 1500 mol ha^{-1} a^{-1}. If flushing of nitrates toward the groundwater has to be limited to a level that keeps the concentration of nitrates below drinking water standards (= 50 mg L^{-1} as NO_3^-) total nitrogen deposition should be limited to 1600 mol ha^{-1} a^{-1} on coniferous forests. In Table 5–2 these critical loads are summarized.

Figure 5–14. Areas in The Netherlands with soils vulnerable to acidification.

Present average deposition of total nitrogen is about 3000 mol ha^{-1} a^{-1}, and the deposition of NH$_x$ is about 1350 mol ha^{-1} a^{-1}. Forest and heathland edges in the eastern parts of the country receive more than twice this amount especially when large ammonia sources are near those receptors.

Table 5–2. Critical loads for deposition of acid and nitrogen in mol (H$^+$ha^{-1} a^{-1}l).

Substance	Criterion	Heathland[a]	Coniferous	Deciduous
NH$_x$	NH$_4$/K ratio	1500[b]	1000	1500
NH$_x$ + NO$_x$	Uptake and nitrate <50	3600	1600	2800
	Nitrate <25 mg l^{-1}	2000	1000	1600
NH$_x$ + NO$_x$	Natural species	400	400	400
NH$_x$ + NO$_x$ + SO$_x$	Ca/Al ratio (acidif.)	1400	1400	1800

[a]In heathlands cutting of sods is not taken into account.

[b]Heathlands considered to be equal to forests in this respect.

It can be concluded that substantial reductions in deposition are needed to prevent further damage to vulnerable systems such as coniferous forests and heathlands. Even less vulnerable systems are severely threatened with the actual deposition levels in The Netherlands. With regard to direct damage to needles and leaves of both agricultural crops and natural vegetation the critical level for ozone concentration is expected to be between 50 and 100 μg m^{-3} (daytime average over the growing season). A peak value for the hourly average critical level is about 200 μg m^{-3}. Both levels are surpassed quite frequently, and ozone concentrations are still rising. An overview of critical levels for some air pollutants is presented in Table 5–3.

In 1980 direct damage by high $SO_2{}^-$ concentrations still occurred. The 1986 concentrations indicated that the damage had lessened. The ongoing research indicates that interactions of changes in the soil environment (both acidification and eutrofication), direct damage to needles and leaves by high ozone concentrations, and natural stress factors such as drought, frost, and pests, are the main contributors to forest decline in The Netherlands. Only when studying "whole tree" systems can the relative importance of each individual factor and the effectiveness of control measures be determined.

It can be concluded that a severe decrease in depositions up to and above 80% is necessary to arrive at the critical levels. Such a decrease cannot be reached within a few years. Most of the initiatives will have to be on a European scale. Measures will be needed to bridge the period from

Table 5–3. Critical concentrations for direct visual effects on vegetation (in μg m^3).

Crop/vegetable component	Agricultural growing season (May–September)	Trees/forests: Annual averages	Low natural vegetations: Annual averages
	Effect/no effect	Effect/no effect	Effect/no effect
SO_2	120 / 30	100 / 25	40 / 10
NO_2*	300 / 100	180 / 60 **	180 / 60
	[300 / 100]	$\rightarrow$ in the absence of SO_2 and O_3	
O_3 daytime average 10–17	100 / 50	100 / 50	100 / 50
NH_3	200 / 80	125 / 50	60 / 25***
O_3 1-hr value during daytime	200 / 150	200 / 150	200 / 150
8-hr value during daytime	75 / 65	75 / 65	75 / 65

*= At concentrations of SO_2 and O_3 within the presented boundaries.

**= Due to a lack of research data the same values as for agricultural crops are taken.

***= Secondary effects, after accumulation and conversion.

now until emission reductions are successful. Further studies of restorative measures will be needed. These might include studies on selective fertilization and on irrigation, in order to delete one or more of the natural stress factors.

References

Alcamo, J., L. Hordijk, J. Kämäri, P. Kauppi, M. Posch, and E. Runca. 1985. Integrated analysis of acidification in Europe. *J Environ Manag* 21:47–61.

Anon. 1986. Luchtkwaliteit, jaarverslag 1984 en 1985 (Air quality, annual report 1984 and 1985). National Institute of Public Health and Environmental Protection, report no. 228216052. Bilthoven, The Netherlands.

Anon. 1987. Verslag van het landelijk vitaliteitsonderzoek (report on the forest vitality survey) 1987. Staatsbosbeheer, sector bosbouw (State Forest Service). Utrecht, The Netherlands.

Anon. 1988. Luchtkwaliteit, jaarverslag 1987 (Air Quality, annual report 1987). National Institute of Public Health and Environmental Protection, report no. 228702009. Bilthoven, The Netherlands.

Asman, W.A.H., and A. Janssen. 1986a. A long-range transport model for ammonia and ammonium for Europe and some model experiments. Institute for Meteorology and Oceanography, University of Utrecht, Report no. R 86-6. Utrecht, The Netherlands.

Asman, W.A.H., and A. Janssen. 1986b. A long-range transport model for ammonia and ammonium for Europe. Paper presented at the Symposium on Interregional Air Pollutant Transport (EURASAP), Budapest, April 22–24, 1986.

Bresser, A.H.M. 1986. Effectiveness of measures to prevent and reduce the effects of acid rain. Report no. 69-01 of Dutch Priority Program on Acidification, National Institute of Public Health and Environmental Protection. Bilthoven, The Netherlands.

Buysman, E. 1986. Historical trend in the ammonia emission in Europe. Institute for Meteorology and Oceanography, University of Utrecht, Report no. R-86-9. Utrecht, The Netherlands.

Denmead, O.T. 1983. Micrometeorological methods for measuring gaseous losses of nitrogen in the field. In J.R. Freney and J.R. Simpson (eds.), *Gaseous Loss of Nitrogen from Plant-Soil Systems*, The Hague, 133–157. The Netherlands: Martinus Nijhof/Dr. W. Junk.

Van der Eerden, L.J., A.E.G. Tonneijck, and J.H.M. Wijnands. 1987. Economische schade door luchtverontreiniging aan gewasteelt in Nederland (Economical damage by air pollution to crops in The Netherlands; in Dutch with summary in English). Publication series Air no. 65, Ministry of Public Housing, Physical Planning and Environment. Leidschendam, The Netherlands.

Van Egmond, N.D. (ed.). 1986. Air pollution by emissions from coal-fired installations. PEO, report no. 20.70-004.11, Stichting Projectbeheerbureau Energieonderzoek. Utrecht, The Netherlands.

Ryden, J.C., and J.E. McNeil. 1984. Application of the micrometeorological mass balance method to the determination of ammonia loss from a grazed sward. *J Sci Food Agric* 35:1297–1310.

Schneider, T., and A.H.M. Bresser. 1986a. Program and Projects. Dutch Priority Program on Acidification, Report no. 00-01 (rev. November 1986). RIVM. Bilthoven, The Netherlands.

Schneider, T., and A.H.M. Bresser. 1986b. Proceedings of the first symposium, Dec. 1985. Dutch Priority Program on Acidification, Report no. 00-02, RIVM. Bilthoven, The Netherlands.

Schneider, T., and A.H.M. Bresser. 1987a. Proceedings of the second symposium, Nov. 1986. Dutch Priority Program on Acidification, Report no. 00-03, RIVM. Bilthoven, The Netherlands.

Schneider, T., and A.H.M. Bresser. 1987b. Verzuringsonderzoek eerste fase; tussentijdse evaluatie aug. 1987 (Acidification research phase 1; interim evaluation Aug. 1987). Dutch Priority Program on Acidification, Report no. 00-04, RIVM. Bilthoven, The Netherlands.

Schneider, T., and A.H.M. Bresser. 1988. Additioneel Programma Verzuringsonderzoek; Evaluatierapport Verzuring (Acidification research phase 1; evaluation report on acidification) Dutch Priority Program on Acidification, Report no. 00-06, RIVM. Bilthoven, The Netherlands.

Schneider, T., and A.H.M. Bresser. 1989. Program and Projects (2nd phase). Dutch Priority Program on Acidification, Report no. 200-01, RIVM. Bilthoven, The Netherlands.

Tamm, C.O. 1985. De skogliga bördighetsförsöken (The Swedish optimal nutrition experiments in forest stands: aims, methods, yield results; in Swedish with English summary). Kungsl. Skogs- o Lantbruksakad. tidskr. suppl. 17:9–29.

Anthropogenic Acidification of Precipitation in the USSR

L.G. Solovyev*

Abstract

Consideration is given to questions of acidic precipitation formation as well as concentration changes of hydrogen ions in cloud water connected with both natural causes and anthropogenic effects of industrial production and cities.

The role of natural and anthropogenic acidic-forming substances is defined in precipitation acidification. The considerable rise of anthropogenic contributions of acidic-forming oxides to pH changes in cloud water and precipitation is noted, as well as the fact that the total contribution of anthropogenic and natural ingredients of acidic-forming substances has almost become equal at a local level. Concentrations of SO_2 (0.5, 10, 40 $\mu g\ m^{-3}$) and NO_2 (0.5, 4, 20 $\mu g\ m^{-3}$) for pure, regionally industrial, and urban zones, respectively, are given that point to a great difference between background and urban atmospheres. The average lifetime of both nitrates and sulphates is estimated at some tens of hours. Therefore, the processes of acidic changes in cloud and rainwater are rather dynamic regarding transient synoptical situations. The importance of speed and direction of air mass travel is evaluated in the change of precipitation acidity in the European part of the USSR.

I. Natural Acidity of Rain

The natural content of hydrogen ions was discovered when determining pH in the glacier waters of both the Antarctic and Greenland for the last two hundred years before the beginning of the twentieth century. It was found that the pH varies between 5.2 and 5.6 and has not changed for 10^2 to 10^3 years.

*Vernadsky Institute of the Academy of Sciences of the USSR, Kosigin 19, Moscow, USSR.

These values coincide with pH determinations of the solution in the atmosphere of which there is about 330ppm of carbon dioxide at the temperature 20°C. The pH of this solution is 5.6 and it, along with the pH value of glacier waters of Greenland and the Antarctic, can be taken as a natural initial point of a determination of acidic changes of atmospheric precipitation brought about by the activity of humans. Note that the pH of chemically pure water is 7.0.

In natural conditions rainfall always has various impurities affecting pH. Both the quantity and composition of impurities depend on regional characteristics where cloud systems form and precipitation occurs. The sea salt contributes most of all to the mineralization of the rain fallout over the ocean and in this case pH of the rainwater rises. As a rule, pH increases occur when precipitation penetrates soil dust in regions with alkaline soils. At the same time in the noncontaminated atmosphere, in addition to a carbon dioxide, there may be a number of acidic-forming matters of a natural origin such as hydrogen sulphide, sulphur dioxide, hydrochloric acid, nitrogen oxides and nitric acid, and organic acids. Their concentrations in the noncontaminated atmosphere is very small. But they can play a role in the formation of hydrogen ions in rainwater.

The solution of an equation on an ion balance for a sample of precipitation relative to proton (H^+) has the following expression at known concentrations of basic cations and anions (Izrael, Rovinsky, and Nazarov, 1983):

$$[H^+] = [Cl^-] + [NO_3^-] + 2[SO_4^{2-}] - [NH_4^+] - [Na^+] - [K^+] - 2[Ca^{2+}] - 2[Mg^{2+}]$$

II. Anthropogenic Emissions into the Atmosphere

A large quantity of matter including such strong acids as sulphuric, nitric, hydrochloric, and hydrofluoric, together with industrial emissions, enters the atmosphere. The emission of hydrogen chloride and hydrogen fluoride as well as many other compounds is tied to specific types of production (e.g., aluminum production), and their effects on atmospheric chemistry have a local character. The emission of sulphur and nitrogen oxides is typical for any branch of industry. Absolute values of the emissions are extremely great. They may be compared with corresponding natural geochemical flows and at the regional level they exceed even them significantly.

In ores and fuels extracted from below the earth's crust there is a considerable amount of sulphur which is oxidized during processing or burning and then enters the atmosphere mainly as sulphur dioxide. About 3% of the latter is oxidized directly into trioxide during the burning process. At present the anthropogenic sulphur emissions into the atmosphere are estimated as 115 million t yr^{-1}, among them: 99 million t of sulphur dioxide, 4 million t of sulphur trioxide, 9 million t of aerosol sulphates, and 3 million t of hydrogen sulphide (Rjboshapko, 1983).

Nitrogen compounds enter the atmosphere both as a result of burning nitrogen-containing fuel and air nitrogen oxidation in a flame at high temperatures. The emissions contain mainly two oxides: nitrogen oxide and nitrogen dioxide. In this case their absolute quantities and ratio depend drastically on both the burning regime and the flame temperature.

The fundamental source of sulphuric dioxide entering the atmosphere is the production of heat and electroenergy; the transport is the source of nitrogen oxides (Table 6–1). The emissions of sulphur dioxide by energy-producing enterprises takes place at 100 to 300 m levels and the emissions of nitrogen oxides occur at ground level.

III. Air Pollution Concentrations

In his research Izrael (1983) singles out three zones of air pollution: pure (background), regionally industrial, and urban (Table 6–2).

From these data it follows that the concentration of all enumerated substances in the atmosphere of urban zones is one or two orders higher

Table 6–1. The emissions of sulphur dioxide and nitrogen oxides (based on 1980 data).

Sources	Sulphur dioxide (%)	Nitrogen oxides (%)
Heat and electroenergy production	55	37
Industry	44	13
Transport	1	50

Table 6–2. Concentrations of sulphur and nitrogen compounds in the atmosphere (fluctuation ranges are shown in parentheses).

	Zones		
Compounds	Urban ($\mu g\ m^{-3}$)	Regionally industrial ($\mu g\ m^{-3}$)	Pure ($\mu g\ m^{-3}$)
SO_2	40(5–200)	10(1–50)	0.5(0.1–3)
H_2S	3(0.5–10)	0.5(0.2–3)	0.3(0.1–1.0)
SO_4^{2-}(particle)	12(3–25)	10(1–20)	2.0(0.5–5)
NO	8(1–30)	1.5(0.3–5)	0.2(0.1–0.5)
NO_2	20(3–100)	4(1.0–15)	0.5(0.2–1.5)
HNO_3(gas)	3(0.5–15)	1.5(0.1–6)	0.2(0.03–2)
NO_3^-(particle)	3(0.5–15)	2.0(0.3–8)	0.4(0.05–4)

than in geochemically pure zones; increased concentrations are observed even at distances more than hundreds of kilometers from the source. The effects of sample acidification and supplies of increased acidic concentration on the underlying surface have the same space scale. In a pure zone carbon dioxide (about 80%) makes a major contribution to the sample acidity while the contribution of sulphuric and nitric acids is about 10%. In the atmosphere of highly industrialized regions the acidity comprises 60% of H_2SO_4, 30% of HNO_3, 5% of HCl, and only about 2% of carbon dioxide solution.

IV. Atmospheric Processes

Because of the presence of oxygen and ozone in the atmosphere, there are molecular and atomic interaction processes of carbon oxide, nitric oxide, and sulphur dioxide with oxygen and then their further conversion to highest oxides and when reacting with atmospheric moisture, the formation of sulphuric, carbonic, and nitric acids.

In his research, Rjboshapko (1983) obtained the experimental dependence of sulphur oxidation rate from the solar radiation (R) and ozone concentration (O_3)

$$S = (0.03 \pm 0.01)(R \cdot h \cdot O_3)$$

where h – height of mixing layer, m.
Oxidation rates resulting from some experiments have the following values:

$0.04^1/_h.$
$0.027^1/_h.$

A lifetime of sulphur dioxide is determined by 30 to 50 hours. There is a profound dependence of oxidation reaction rate of sulphur dioxide on the air moisture and the time of day. Being numerous, oxidation reactions of nitric oxides result in equilibrium states between nitric oxides and dioxides. The relationship between these two nitrogen forms fluctuates from 0.3 to 0.9. Average daily conversion rate of nitric oxides to nitric acid for middle latitudes and normal temperatures is $0.1^1/_h$ and their average lifetime is about 10 hours. In this case the lifetime of sulphates and nitrates makes up 100 hours (Izrael, Rovinsky, and Nazarov, 1983). The total concentration of hydrogen ions is equal to the sum of their concentrations formed during a dissociation of sulphuric and nitric acids and their natural content (Izrael et al., 1983).

$$C^{H+} = C^{H+(SO_4)} + C^{H+(NO_3)} + C^{H+(nat.)}$$

From this, pH = Log(C^{H+}).

Concentrations of nitrogen and sulphur compounds in the air of background (pure) regions of the USSR are the following:

$NO_2 = 0.1-18\mu g\ m^{-3}$ $SO_4^{2-} = 0.5-10\mu g\ m^{-3}$
$SO_2 = 0.05-4\mu g\ m^{-3}$ $H_2S = 0.02-1\mu g\ m^{-3}$

V. Air Pollution with Heavy Metals (Background)

Concentrations of heavy elements in background regions of the USSR forming poorly soluble compounds with sulphate ion are as follows:

$Pb - 2-4\cdot10^{-9}\ g\ m^{-3}$ $Cd - 0.2-0.3\cdot10^{-9}\ g\ m^{-3}$
$As - 1.2-2.2\cdot10^{-9}\ g\ m^{-3}$ $Hg - 5-15\cdot\ 10^{-9}\ g\ m^{-3}$

Average concentrations in precipitations of background regions are $Pb - 15\ \mu g\ l^{-1}$, $Hg - 1.2\ \mu g\ l^{-3}$, $As - 1.6\ \mu g\ l^{-1}$.

VI. Acidic Deposition in the USSR

The transfer of sulphur and nitrogen compounds from the atmosphere to the soil occurs both with atmospheric precipitation and by aerosol particle sedimentation. Dry deposition may account for 40% of the total oxide flow from the atmosphere to the soil.

The trajectory of air mass movement and their transfer of clouds has a great influence on the content of sulphates and chlorides. Transferred from sea regions deep into a continent, the precipitation, as a rule, is enriched in chlorides, bromides, sulphates, and sodium. This is a good indicator for a determination of precipitation origin. Petrenchuk (1979) studied the content change of chlorine, sulphates, sodium, and calcium as well as the relationship of chlorine to sodium as an indicator for a determination of precipitation origin (a movement trajectory) at the stations located in different regions of the USSR.

Observations performed at Voeykovo (Leningrad), Valday (Novgorod region), Mudyug (Arkhangelsk), and High Dubrava (Urals) stations have shown that through the travel of air masses from oceanic regions the precipitation at the stations are enriched with sulphates and chlorine ion and at the same time while traveling along the continent they are enriched with calcium (Table 6–3) and chlorine ion content is decreased.

The effects of the transfer of chemical compounds such as NO_2 and SO_2 playing a significant role in the change of precipitation acidity have been considered (e.g., Petrenchuk, 1979). Researchers established background concentrations of NO_2 equal to 2 to 7 $\mu g\ m^{-3}$ according to determinations

Table 6–3. The concentration of sulphates, chlorides, sodium, and calcium in the precipitations over the European part of the USSR.

Direction of air mass travel	Observation stations							
	Voeykovo				Mudyug			
	SO_4^{2-}	Cl^-	Na^+	Ca^{2+}	SO^{2-}	Cl^-	NA^+	Ca^{2+}
North	5.97	1.33	1.54	0.85	3.20	2.79	1.93	0.45
	Voeykovo				Mudyug			
	SO_4^{2-}	Cl^-	Na^+	Ca^{2+}	SO^{2-}	Cl^-	NA^+	Ca^{2+}
West	6.18	1.59	1.61	1.16	4.05	2.55	1.29	0.53
South	5.09	1.11	1.42	1.29	3.96	1.80	1.14	0.45
	Valday				High Dubrava			
North	2.77	0.75	1.31	0.32	4.01	1.02	1.18	0.75
West	3.89	1.15	1.45	0.52	4.89	1.65	1.59	0.86
South	4.0	0.80	1.23	0.45	4.65	0.98	1.13	0.91

in reservations of the South Baltic and the Caucasus. However, they marked the momentary high concentrations of NO_2 as great as 37 to 400 μg m^{-3}. A daily change of NO_2 concentrations in natural conditions varies very substantially (Figure 6–1), which is essential to take into consideration when determining anthropogenic contributions to changes of NO_2 content in the atmosphere (see Figure 6–1).

The effect of the direction change of air mass travel on NO_2 and SO_2 concentrations is shown in Figure 6–2 illustrating the data on SO_2 and NO_2 concentration changes in Varenska Puscha Reservation when changing the air travel direction available from the Atlantic Ocean up to 6 P.M. on July 29, and then from the southwest European regions beginning 6 A.M. on July 30. These data show that it is necessary to take into account the daytime and the transboundary transfer when calculating the concentrations of NO_2 and SO_2 of anthropogenic origin as well as an increase of precipitation acidity associated with them.

The changes of sulphate and nitrate concentrations as well as changes induced by them of precipitation acidity occur also due to anthropogenic sources. It is evident from the trend of SO_4^{2-} concentration changes registered by observers at Kemeri, Shilutee, and Beresino stations (Figure 6–3) from 1958 to 1976. Here as a rule the precipitation acidity is greater in winter months than in summer.

The determination of the precipitation acidity in the Baltic, in the northern European section, Ukraine, Zavolje, East Siberia, and other territories of the USSR show that the average pH of precipitation and cloud water ranges from 5.2 to 5.7 indicating a small anthropogenic effect in these regions (Table 6–4).

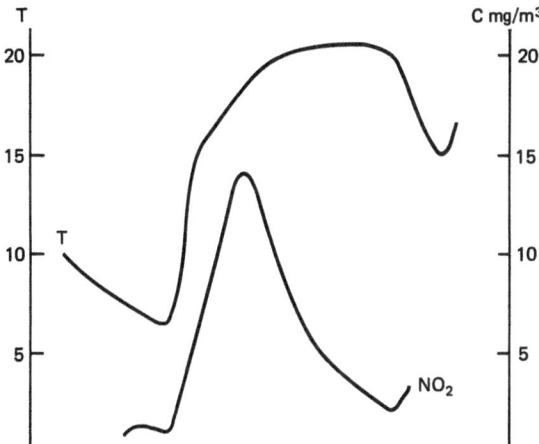

Figure 6–1. A Daily course of concentration (NO$_2$) and temperature (T), Varenska Puscha, July 14, 1976.

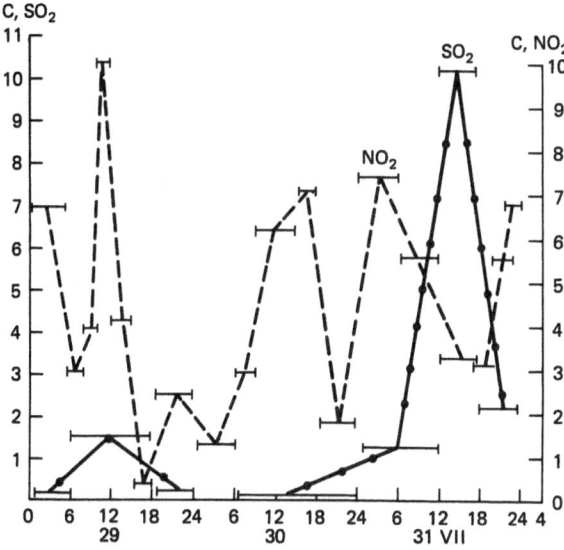

Figure 6–2. Change of air masses and changing SO$_2$ and NO$_2$ concentrations, Varenska Puscha, July 30, 1976.

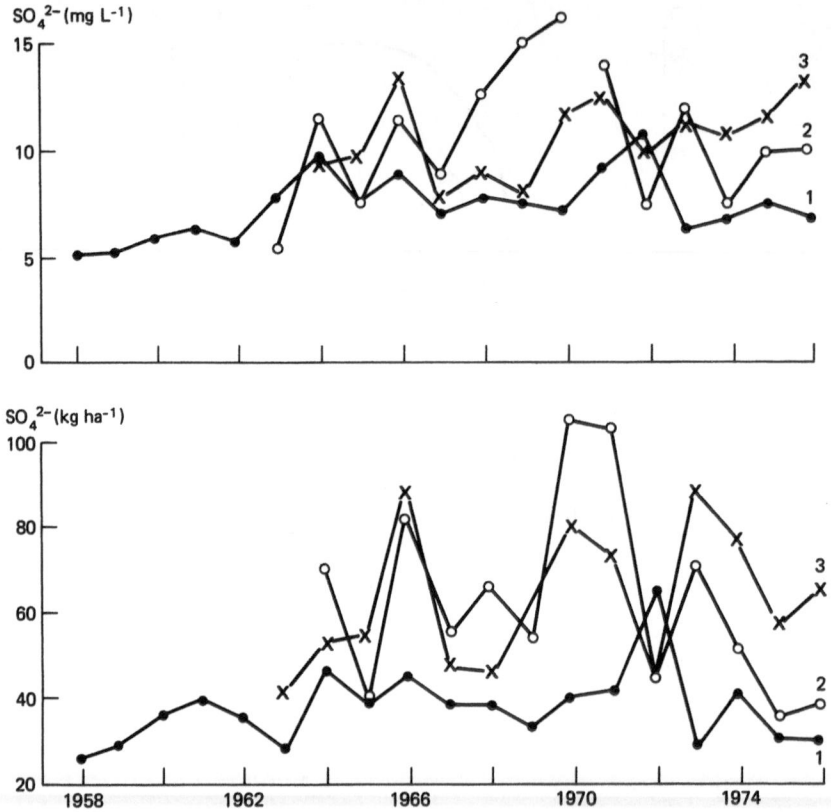

Figure 6–3. Concentration of SO_4^{2-} (mg L^{-1}) in precipitation and fallout of SO_4^{2-} (kg ha^{-1}) with precipitation at Kemeri (1), Shilute (2), and Bersino (3) stations.

Table 6–4. The acidity of cloud water in different regions of the USSR.

Regions	pH
The North European section	5.4
Baltic	5.2
Ukraine	5.7
Zavolje	6.0
East Siberia	5.7

In this case it should be noted that there are changes of sulphate and nitrate concentrations as well as changes of pH in precipitation caused by industrial centers. For example, in the neighborhood of such towns as Kazan, Chistopol, Kanash, and so on, significant precipitation acidification has been observed (Table 6–5).

Table 6–5. The acidity of cloud water in the regions of some industrial centers.

Town	pH
Kazan	3.3–3.8
Leninogorsk	4.1
Chistopol	3.8
Mamadish	3.1
Shumerla	4.6
Kanash	4.4
Kiev	4.0
Dnepropetrovsk	5.5
Odessa	5.9
Leningrad	3.7–4.8

The effect of industrial centers on SO_4^{2-} content and hence the precipitation acidity is well illustrated in the research of where they reveal the changes of sulpate ion and mineralization in the direction Dnepropetrovsk to Odessa.

Many years of changes of concentrations of acidic-forming ions SO_4^{2-}, NO_3^-, and pH values at Kemeri station (Latvia) demonstrate that in spite of an increase of SO_4^{2-} concentrations in precipitations of approximately 1.5 times over 18 years (1958 to 1976), the value of pH remained in the range of 5.2 to 6.7 and acidification of precipitation did not occur.

A synchronic increase of sulphate concentrations in precipitation occurring at various stations in the western part of the European territory of the USSR makes us think about the existence of a single mechanism causing this increase. Petrenchuk (1979) does not exclude the possibility of sulphur compound transfer over the western border of the USSR.

However, it should be noted that in spite of intensive industrial development in European countries, the most prominent acidification of precipitation in the western regions has not been observed, although the sulphate depositions here are 2.5 times more than in Norway. Mid-annual pH values are 5.0 to 6.0 and only in rare cases are they reduced to 4.3 to 4.6.

Petrenchuk (1963) offers a very interesting example of the dependence of the chemical composition and the pH value of cloud water in a synoptical situation over Krasnoyarsk (Table 6–6).

So the influence of an industrial center on the chemical composition and pH value of precipitation depends to a large extent on air mass travel and transferred atmospheric moisture, in other words, on the time of the atmospheric moisture contact with gaseous and solid wastes of industrial enterprises in the region's atmosphere.

The time of precipitation fallout (precipitation amounts) (e.g., Petrenchuk, 1963) has a rather severe influence on pH value and on mineralization (Table 6–7).

Table 6–6. Dependence of the chemical composition and pH value of cloud water in a synoptical situation.

Synoptical situation	pH	Electrical conductivity $(x \cdot \mu s \cdot cm^{-3})$
Low advection of air masses in the center of anticyclone	3.8–4.9	65
Cyclone development	5.0–5.2	8–13
Distinct advection of air masses from the North	4.8–5.5	16–27

Table 6–7. Changes of pH and ion sums in precipitation during their time fallout.

Time	pH	Σ of ions
3 P.M. 34 min.	5.30	11.71
4 P.M. 55 min.	6.20	6.73
8 P.M. 35 min.	5.89	4.91

The study of data on pH value, mineralization, and ion composition of precipitation in the USSR shows that the changes of pH values are observed and localized near industry centers; while over most of the USSR, the pH values are 5.0 to 6.0. However, in connection with industrial production growth consideration must be given to a study of the possibility of precipitation acidification and its effects on plants, animals, and humans.

References

Izrael, Y.A., F.J. Rovinsky, and I.M. Nazarov. 1983. *Acid rains.* Moscow: Gidrometeoizdat. 273 pp.

Petrenchuk, O.P. 1963. *Changing precipitation composition in Sverdlovsk region in accordance with meteorological conditions.* Leningrad: GGO Collections. 136 pp.

Petrenchuk, O.P. 1979. *Experimental study of atmospheric aerosol.* Leningrad: Gydrometeoizdat. 263 pp.

Rjboshapko, A.G. 1983. *Atmospheric circle of sulphur.* Moscow: Science.

Atmospheric Acidic Deposition and Its Environmental Effect in Hungary

I. Pais* and L. Horváth†

Abstract

The data concerning atmospheric acidic deposition and its effect on the environment in Hungary are briefly summarized. The yearly SO_2 and NO_x emissions are between 0.5 and 0.8 million tS year^{-1} and 0.05 to 0.10 million tN yr^{-1}, respectively. The amount of sulphur compounds emitted exceeds the rate of dry plus wet deposition on a yearly basis.

The dry deposition rate of atmospheric S and N compounds expressed in H$^+$ equivalents is 146 meq m^{-2} yr^{-1}. Sulphur dioxide, NO, and HNO$_3$ are responsible for 95% of dry deposition; the effect of SO_4^{2-}, NO_3^-, and NH_4^+ particles as well as NH_3 gas is negligible in this process.

Wet deposition of acidifying compounds (SO_4^{2-}, NO_3^- and NH_4^+) is 128 meq m^{-2} yr^{-1}, determined from results of the Hungarian precipitation chemistry network. The sum of wet-and-dry deposition amounts to 274 meq m^{-2} yr^{-1}.

During the last few decades the pH of some forest soils has decreased as a consequence of acidic deposition. The pH value of agricultural soils has also decreased, but it has been demonstrated that the main reason for the acidification is the use of fertilizers. Acidic deposition causes mostly direct damages (physiological effects) on plants, rather than indirect effects through soil acidification.

The corrosive effect of acidic pollutants is also a great problem in Hungary. The total cost of damage caused by corrosion amounts to nearly $1 billion annually. Damage of buildings and sculptures made of limestone is also observed. It has been calculated, however, that most surface waters tolerate the atmospheric acidic input well because of their high buffer capacity.

*University of Horticulture and Food Industry, H-1114 Budapest, Villányi u. 29/31, Hungary.
†Institute for Atmospheric Physics, H-1675 Budapest, P.O. Box 39, Hungary.

I. Introduction

In Hungary, similar to other central eastern European countries, damaging effects of acidic deposition can be observed. It has been demonstrated that sulphur and nitrogen oxides, called primary air pollutants, are responsible for the damage. Primary and secondary pollutants (HNO_3, H_2SO_4, etc., transformed from NO and SO_2) deposited in Hungary are coming partly from domestic, partly from foreign sources. Simultaneously part of the acidic air pollutants emitted in Hungary are deposited abroad.

A sampling network has been monitoring the deposition of air pollutants since the mid-1970s. Systematic observation of consequences of the acidic deposition has been carried out for years as well.

II. Emission of Sulphur and Nitrogen Oxides

In Hungarian energy production nuclear energy has a share of about 30% to 35%, and most of the other energy is produced by burning Hungarian coal and oil containing relatively high proportions of sulphur (see Table 7–1).

Accurate estimates of anthropogenic SO_2 emissions are rather uncertain (Bede and Gács, 1986). According to the model calculation of the European Monitoring and Evaluation Program of the Economic Commission for Europe (EMEP, 1986), the Hungarian SO_2 emissions were 0.55 million tS in 1984. Grennfelt (1981) estimated 0.75 million tS for yearly SO_2 emissions. On the basis of a Hungarian survey (MTA/OKTH, 1985) the strength of domestic SO_2 sources was calculated at between 0.5 and 0.8 million tS yr^{-1}. It should be mentioned that the rate of natural emissions is negligible as compared to the strength of anthropogenic sources (Várhelyi, 1982).

The materials listed in Table 7–1 were previously burned without special purification, but there are plans to reduce the amount of sulphur

Table 7–1. The sulphur content of Hungarian fuels (Horváth, 1986).

Used material	Average sulphur content (%)	Used quantity (million t yr^{-1})
Lignite	1.4	8.5
Brown coal	2.3	14.2
Stone coal	1.4	5.0
Mineral oil	2.1	9.8
Natural gas	Negligible	9.7 million m^3

emitted to the atmosphere. The Hungarian government, in agreement with other European countries, has joined the program to reduce sulphur emissions (or exports) by 30% by 1993.

As for the Hungarian nitrogen oxide emission, Bónis (1981) calculated 0.047 to 0.102 million tN yr^{-1} for the strength of natural plus anthropogenic sources. According to calculations of Horváth (1986) this figure is about 0.080 million tN yr^{-1}. Main NO_x sources are coal burning, oil consumption, and motor vehicles. Contribution rates of these factors to the total NO_x emissions are practically the same. Natural emissions, as in the case of SO_2, are negligible as compared to man-made sources.

III. Chemical Transformation of Primary Air Pollutants

As we can see, a large amount of anthropogenic sulphur and nitrogen oxides are emitted into the air in Hungary year after year. Considering the oxidizing feature of the air these materials can be oxidized in the atmosphere to form acids (HNO_3, H_2SO_4). Oxidation occurs mainly by tropospheric ozone and photochemical oxidants (e.g., free radicals such as OH and HO_2). This is supported by some Hungarian observations. In the summer the ratio of SO_4^{2-}/SO_2 concentrations, characterizing the chemical transformation, correlates well with solar radiation intensity, which controls the atmospheric photochemical reactions (Mészáros, 1973). This phenomenon cannot be observed in the winter when the concentration of SO_4^{2-} correlates first of all with the mass concentration (and surface) of aerosol particles (Mészáros, 1974a). In the winter the $SO_2 \rightarrow SO_4^{2-}$ transformation probably takes place on the surface of particles with heterogeneous processes. It has recently been shown that elemental carbon particles (soot) play an essential role in controlling the transformation of SO_2. This is supported by electron microscopic observations of Á. Mészáros (1986).

There were some attempts to estimate the kinetic parameters of $SO_2 \rightarrow SO_4^{2-}$ transformation in Hungary. According to these results, in the plume of Budapest, 15 km from the middle of city (the supposed center of sources), the half-life of SO_2 is 2 to 10 hours or 1.8 hours (Mészáros, Moore, and Lodge, 1977; Horváth and Mészáros, 1978, respectively). On the basis of another field measurement (Horváth and Bónis, 1980) in the urban plume moving from Budapest to one of our regional background air pollution stations (approximately 60 km from the city) the estimated half-life of SO_2 is 7 hours.

Chemical transformation of atmospheric nitrogen oxides (NO, NO_2) to HNO_3 was studied by means of model calculations (Haszpra and Turányi, 1986). According to this investigation the transformations of the nitrogen compounds are controlled by complicated nonlinear interactions. The calculations prove that there is no linear relation between the rate of the nitric acid production and nitrogen oxide concentration. The production rate and the concentration of nitric acid are influenced by the hydrocarbon emission/concentration to a large extent.

IV. Transport of Acidic Atmospheric Air Pollutants

In connection with the transport of air pollutants, one of the most important problems is to estimate the flux of sulphur and nitrogen compounds across the border. The result of an international model calculation of sulphur budgets for Europe shows that in Hungary the SO_2 emission was 0.552 million tS yr^{-1}; total sulphur deposition was 0.351 million tS yr^{-1} in 1984 (EMEP, 1986). This means that Hungary is a significant sulphur exporter. From foreign sources around 0.125 million tS yr^{-1} sulphur arrives above Hungary, and 0.326 million tS yr^{-1} leaves the country. The situation is more unfavorable when we take into account the results of budget calculations. According to Várhelyi (1982) sulphur dioxide emissions are about 0.75 million tS yr^{-1}, whereas total deposition amounts to only 0.38 million tS yr^{-1} in Hungary.

The Hungarian nitrogen budget was calculated by Bónis (1981). She demonstrated that sources and sinks of nitrogen compounds (NO_x, NH_x) are practically balanced.

A regional scale transmission model was constructed by Fekete and Szepesi (1987) for Hungary taking into account both sulphur- and nitrogen-containing species. Researchers in their model calculation considered the effects of domestic area sources and tall stacks as well as that of foreign sources separately. Transmission of pollutants was partly calculated by means of an air trajectory moving box model and partly by a Gaussian model. The calculated average maximum acidic deposition (without NH_3 and HN_4^+) is 229 meq m^{-2} yr^{-1} (expressed in H^+ equivalents), which is in strong agreement with the corresponding figure from Table 7–5. (234 meq m^{-2} yr^{-1}) calculated on the basis of direct measurements.

V. Acidic Deposition

A. Dry Deposition

During dry weather situations the atmospheric acidifying constituents are deposited onto surfaces by turbulent diffusion processes (gases) as well as by turbulent and gravitational mechanisms (aerosol particles). In order to determine the rate of dry deposition, we have to know, on one hand, the surface-level concentrations of acidic materials and on the other, the so-called dry deposition velocity of the given compound.

Background atmospheric concentration of acidic sulphur and nitrogen compounds are measured at three background air pollution monitoring stations in Hungary. The location of the stations (Farkasfa, K-puszta, Szarvas) can be seen in Figure 7–1. The siting criteria of these stations are practically similar to those proposed by WMO (1978). This means that the background concentrations measured at these stations are representative of Hungary and are not influenced by direct effect of larger emission sources. Since the ratio between the area of directly polluted regions and the total surface area in Hungary is relatively small (Várkonyi and Cziczó, 1980), background concentrations are practically representative of the whole country.

In terms of acidification the following sulphur and nitrogen compounds have considerable high atmospheric levels in Hungary: SO_2 and NO_2 gases, HNO_3 vapor as well as SO_4^{2-} and NO_3^- containing aerosol particles. Furthermore, it has been demonstrated (e.g., Asman, Jonker, Slanina, and Baard, 1982) that NH_3 gas and NH_4^+ particles play an important role in acidification, since these materials can take part in nitrification processes resulting in nitric acid in the soil. Therefore, for an estimation of the measure of wet-and-dry acidic deposition we should take into account the NH_x compounds as well.

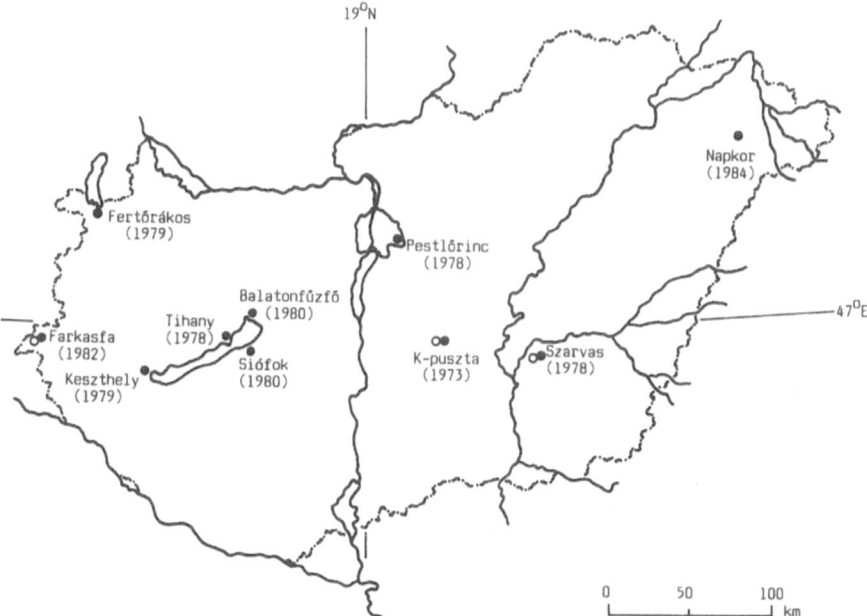

Figure 7–1. Background air pollution monitoring stations (o) and wet-only collectors (•) in Hungary. Years in parentheses show the first year of operation period.

The sampling program of Hungarian background air pollution monitoring stations can be seen in Table 7-2. The K-puszta station with extended programs supports data for WMO and ECE/EMEP networks. Sulphur dioxide is sampled and analyzed according to the West and Gaeke (1956) method. Nitrogen oxide is measured by using a modified version of Saltzman's method (Levaggi, Siu, and Feldstein, 1973). The sampling and analysis for ammonia measurement is carried out on the basis of Ferm (1979) and Koroleff (1970), respectively.

Aerosol particles for NH_4^+, NO_3^-, and SO_4^{2-} analyses are captured by a hydrophobic teflon filter. Sulphate is identified by isotope dilution technique (Klockow, Denzinger, and Rönicke, 1974). Ammonium ions are analyzed with the method mentioned for NH_3; NO_3^- is determined by the nitration of salicylic acid with nitric acid in a medium acidified with sulphuric acid.

Nitric acid vapor sampling is done by using two prepared Whatman 41 filters after teflon prefilter (Forrest, Tanner, Spandau, D'Ottavio, and Newman, 1980). Chemical analysis of HNO_3 is carried out with the same method as for NO_3^-. The daily 24-hr measurements are performed 1.5 m above the surface.

Average concentrations of acidifying sulphur and nitrogen compounds can be seen in the second column of Table 7-3. Further information concerning the concentration of sulphur and nitrogen compounds in Hungary can be found in the research of Mészáros and Horváth (1984).

For estimating dry deposition it is necessary to know the dry deposition velocity of a given compound in addition to the atmospheric concentrations. Dry deposition velocities for gases (SO_2, NO_2, NH_3) were determined by means of the so-called gradient method, above grass-

Table 7-2. Time period of the sampling at Hungarian background air pollution monitoring stations for nitrogen and sulphur compounds.

Station	Component	Operation period	Average concentration ($\mu g\ m^{-3}$)
K-puszta	SO_2	1974–1986	15.6
	NO_2	1974–1986	7.10
	NH_3	1981–1986	See Table 7-3
	HNO_3 and NH_4^+, NO_3^-, SO_4^{2-} particles	1982–1986	See Table 7-3
Szarvas	SO_2	1981–1986	17.8
	NO_2	1981–1986	10.5
Farkasfa	SO_2	1983–1986	11.5
	NO_2	1983–1986	6.90

Table 7-3. Average background concentrations, deposition velocity, and dry deposition rate of sulphur and nitrogen compounds.

Species	Average concentration ($\mu g\ m^{-3}$)	Deposition velocity ($cm\ s^{-1}$)	Dry deposition ($g\ m^{-2}\ yr^{-1}$)
SO_2	15.0[a]	0.6	2.84
SO_4^{2-}	6.60	0.024	0.05
NO_2	8.17[a]	0.5	1.29
HNO_3	2.14	2.0	1.35
NO_3^-	2.10	0.030	0.02
NO_3	1.00	0.33	0.09
NH_4^+	2.00	0.022	0.01

[a] average of three stations

covered surfaces (e.g., see Horváth, 1983a). Dry deposition velocities for SO_2 and NO_2 were adapted from the work of Várhelyi (1980).

It should be mentioned that in the case of NH_3 gas different kind of surfaces can be source or sink for ammonia depending on the physical and chemical state of surfaces (e.g., soil or water) and that of the air. The sign of the ammonia flux is controlled by the difference between the actual atmospheric NH_3 level and the equilibrium ammonia concentration determined by some physical and chemical factors in soil or water (pH, temperature, NH_3/NH_4^+ concentrations, etc.). Maximum deposition velocity of NH_3 is 1 cm s^{-1} both above water and soil surfaces (Horváth, 1982, 1983a). For our calculations a value of 0.33 cm s^{-1} was accepted as an average for Hungary (Horváth, 1983a). Finally, because of the lack of Hungarian measurement, dry deposition velocity for HNO_3 was accepted from the literature (Huebert, 1983).

In order to determine the dry deposition velocity of NO_3^-, NH_4^+, and SO_4^{2-} containing particles, mass median diameters were taken from size distribution data of Mészáros (1972), while dry deposition velocities for these diameters were determined by means of Chamberlain's curve as presented by Hidy (1973). The dry deposition velocity figures can be seen in the third column of Table 7-3. The last column shows the yearly dry deposition figures calculated as a product of concentration and deposition velocity. It can be seen that only SO_2, NO_2, and HNO_3 play an essential role in dry acidic deposition; the effect of other S and N compounds is practically negligible.

B. Temporal Variation of the Rate of Dry Deposition

Most of the atmospheric background concentration (and dry deposition) of air pollutants has a characteristic yearly variation (Mészáros and Horváth, 1984). This phenomena is primarily due to the yearly variation of emission strengths. Concentrations of the most important primary air pollutants (SO_2, NO_2) have a definite maximum during winter months (see Figures 7–2a and 7–2b). They are due without a doubt to the increased energy consumption and emission in the winter months.

Of course, beside emission other factors may also control the yearly variation of the concentration (and deposition) of air pollutants, such as chemical transformation rates, temporal variation of meteorological factors, and so on.

The concentration of HNO_3 (which is the most important of air pollutants besides SO_2 and NO_2 shows a less pronounced pattern than SO_2 and NO_2 (see Figure 7–2c). Since HNO_3 is a so-called secondary air pollutant (formed from NO_x; see Section III), yearly variation of the concentration of nitric acid vapor is controlled mainly by yearly variation of rates of chemical transformations and by atmospheric concentrations of hydrocarbons (Haszpra and Turányi, 1986).

Another important question is the long-term trend of the concentration (and dry deposition) of acidic air pollutants. To observe the trend we have a long measuring period for SO_2 and NO_2 only at the station K-puszta (Figures 7–3a and 7–3b). Supposing that meteorological factors do not have systematic long-term variations, the concentrations of SO_2 and NO_2 (in accordance with Section IV) are controlled first by the change of Hungarian emission rates, second by the foreign sources. The

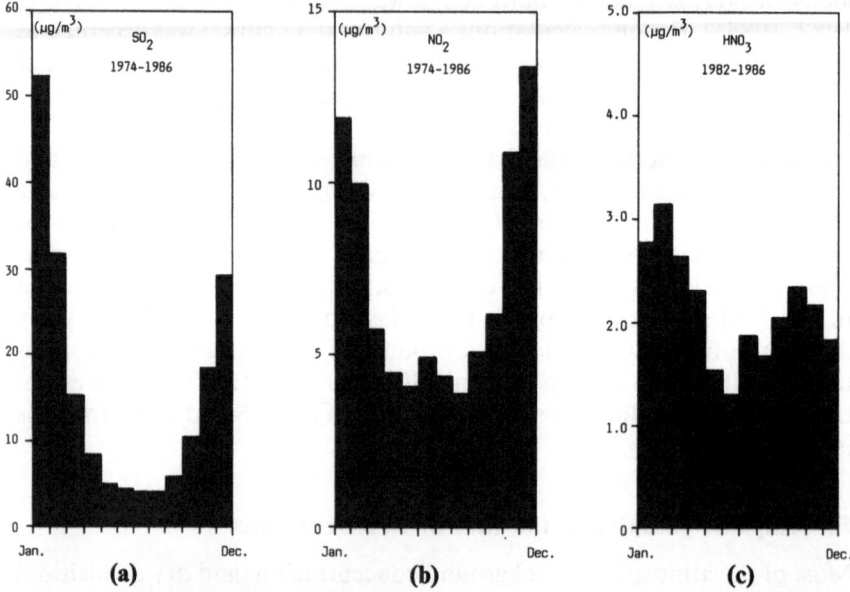

Figure 7-2. Yearly variation of concentrations of SO_2(a), NO_2(b), and HNO_3(c) at K-puszta station.

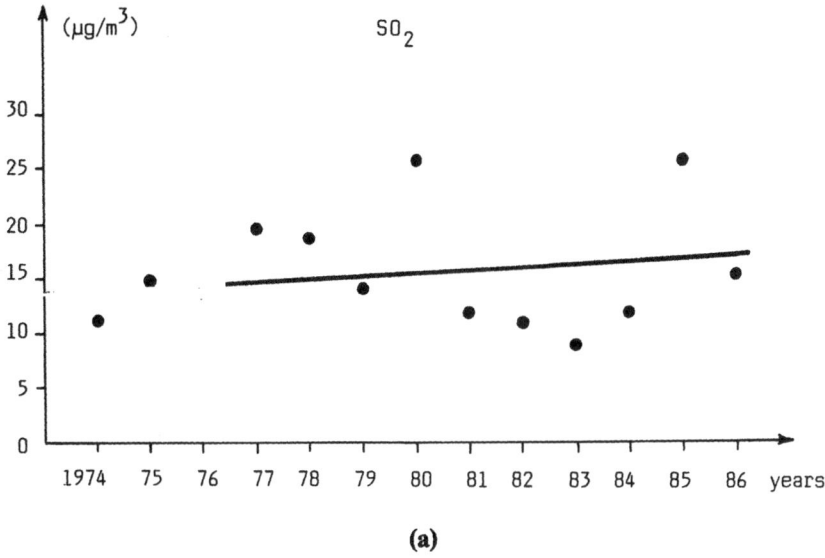

(a)

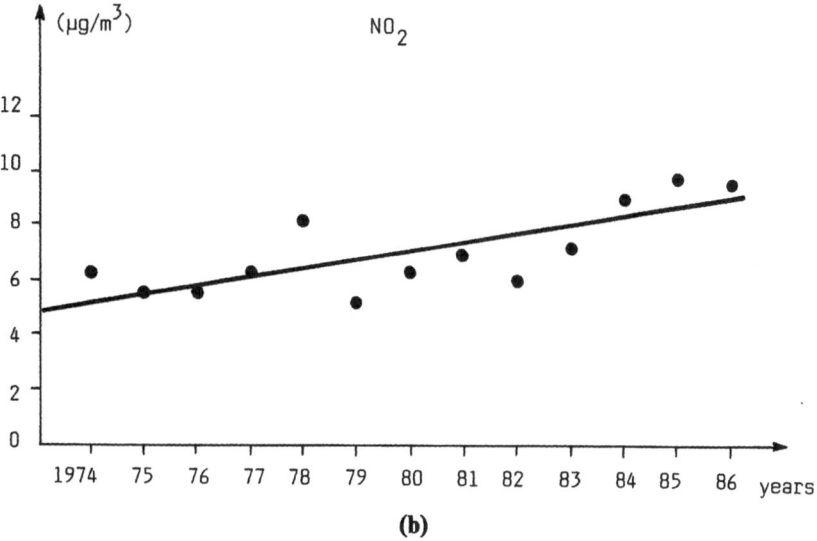

(b)

Figure 7–3. Trends of SO_2(a) and NO_2(b) concentrations measured at K-puszta station.

concentration of SO_2 practically has no long-term trend (Figure 7–3a). The unusually high figures in 1980 and 1985 can be attributed to accidental variation of meteorological conditions. For example, the extremely high SO_2 level in 1985 was due to the "episode" event in January, when the background SO_2 concentration had been increasing for weeks in Europe owing to meteorological reasons (Mészáros, Mersich, and Szentimrey, 1987).

Maximum SO_2 concentration had nearly reached the level of 600 μg m^{-3} at K-puszta station which is around two orders of magnitude higher than the yearly average figure. Apart from these points (in 1980 and 1985) one can say that background concentration (and dry deposition) of SO_2 is practically constant or has slightly decreased during recent years. It is in accordance with the emission trend, which has approximately been constant (or has decreased) since 1970 (Horváth and Mészáros, 1984). Because Hungary joined the program to reduce SO_2 emissions, we can expect further decreases in SO_2 concentration and dry deposition in the future. However, it may be expected that a 30% decrease in emission would cause less (approximately 14%) of a decrease of total acidic deposition according to calculations of Horváth and Möller (1987).

Concentrations of NO_2 showed a regular increase from 1974 to 1986 (Figure 7–3b). One of the most important sources of nitrogen oxides in Hungary is the motor traffic. Since the number of private cars in the mid-1980s (around 1.3 million) was around five times the figure in 1970 (around 0.24 million), the NO_x emission of cars has proportionately increased (Horváth and Mészáros, 1984). In the light of this fact, it is not surprising that atmospheric concentration (and dry deposition) of NO_2 has nearly doubled in the last 13 years.

A decrease of deposition of SO_2 and an increase of deposition of NO_2 and HNO_3 can be expected in the future. Therefore, the relative importance of N to S compounds will probably be increasing in Hungary.

C. Wet Deposition

For the calculation of wet deposition rates of acidic air pollutants we used the results of chemical analyses of precipitation samples. In the Hungarian precipitation chemistry network 10 wet-only precipitation collectors are in operation, as it can be seen in Figure 7–1. The main results of the Hungarian precipitation network up until 1980 is summarized by Horváth and Mészáros (1984). At each station monthly precipitation samples are taken, except K-puszta station where daily and weekly samples are also collected, as proposed by WMO and ECE/EMEP.

Concentrations of important ions, with respect to acidification (SO_4^{2-}, NO_3^-, NH_4^+), are determined by means of analytical methods described in Section VA; pH of samples are measured with glass electrode.

The weighted mean pH values and concentration of SO_4^{2-}, NO_3^-, and NH_4^+ ions for the 10 Hungarian stations can be found in Table 7-4.

The mean pH value of precipitation in Hungary is around 4.8, an acidity nearly one order of magnitude higher than the reference value (pH = 5.6) controlled by the absorption and hydrolysis of atmospheric CO_2. There are large deviations, however, from the mean value. For example, the frequency distribution of daily precipitation samples collected in K-puszta from 1978 to 1982 shows that H^+ concentration of samples varies within 6 orders of magnitude (Horváth and Mészáros, 1986). In 6% of the cases precipitation samples were strongly acidic (3<pH<4). The number of moderately acidic samples (4<pH<5) was 31%; in 41% of the cases the precipitation was weakly acidic or neutral (5<pH<6). Twenty-two percent of the samples were alkalinic because of the washing out of CO_3^{2-} and HCO_3^- containing soil-derived alkalinic aerosol particles (Horváth, 1981).

It should be noted, however, that in the evaluation of wet acidic deposition pH is not a suitable parameter. Concerning pH of precipitation we should take into account at least three disturbing factors (Horváth and Mészáros, 1986).

1. In Hungary the pH of precipitation water is affected to a great extent by the soil-derived particles. There is, for example, a good correlation between pH and Ca^{2+} content of precipitation (Horváth, 1981). Owing to air pollutants, spread as gases or in the form of fine aerosol particles, an acidity is developing in clouds and in falling precipitation, which is representative of the pollution of a given region (e.g., of

Table 7-4. Yearly averaged pH and concentrations of acidifying ions in precipitation water.

| Station | pH | SO_4^{2-} | NO_3^- | NH_4^+ | Precipitation amount |
			(mg L⁻¹)		(mm yr⁻¹)
Balatonfűzfő	5.44	6.40	2.72	1.36	585
Farkasfa	4.83	4.93	2.17	0.62	691
Fertőrákos	4.88	7.25	3.48	1.24	547
Keszthely	4.88	4.86	3.76	0.64	597
K-puszta	4.39	5.96	2.48	1.59	528
Napkor	5.19	6.29	3.21	1.13	376
Siófok	5.52	7.04	2.92	1.21	516
Szarvas	4.73	5.73	2.65	1.21	435
Pestlőrinc	4.97	9.31	2.83	1.48	470
Tihany	4.58	6.20	3.13	1.19	542
Weighted mean	4.82	6.32	2.92	1.14	529

Hungary). This "background" acidity is affected by the local effect of large alkaline particles ($r > 1 \mu m$) with a short lifetime.

2. During monthly sampling acidic and alkaline samples may get neutralized. Without this effect frequency distribution of samples would be more variable resulting in more acidic mean pH because of the logarithmic relationship between the pH and H^+.

3. The pH may change during sampling (storage) of precipitation water. In general, both the pH of acidic and alkaline samples tend to alter to neutral direction (Horváth and Mészáros, 1986). This leads to homogenization of samples as well.

It follows that without these disturbing factors the average pH, representative of the region of Hungary, would be more acidic than the measured value. Theoretically, the upper limit of acidity of precipitation water (without local neutralization) can be estimated with the sum of concentrations of SO_4^{2-} and NO_3^-, expressed in H^+ equivalents. Namely, concentration of these ions does not change during neutralization. The concentration of sulfate and nitrate ions according to Table 7-4 are 6.32 and 2.92 mg L^{-1}, respectively. The corresponding values, expressed in H^+ equivalents are 132 and 47 meq L^{-1}. The initial pH value of 3.75, corresponding to this figure, can be regarded as the potentially highest limit of acidity.

Taking into account the mean concentrations in Table 7-4, the calculated wet deposition figures can be found in Table 7-5.

D. Temporal Variation of the Chemical Composition of Precipitation Water

In Hungary concentrations of most of the inorganic ions in precipitation have a characteristic yearly variation with a pronounced spring maximum. A reasonable explanation of this phenomenon might be the effect of tropospheric ozone, which also has a maximum level in the atmosphere in the spring. Ozone molecules may promote the liquid-phase oxidation and neutralization processes in cloud and precipitation water (Mészáros, 1974b).

On the other hand, it is worth mentioning that whereas concentrations of SO_4^{2-}, NH_4^+, and H^+ ions in precipitation show no long-term variation, nitrate has a positive trend. Relying on the results of earlier research, Horváth (1983b) demonstrated that concentration of NO_3^- in precipitation had increased 6.6 times by the 1980s as compared to the figure measured in 1902 in Hungary. At the same time, concentration of NH_4^+ had remained at practically the same value.

E. Comparison of the Rate of Dry and Wet Deposition in Hungary

Table 7-5 shows the comparison of dry and wet deposition in Hungary expressed in H^+ equivalents. The most important information in the

Table 7-5. Comparison of dry and wet acidic deposition in Hungary expressed in H+ equivalents.

Components		Dry deposition	Wet deposition	Total deposition
		meq m^{-2} yr^{-1}		
Sulphur	SO_2	89	—	89
compounds	SO_4^{2-}	1	70	71
Subtotal	(S)	90	70	160
	NO_2	28	—	28
	HNO_3	21	—	21
Nitrogen				
	NO_3^-	−1	25	25
Compounds				
	NH_4^+	1	33	34
	NH_3	6	—	6
Subtotal	(N)	56	58	114
Total	(S+N)	146	128	274

table is that total (dry plus wet) deposition of different S and N species amounts to 274 meq m^{-2} yr^{-1} on an average. In the calculation of wet deposition the potentially highest acidity (including NH_4^+) was taken into account (see Section VC). It can be seen that the rate of dry deposition exceeds the effect of wet removal processes. Contribution of S and N compounds to total deposition are 58% and 42%, respectively. However, the relative importance of sulphur compounds is continuously decreasing for the reasons mentioned in Section VB.

VI. Effect of Acidic Deposition on the Environment in Hungary

A. Acidification of Forests

Stefanovits (1986) pointed out that in some Hungarian forest soils (podzole, brown forest soil) the pH value has decreased to a great extent during the last 25 to 30 years. Because there is no fertilization in the forests, the changes are the consequences of acidic deposition. These data are compiled in Table 7-6. Stefanovits (1986) proposed that in Hungary we ought to start with a supply of $CaCO_3$ from the air to balance the acidification of forest soils.

Solymos (1986) pointed out some consequences of acidic deposition for the health of Hungarian forests. We can estimate the forest-covered

Table 7-6. Change of pH values of some Hungarian forest soils.

Forest (county in Figure 7-4)	Depth (cm)	pH	
		In 1955–1960	In 1985
Pilis (2,10)	0–8	5.1	4.5
Mátra (3,4)	0–15	6.6	4.7
Zala (13)	0–6	6.0	5.6
Veszprém (8)	0–5	6.6	5.3

area in Hungary to be 18% of the total territory. Hungarian scientists have observed serious problems in sessile oaks mostly in the northern mountainous area. The main sources of this damage are not quite clear today; results do not prove unambiguously the primary role of acidic deposits in the initiation of damage.

Jakucs (1986) prepared a review on the effect of acidic deposition on the natural living ecosystems in Hungary. He wrote that these effects may be direct or indirect. In the case of direct impact, air pollutants initiate biochemical processes resulting in toxic effects. A large part of the direct effect is the so-called free radical pathology. For example, on the influence of SO_2, free-radical processes induce producing hydrogen peroxide or organic peroxides. These products may oxidize the unsaturated fatty acids of cell membranes. In this way the functional regulation becomes less effective. This effect is called *radical pathology.*

Among the indirect effects we can find the destruction of mycorrhizal fungi which causes the decrease of the uptake capacity of root systems and decay of trees.

B. Effects on Soils and Plants

One of the most serious environmental problems in Hungary is the acidification of soils. Várallyay, Rédly, and Murányi (1986) studied the influence of acidic deposition on the state of soils, and classified the Hungarian soils into six groups according to their susceptibility to acidification.

1. Strongly acidic soils which form about 13% of the total area of Hungary and whose further acidification is practically negligible.
2. Highly susceptible soils with light texture having low buffer capacity (about 14%).
3. Medium buffer-capacity soils with medium organic matter content which are susceptible to acidification and represent only 5%.
4. Slightly acidic soils with heavy texture, moderately susceptible to acidification (about 23%).
5. Slightly susceptible soils (4%, a relatively low percentage).

6. Nonsusceptible soils (about 41%), whose pH value cannot be easily changed because their $CaCO_3$ content is high enough.

According to these authors, acidic deposition has had a minimal effect on the acidification of soils up to now; however, some artificial fertilizers, like NH_4NO_3, are responsible for this effect. The main reason for the acidification of soils in Hungary is fertilization. As Kiss (1986) and Stefanovits (1986) pointed out, in most of the Hungarian soils the pH value has significantly decreased during the last 25 to 30 years. Kiss (1986) supported his finding with international data showing that the yield of some plants becomes very low at lower pH values (see Table 7–7).

One of the most serious consequences of soil acidification is the mobilization of different toxic metals, like aluminum, manganese, and so on. Pais (1984) dealt briefly in his book with the changes in environment and their effects on the trace element status of agricultural production.

Várallyay et al. (1986) concluded that the effect of atmospheric acidic deposition has direct effects on plants and microorganisms much more than through the soil acidification. In other words, negative consequences are usually physiological effects and very rarely direct pH changes.

The Agrochemical and Plant Protection Organization (MÉM-NAK), a state organization under the control of the Ministry of Agriculture and Food, has 15 research stations in most of the county centers. Specialists are dealing with soil and plant analysis including the measurement of residues of plant protection chemicals and toxic compounds in an organization called the Agrochemical Service Organization. Its main task is to give agrochemical advice to big state and cooperative farms (in Hungary big farms represent 99% of the agricultural production). In the 1970s this organization collected measurements of the macro- and microelement content of soils, the pH values, and other parameters of different soils in Hungary.

In the following section we summarize the pH values measured before 1981 (first period) and in 1984 to 1985 (second period), from which it is

Table 7–7. The yield of some plants at different soil pH values (after Kiss, 1986).

| Plant type | pH | | | | |
	4.7	5.0	5.7	6.8	7.5
	Yield (in percentages)				
Alfalfa	2	9	42	100	100
Barley	0	23	80	95	100
Corn	34	73	83	100	85
Red clover	12	21	52	98	100
Sweet melilot	0	2	49	89	100

possible to evaluate the changes during recent years. For this purpose researchers have classified the soils in Hungary into six groups: I. Chernozem soils; II. Brown forest soils; III. Meadow soils; IV. Sandy soils; V. Sodic soils (high sodium-carbonate content) and VI. Textured soils. In Table 7–8 we can see the percentage distribution of different pH values in the six different groups of soils.

The degrees of pH decrease of different soil types are as follows (in percentages): I. = 5.9, II. = 4.5, III. = 6.8, IV. = 8.5, V. = 6.3, and VI. = 3.4. As the data show, the pH decrease is higher in lower pH range in the third and fifth type of soils, while in the fourth one it is higher in higher pH range. The lowest values can be found in the textured soils, because these have relatively high buffer capacity. Because these data concern agriculturally used land, the pH decrease may be attributed mostly to the effect of fertilizers and only in smaller part to acidic deposition.

The MÉM-NAK laboratories have analyzed the pH values in the whole country. It can be seen from the data collected in Tables 7–9 and 7–10 that in the north and the southwest counties of Hungary the pH of soils is relatively low. We can generally find the highest values between the two big Hungarian rivers in the Great Hungarian Plain.

If we calculate the need for $CaCO_3$ in Hungarian soils, it can be estimated as 0.4 to 0.5 t ha^{-1} yr^{-1}. This quantity is only enough to hinder the further decrease of the pH in Hungarian soils.

In Hungary about 35% of agricultural lands (2.3 million ha) have relatively low pH and in some counties (3, 5, 6, 13, and 14) this percentage is over 40. A much greater problem is that in these counties the soils usually

Table 7–8. Changes in the distribution of pH values in Hungarian soils.

Type of soil	Period	pH Intervals				
		<4.5	4.5–5.5	5.5–6.5	6.5–7.5	>7.5
		Distribution in percentages				
I. Chernozem	1.	1.0	6.5	14.6	66.5	11.4
	2.	1.2	9.2	17.6	65.7	6.3
II. Brown forest	1.	13.7	26.0	26.4	31.4	2.5
	2.	15.4	27.5	27.7	28.8	0.6
III. Meadow	1.	5.8	19.1	27.2	39.6	8.3
	2.	7.2	24.5	26.8	37.3	4.2
IV. Sandy	1.	18.2	17.7	10.3	24.7	29.1
	2.	21.7	17.8	12.8	27.1	20.6
V. Sodic	1.	3.6	15.0	34.2	39.6	7.6
	2.	5.6	19.3	33.7	38.2	3.2
VI. Textured	1.	9.5	17.5	20.5	49.6	2.9
	2.	11.7	18.5	20.6	46.2	3.0

Table 7–9. Soil pH values in different counties of Hungary from 1979 to 1981.

County (see Figure 7–4)	Distribution of pH values in percentages					
	<4.5	4.5–5.5	5.5–6.5	6.5–7.5	7.5–8.5	>8.5
1.	3.2	11.8	15.5	31.4	37.8	0.3
2.	0.5	4.1	9.3	69.1	16.7	0.3
3.	14.1	29.5	28.2	27.7	0.1	0.4
4.	6.7	31.1	27.0	34.5	0.4	0.4
5.	15.7	43.1	24.5	16.5	0.2	—
6.	24.3	34.5	19.4	18.3	3.5	—
7.	10.5	28.3	31.5	26.6	3.1	—
8.	6.7	16.1	17.2	49.8	10.0	0.2
9.	0.1	0.5	3.8	80.8	14.0	0.8
10.	0.8	4.6	7.5	55.2	31.4	0.5
11.	1.8	16.8	25.6	53.6	2.2	—
12.	5.2	20.8	30.7	40.8	2.3	0.2
13.	16.8	28.8	24.2	26.9	2.9	0.4
14.	19.7	20.6	16.7	39.9	3.1	—
15.	0.8	3.3	10.9	80.0	4.9	0.1
16.	7.6	19.3	18.6	51.9	2.6	—
17.	0.5	0.4	0.7	55.3	12.5	0.6
18.	0.1	5.6	11.2	58.2	24.8	0.1
19.	0.4	7.2	19.4	65.9	7.1	—

Table 7–10. Soil pH values in different counties of Hungary from 1984 to 1985.

County (see Figure 7–4)	Distribution of pH values in percentages					
	<4.5	4.5–5.5	5.5–6.5	6.5–7.5	7.5–8.5	>8.5
1.	2.0	16.4	28.5	36.1	17.0	—
2.	0.6	3.7	11.0	75.6	9.1	—
3.	15.6	31.0	42.0	11.4	—	—
4.	4.3	29.2	41.2	25.3	—	—
5.	10.2	45.9	34.8	9.0	0.1	—
6.	24.1	35.2	22.4	15.0	1.8	1.5
7.	9.2	36.2	38.9	15.6	0.1	—
8.	3.2	18.0	29.7	46.7	0.9	1.5
9.	2.0	16.4	28.5	36.1	17.0	—
10.	0.8	3.4	7.9	64.2	23.7	—
11.	0.6	10.5	42.9	45.8	0.2	—
12.	3.3	24.1	44.4	27.8	0.4	—
13.	11.7	37.4	36.0	14.9	—	—
14.	16.5	28.7	22.1	32.6	0.1	—
15.	—	1.3	8.5	71.3	18.9	—
16.	3.6	16.0	32.7	47.3	0.4	—
17.	—	0.2	1.0	80.3	18.5	—
18.	3.1	4.0	15.9	68.1	8.9	—
19.	0.4	10.7	28.9	59.3	0.7	—

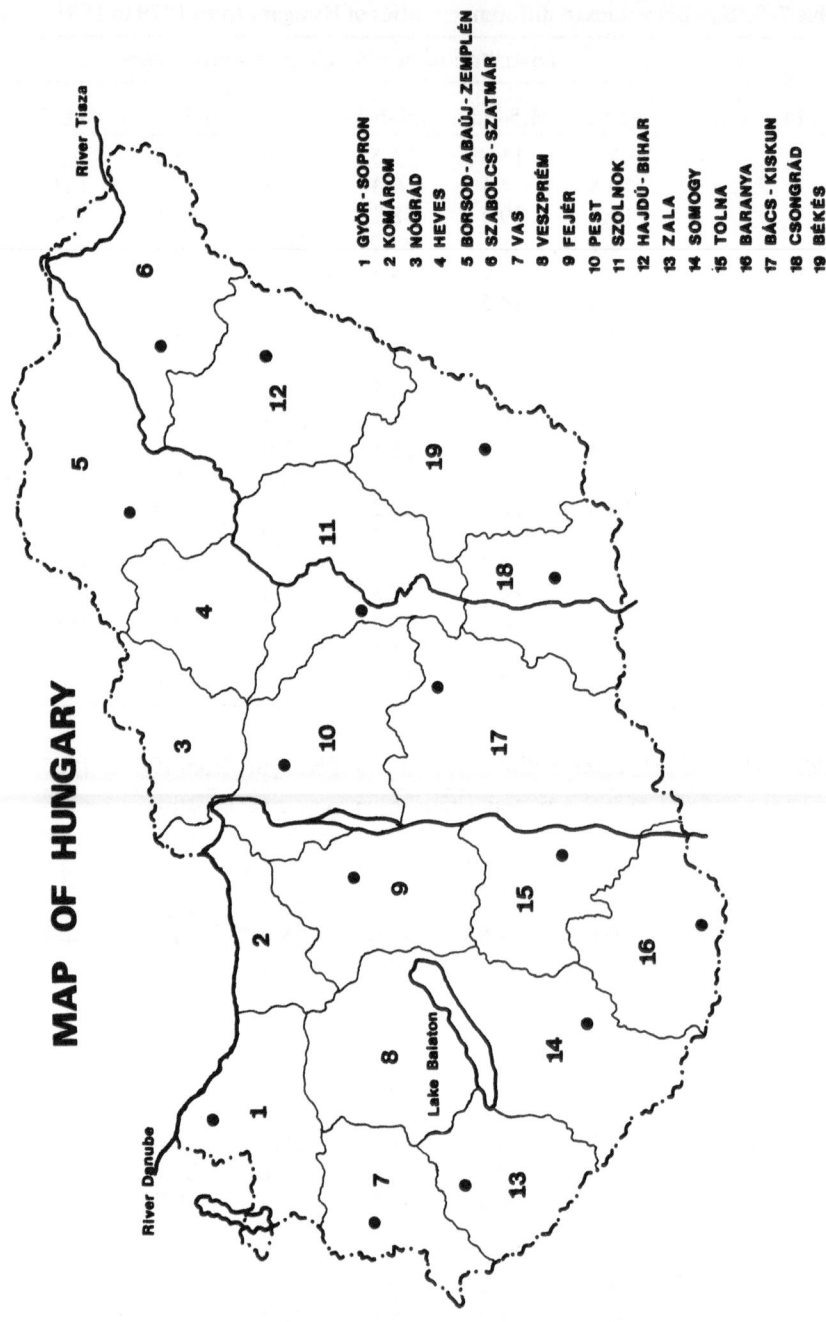

Figure 7-4. Location of counties in Hungary.

have a low $CaCO_3$ level, therefore their buffer capacity is also very low. Results of measurements of MÉM-NAK show that the $CaCO_3$ content of the soil has decreased since the beginning of the 1980s to the present.

It is well known that the uptake of different nutritive elements by plants is in strong correlation with the pH of the soil. When the soil pH values begin to decrease, the mobilization of aluminum and manganese is stronger. Generally, aluminum itself is not toxic for plants, but with the increase of Al concentration the uptake of calcium and other essential nutritive elements becomes lower and the plants show deficiency symptoms.

In like manner, with the decrease of pH, the uptake of manganese, cadmium, and some other toxic heavy metals increases, and these unusually high concentrations are not tolerated by most cultivated plants.

Bokori and Tölgyesi (1982) dealt in their research with the mineral elements in *Brassica rapa*. According to their measurements manganese content of *Brassica rapa* was 54.8 and 164.3 mg kg^{-1} of soil, pH values of 7.0 and 5.0, respectively. As it can be seen, at lower pH the manganese content is about 3 times higher than in neutral soil.

C. Other Damaging Effects of Acidic Air Pollutants in Hungary

Somlyódi and Zotter (1986) studied the expected effect of acidic deposition on the quality of water resources in Hungary. They concluded that the great majority of Hungarian surface waters, as a consequence of their high buffer capacity, tolerate the atmospheric acidic inputs well. However, in the northern mountain area (especially in Mátra Mountain) one can expect acidification of drinking water reservoirs due to acidic deposition.

The corrosive effect of acidic materials on machinery, instruments, buildings, and sculpture is also a great problem. Machines in the open air that are exposed to the acidic atmosphere show a relatively high amount of corrosion damage. In Hungary it is estimated to be $100 million a year (1987 exchange rate). The damage to buildings and sculpture has the same magnitude: We can calculate extra expenses of $100 to $120 million a year. The total cost due to corrosion effects including the expenses of prevention amounts to nearly $1 billion a year in Hungary (MTA/OKTH, 1985).

The building that houses the Hungarian Parliament, one of the biggest, most beautiful, and famous buildings in Europe, is continuously being damaged by the acidic components of air in Budapest. This building was decorated with structural elements made from relatively soft limestone, whose parts now have to be replaced gradually by more resistant stone.

As for human health problems, we can also cite very significant problems. For example, the number of SIDS deaths (Sudden Infant Death

Syndrome) is increasing steadily. This disease is probably a consequence of the high atmospheric level of acidic pollutants (e.g., SO_2, PAN, etc.). The infants who are more sensitive to these compounds die very rapidly. Bone development also exhibits different problems as a consequence of acidic materials. The increase of the NO_3^- content of some foods, which has dangerous effects on the health of the population, can be related to acidic pollution of the atmosphere.

VII. Conclusion

It is obvious that, similar to other European countries, damaging effects of atmospheric acidic deposition have appeared in Hungary as well. For example, forest soils in this country are continuously acidifying, and corrosion costs amount to several hundred million dollars annually. Although the main reason for acidification of cultivated land is the use of fertilizers, atmospheric acidic deposition is potentially dangerous for agricultural soils.

Since the effects of acidic deposition are cumulative, appearance of damages may be expected in more and more places. To prevent or reduce further damages we should concentrate our forces internationally. A first step may be to reduce SO_2 emissions in Europe by 30%. But it is self-evident that this reduction alone will not solve the problem of acidification. Only radical changes in the structure of energy production can lead to permanent improvement.

References

Asman, W.A.H., P.J. Jonker, J. Slanina, and J.H. Baard. 1982. Neutralization of acid in precipitation and some results of sequential sampling. *In* H.W. Georgii and J. Pankrath, eds. *Deposition of Atmospheric Pollutants*, 115–123. Dordrecht: Reidel.

Bede, G., and I. Gács. 1986. Present and future SO_2 and NO_x emissions from sources in Hungary [in Hungarian]. *Időjárás* 90, 77–83.

Bokori, J., and Gy. Tölgyesi. 1982. Newer green forage plants, on the forage rapes, with special reference to their mineral substance content [in Hungarian]. *Magyar Állatorvosok Lapja* 37, 110–114.

Bónis, K. 1981. The atmospheric budget of nitrogen compounds over Hungary [in Hungarian]. *Időjárás* 85, 149–156.

EMEP. 1986. *Sulfur budgets for Europe for 1979, 1980, 1983 and 1984.* EMEP/ MSC-W report, January 1986.

Fekete, K., and D. Szepesi. 1987. Simulation of atmospheric acid deposition on a regional scale. *J Environ Management* 24, 17–28.

Ferm, M. 1979. Method for determination of atmospheric ammonia. *Atmospheric Environ* 13, 1385–1393.

Forrest, J., R.L. Tanner, D. Spandau, T. D'Ottavio, and L. Newman. 1980. Determination of total inorganic nitrate utilizing collection of nitric acid on NaCl-impregnated filters. *Atmospheric Environ* 14, 137–144.

Grennfelt, P. 1981. *Sources and sinks of acidifying compounds.* Report of the European Conference of Acid Rain, 31–47, Göteborg, May 9–12, 1981.

Haszpra, L., and T. Turányi. 1986. Production of nitric acid in the atmosphere under different conditions. *Időjárás* 90, 332–338.

Hidy, G.M. 1973. Removal processes of gaseous and particulate pollutants. In S.I. Rasool (ed.), *Chemistry of the Lower Atmosphere,* 121–176. New York: Plenum.

Horváth, L. 1981. Chemical composition of precipitation over Hungary [in Hungarian]. *Időjárás* 85, 201–212.

Horváth, L. 1982. On the vertical flux of gaseous ammonia above water and soil surfaces. *In* H.W. Georgii and J. Pankrath, eds. *Deposition of Atmospheric Pollutants,* 17–22. Dordrecht: Reidel.

Horváth, L. 1983a. Concentration and near vertical flux of ammonia in the air in Hungary. *Időjárás* 87, 65–70.

Horváth, L. 1983b. Trend of the nitrate and ammonium content of precipitation water in Hungary for the last 80 years. *Tellus* 35B, 304–308.

Horváth, L. 1986. *Acid rain* [in Hungarian]. Budapest: Gondolat Könyvkiadó.

Horváth, L., and K. Bónis. 1980. An attempt to estimate the rate constant of sulfur dioxide-sulfate conversion in the urban plume of Budapest. *Időjárás* 84, 190–195.

Horváth, L., and E. Mészáros. 1978. Determination of the kinetic parameters of sulfur dioxide-sulfate conversion on the basis of atmospheric measurements. *Időjárás* 82, 58–62.

Horváth, L., and E. Mészáros. 1984. The composition and acidity of precipitation in Hungary. *Atmospheric Environ* 18, 1843–1848.

Horváth, L., and E. Mészáros. 1986. Acid deposition in Hungary [in Hungarian]. *Időjárás* 90, 143–149.

Horváth, L., and D. Möller. 1987. On the "natural" acid deposition and the possible consequences of decreased SO_2 and NO_2 emission in Europe. Accepted for publication in *Időjárás* 91.

Huebert, B.J. 1983. *The dry deposition of HNO_3 vapour as sink for NO_y.* Presented at the CACGP Symposium of Tropospheric Chemistry, Oxford, U.K.

Jakucs, P. 1986. The impact of atmospheric acidification on living organisms [in Hungarian]. *Időjárás* 90, 150–158.

Kiss, A.S. 1986. The importance of the inhibition of acidification of soils [in Hungarian]. *Melioráció Iöntözés és tápanyaggazdálkodás 1,* 63–70.

Klockow, D., H. Denzinger, and G. Rönicke. 1974. Anwendung der substöchiometrischen Isotopenverdünnungsanalyse auf die Bestimmung von atmosphärischen Sulfate und Chloride in "Background" Luft. *Chemie-Ingenieur-Technik* 46, 831.

Koroleff, F. 1970. Direct determination of ammonia in natural waters as indophenol blue. In *Information on Techniques and Methods for Seawater Analysis. Int. Courc. Exploration of the Sea.* Report No. 3. Copenhagen: Charlottenlund Castle.

Levaggi, D.A., W. Siu, and M. Feldstein. 1973. A new method for measuring average 24-hour NO_2 concentration in the atmosphere. *J Air Pollut Control Association* 23, 30–33.

Mészáros, Á. 1986. Sulfate formation on elemental carbon particles [in Hungarian]. *Időjárás* 90, 122–130.

Mészáros, E. 1971. The size distribution of water soluble particles in the atmosphere. *Időjárás* 75, 308–314.

Mészáros, E. 1973. Evidence of the role of indirect photochemical processes in the formation of atmospheric sulphate particulate. *J Aerosol Science* 4, 449–434.

Mészáros, E. 1974a. On the formation of atmospheric sulphate particulate in the winter months. *J Aerosol Science* 5, 483–485.

Mészáros, E. 1974b. On the spring maximum of the concentration of trace constituents in atmospheric precipitation. *Tellus* 24, 402–407.

Mészáros, E., and L. Horváth. 1984. Concentration and dry deposition of atmospheric sulfur and nitrogen compounds. *Atmospheric Environ* 18, 1725–1730.

Mészáros, E., I. Mersich, and T. Szentimrey. 1987. The air pollution episode of January 1985 as revealed by background data measured in Hungary. Accepted for publication in *Atmospheric Environ* 21.

Mészáros, E., D.J. Moore, and J.P. Lodge. 1977. Sulfur dioxide-sulfate relationships in Budapest. *Atmospheric Environ* 11, 345–349.

MTA/OKTH. 1985. On the increasing acidification of the environment and its effect [in Hungarian]. *MTA/OKTH Report,* August 1985.

Pais, I. 1984. *The importance of trace elements in the agricultural production. Their research situation in the world* [in Hungarian]. Edition of University of Horticulture, 1–224.

Solymos, R. 1986. The health of Hungarian forests and the acid deposition [in Hungarian]. *Időjárás* 90, 181–191.

Somlyódi, L., and K. Zotter. 1986. The expected effects of acid precipitation on the quality of water resources in Hungary [in Hungarian]. *Időjárás* 90, 159–168.

Stefanovits, P. 1986. Some new data on soil acidification [in Hungarian]. *Magyar Tudomány* 93, 339–341.

Várallyay, Gy., L. Rédly, and A. Murányi. 1986. The influence of acid deposition on soils in Hungary [in Hungarian]. *Időjárás* 90, 169–180.

Várhelyi, G. 1980. Dry deposition of atmospheric sulphur and nitrogen compounds. *Időjárás* 84, 15–20.

Várhelyi, G. 1982. On the atmospheric sulphur budget over Hungary. *Időjárás* 86, 333–337.

Várkonyi, T., and T. Cziczó. 1980. *Test of the air quality* [in Hungarian]. Budapest: Műszaki Könyvkiadó.

West, P.W., and G.C. Gaeke. 1956. Fixation of sulfur dioxide as disulfitomercurate (II) and subsequent colorimetric estimation. *Analytical Chem* 28, 1816–1819.

WMO. 1978. *International Operations Handbook for Measurement of Background Atmospheric Pollution.* No. 491, Geneva.

Acidic Precipitation Research in Poland

Z. Strzyszcz*

Abstract

Investigations of the chemical composition of precipitation were carried out in Poland rather early at 22 sites (1964 to 1966), although the studies were not connected with the occurrence of acidic rain. In some regions with heavy air pollution (e.g., Katowice district), the investigations of the chemical composition of rainfall were carried out from 1965 to 1968, however, attention was mainly on the chemical composition of the solid phase of rainfall, connected with the air pollution in that region. Investigations connected with the dependence of air pollution on the acidity of precipitation and the contents of sulphates and ether components in it was initiated in 1976. Later, from 1980 to 1986, tests were carried out to define the impact of the dry and wet depositions on such elements of the ecosystem as soils and plant life, mainly forests.

I. Introduction

The yearly output of coal in Poland is approximately 200 million ts of pit coal and 60 million ts of brown coal. The amount of pit coal will probably not increase until the year 2000, but a probable increase of some 100 to 120 million ts of brown coal output is expected until the year 2000.

About 165 million ts of pit coal is used for domestic production and the needs of power production as well as the entire amount of brown coal. In 1985, power production used up about 52 million ts of pit coal and 53 million ts of brown coal, emitting into the atmosphere in 1982 to 1983 an amount of SO_2 in the range of 3.38 to 4.08 million ts and an amount of nitric oxides from 1.57 to 1.97 million ts. It has been assumed that the global SO_2 emissions in 1985 amounted to between 4.74 and 6.12 million ts,

*Polish Academy of Sciences, Institute of Environmental Engineering, 41-800 Zabrze, Sklodowskiej Curie 34, Poland.

from which about 35% is contributed by power production, about 18% by metallurgy, about 30% by transport, and about 18% by the municipal and living sector (Ochrona and Gospodarka, 1986).

The energy needs of neighboring countries now and in the future will be met by the burning of pit coal and brown coal, emitting simultaneously considerable amounts of gaseous pollution into Poland. The derivation of the SO_2 emissions over Poland is as follows: Our own contribution is equal to 42%, foreign contribution is equal to 52%, and the unidentified part is equal to 6%. Our largest pollution contributors are E. Germany with 21.3 thousand ts, Czechoslovakia with 13.6 thousand ts, and W. Germany with 8.0 thousand ts. Poland emits SO_2 to other countries, too. Reducing it to sulphur we emit 130 thousand ts of pollution every month, from which 56.5 thousand ts remain in our country, 38.6 thousand ts go to the Soviet Union, 9.5 thousand ts to Czechoslovakia, and 3.1 thousand ts to Sweden (Chudzyńska, 1985). The balance is at once credit and debit, but on the whole we "import as much as export."

It is obvious that a considerable amount of SO_2 and NO_x has a detrimental effect on plants and soils as well as on other elements of the environment, although opinion differs whether the gaseous air pollution or the acidic precipitation are the real cause. The opinion prevails that the impact of the gaseous pollution is much greater than that of the acidic precipitation, although until now no complex investigations have been carried out within the impact of the acidic precipitation for the whole country.

The following items were included in the sphere of research connected with the impact of acidic precipitation on the natural environment in Poland:

the chemical composition of precipitation and the degree of its acidification,
the impact of acidic precipitation on cultivated and forest soils,
the impact of acidic precipitation on vegetation with special regard to the impact on forests.

II. The Chemical Composition of Rainfall and the Degree of Its Acidification

Rainfall has been investigated by Chojnacki since 1964 (Chojnacki, 1967), when 22 sites in Poland were used for research. He stated, based on the sulphate content in the rainfalls, that within Poland 10 to 75 kgS ha^{-1} yr^{-1} falls down in wet form. Considering that these investigations refer to the years 1964 to 1966, the results should be regarded as very accurate. At that time the amount of sulphur was not investigated when the source was the so-called dry deposition.

The investigation of the chemical reaction, conductivity, and also chemical composition of rainfalls was initiated in Silesia in 1964. (Czyż, Gadżik, Geszta, and Olszowski, 1968). These investigations proved that the rainwater acidity ranged from pH 3.7 to pH 7.7, whereby no dependence between the pH and SO$_2$ contents in the air was found as well as between the contents of sulphates in rainwater and pH. No dependence of dust precipitation on the contents of sulphates in rainwater was recorded either.

The fact that no relationship between the investigated parameters was discovered probably resulted from the specific character of the emitted pollution (power production industry, iron and noniron metallurgy, cement industry, chemical coke industry) as well as from certain method difficulties which prevailed in those years. However, it was noticed that the acidic reaction occurred mainly on the borders of the upper Silesian industrial region where the emission of air pollution reached the limits of 150 t km^{-2}. However, the alkaline reaction was found in regions where pollution exceeded 250 t km^{-2}.

The results of investigations connected with a higher pH in regions with greater pollution were confirmed in the following years when investigations of rainfall began in the vicinity of the Institute at Zabrze (see Tables 8–1 and 8–2).

Table 8–1. Mean results of measurements of concentration of air contamination between July 1964 and January 1965 (Czyż, 1968).

Number of sample area	Distance from source of air pollution km	Water from the sedimentary vessels			
		mg cm^{-3} d^{-1}	pH	Dry remains mg m^{-2} d^{-1}	SO$_4$ mg^{-1} m^{-2} d^{-1}
I	2.0 SW	0.440	5.1	726.1	10.7
II	3.4 NW	0.566	4.8	755.0	19.21
III	5.2 W	0.719	6.0	150.0	55.31
IV	4.0 S	0.533	5.6	692.1	16.74
V	2.0 SE	0.585	5.3	726.2	10.98
VI	6.0 SE	0.584	5.8	221.3	7.63
VIIa	6.6 S	0.655	4.5	470.5	9.90
VIIb		0.999	3.9	372.5	11.80
VIII	5.8 E	0.448	4.7	410.3	10.51
IV	3.5 N	0.584	4.9	604.3	13.66
X	7.2 E	0.386	5.2	471.8	12.50

[a] measurement equipment in canopy

[b] above canopy

Table 8–2. The average monthly contents of some substances in rainwater: 1978 to 1979 at Zabrze.

Month/year	Precipitation (mm)	pH	μS^{cm-1}	SO_4^{2-}	NO_3^- mg dcm^{-3}	NH_4^+
May/1978	28.45	7.03	133.0	38.45	0.744	—
June	39.87	5.59	87.5	17.18	1.200	—
July	50.80	4.37	69.5	21.14	2.120	4.980
August	74.50	6.96	—	14.47	0.938	1.855
September	24.85	5.39	78.2	12.86	0.958	3.466
October	57.80	5.56	59.1	11.70	0.797	2.716
November	22.21	5.50	48.3	11.61	0.430	1.968
December	37.36	5.70	96.5	28.55	0.474	2.177
January/1979	53.63	6.50	60.6	13.20	0.656	1.194
February	21.48	6.17	11.9	20.80	0.842	3.720
March	33.90	6.30	99.8	32.36	0.912	3.473
April	51.85	5.63	52.0	15.43	0.986	1.938
May	68.75	5.50	45.2	11.57	0.727	1.515
June	55.65	6.02	73.6	17.10	0.726	2.501
July	44.95	4.73	59.6	11.74	0.504	1.534
August	45.20	5.81	59.4	16.17	0.569	1.381

The investigations of rainfalls carried out in South Poland by Kasina were of a more complex character (Kasina, 1980; Kasina, Kwiek, and Lewinska, 1984). The investigations were conducted from April 1977 to March 1978 with the aim, among others, to define the pH of rainfall in cities and industrial regions as well as in the mountain area of Zakopane. Besides pH the contents of sulphates and the so-called wet deposition was examined. These investigations were increased in the following years by including more testing points and also enlarging the number of specific parameters (see Table 8–3).

The Institute of Meteorology and Water Management in Warsaw, thanks to two international programs (of the World Meteorology Organization and of the European Economic Committee), investigates the chemical composition of precipitation at two sites in Poland—in Suwał ki (some 250 km northeast of Warsaw) and on the mountain Snieżka (about 400 km southwest of Warsaw). The investigations at Suwałki have been carried out since 1975 and on the mountain Snieżka since 1981. In 1985 the investigations have been increased with testing points in Warsaw and Jarczewo (about 75 km southeast of Warsaw). Also, the Institute of Environmental Engineering of the Polish Academy of Sciences at Zabrze started investigations on precipitation in the Beskid region near the Polish-Czechoslovakian border in 1986 and also in Zabrze within the institute, where the chemical composition of precipitation was investigated from 1978 to 1980 (Table 8–2; Wpływ, 1979).

According to long-term data of the Institute of Meteorology and Water Management in Warsaw (Chudzyński after Hryniewicz, 1985) the average yearly pH in Suwałki was equal to 4.4; the average monthly pH on the Snieżka mountain was lower than 4.0. The reason for that is the vicinity of the power station at Turoszów (70 km) and industrial plants (mainly power stations) in E. Germany and Czechoslovakia. High dynamics of pH in rainfall ranging from pH 2.8 to pH 8.0 with the average of 4.8 were found in Warsaw and its vicinity. It is a rather important fact that at the beginning of investigations a pH lower than 4.0 occurred once or twice during the year. But nowadays such a pH occurs once a month in that region.

Table 8–3. Annual mean weighted pH values and sulphate concentrations in precipitation. Deposition of sulphate from precipitation.

Station numbers refer to Fig. 8–1	pH	Sulphate concentration $\mu g\ S\ cm^{-3}$	Wet deposition $gS\ m^{-2}\ yr^{-1}$
City-industrial area Period: April 1975– March 1979			
1	5.30	5.04	3.55
2		Incomplete data	
3	5.25	5.76	2.97
4	5.10	4.30	2.77
Rural area Period: April 1975– March 1977			
6	4.64	2.98	1.78
7	4.65	2.40	1.48
8	4.40	2.35	1.22
9	4.85	2.91	1.73
10	4.50	2.41	1.20
11	4.95	3.01	1.40
12	4.30	2.67	1.61
13	4.50	2.59	1.79
14		Incomplete data	
15	4.30	2.34	1.33
Period: April 1977– March 1979			
16	4.92	1.56	0.80
Mountain area Period: April 1975– March 1979			
5	4.45	0.55	0.88

According to estimated calculations by the Institute of Meteorology and Water Management, about one-third of Poland's area is situated within the area of the most acidic rainfall in Europe (pH 4.1), and the remaining area has an average pH of rainwater of about 4.3.

III. The Impact of Acidic Precipitation on Soil Acidification

The investigations carried out in some regions of Poland connected with acidity and the chemical composition of precipitation do not give a direct answer to the degree of the acidic rain hazard to soils and plants. That results mainly from method difficulties and also from lack of suitable apparatus which would make possible a separation of the dry and wet deposition. The dry deposition, especially in some parts of our country, could play a decisive role, the same one which wet deposition plays in other countries. That appertains mainly to heavy metals (Nürnberg, Nguyen, and Valenta, 1983).

Having no sufficient data dealing with acidity of precipitation and its impact on the environment, tests were undertaken to assess the impact of acidic rain by assessing the acidification of cultivated and forest soils. That course of action also has been applied in other countries (Stöhr, 1984). To the choice of soil as the indirect factor for assessing the impact of acidic rains on the environment, previous investigations also contributed by proving that there was a rather close relationship between the general sulphur and sulphate content in soils and the air pollution by SO_2. The correlation factor between the first relationship was 0.93 and the second was 0.87. (Warteresiewicz, 1979).

Investigations carried out by 19 Regional Chemical and Agricultural Stations (Figure 8–1) connected with acidification of cultivated soils proved after the first rotation of tests (1954 to 1965) that 58% of cultivated soils should be regarded as very acidic and acidic soils (pH 1 n KCl < 5.5), 25% as weakly acidic ones (pH 5.6 to 6.5), and 17% as neutral and basic soils (pH > 6.5). The second rotation of tests, carried out from 1966 to 1977 did not show essential changes in that range although the average yearly dose of lime increased from 12.4 from 1959 to 1960 to 125 kg/ha^{-1} CaO from 1976 to 1977 (see Figure 8–3). It shows that besides known natural causes of soil acidification, unknown factors, to which the air pollution by SO_2 and NO_x belongs, have impact on that phenomenon. Unfortunately, systematic soil tests were not associated with measurements of air pollution caused by factors mentioned earlier. These measurements started in Poland but did not encompass the whole country before 1975. Due to investigations by Chojnacki (Chojnacki, 1967) referring to the sulphate content in precipitation and data by OECD (OECD, 1977) it can be assumed that in that time the level of air pollution by SO_2 in many parts of our country already was higher than 25 μg m^{-3} (Figure

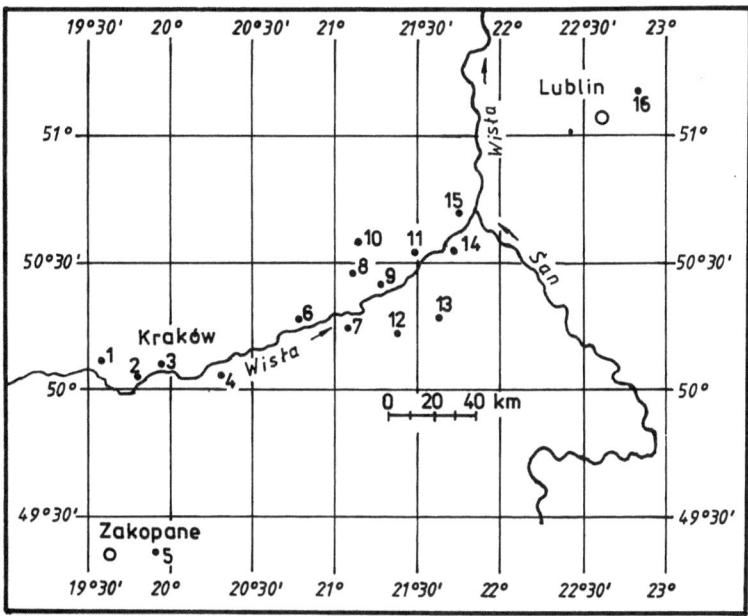

Figure 8–1. Location of precipitation chemistry stations in southeast Poland (Kasina, 1980).

8–2), which taking into account the factor of time could be one of the reasons for heavy acidification (Strzyszcz, 1982, 1983). Comparing the obtained data connected with air pollution by SO_2 with the proportional share of acidic soils (Figures 8–2 and 8–3), a convergence can be noticed in many cases. The observed differences are caused not only by the level of agriculture but also by the buffer capacity of soils and by the local specificity of air pollution. The Opele district is typical with a very high level of agriculture and where the amount of lime applied exceeds 330 kg ha[-1] CaO, nevertheless the proportion of acidic soils exceeds 50%. The small soil acidity in the regions of Chełm and Zamość is caused on the one hand by low air pollution by SO_2 and on the other by the high buffer capacity of soils (humus, loess formation). The share of acidic soils in the Suwałki district is low too, while there the smallest amount of air pollution occurs. The soils in that region are much more vulnerable, which can be seen already in points of acidic reaction by soils. As has been shown by the Chemical and Agricultural Station at Białystok, within communes there are villages where the proportional share of soils with a reaction below pH 5.5 is as follows: in the commune Puniek, 4% to 81%; Winiajny, 0% to 99%; Jeleniowe, 1% to 67%; Suwałki, 0% to 74%; Szypliczki, 2% to 74% (Strzyszcz, 1982).

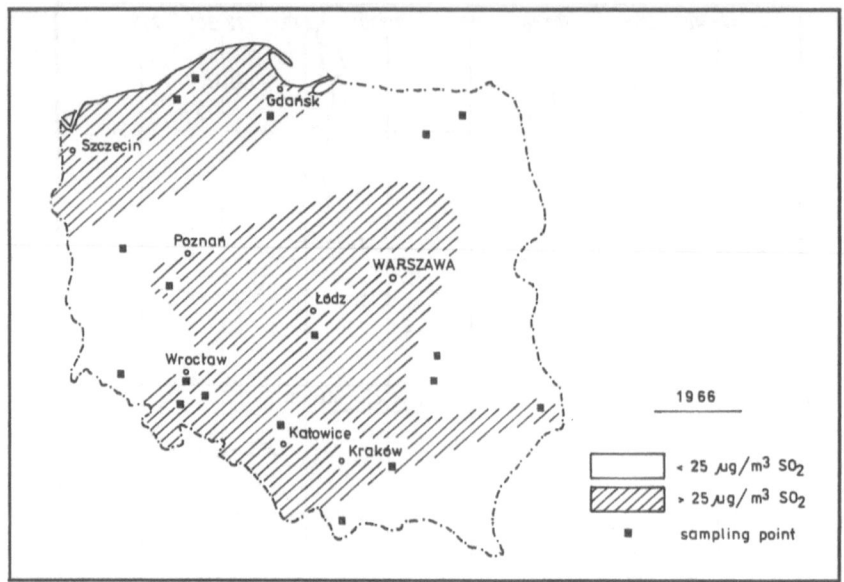

Figure 8–2. Contents of SO$_2$ in the air in 1966 calculated from OECD data and contents of SO$_4$ in atmospheric precipitation.

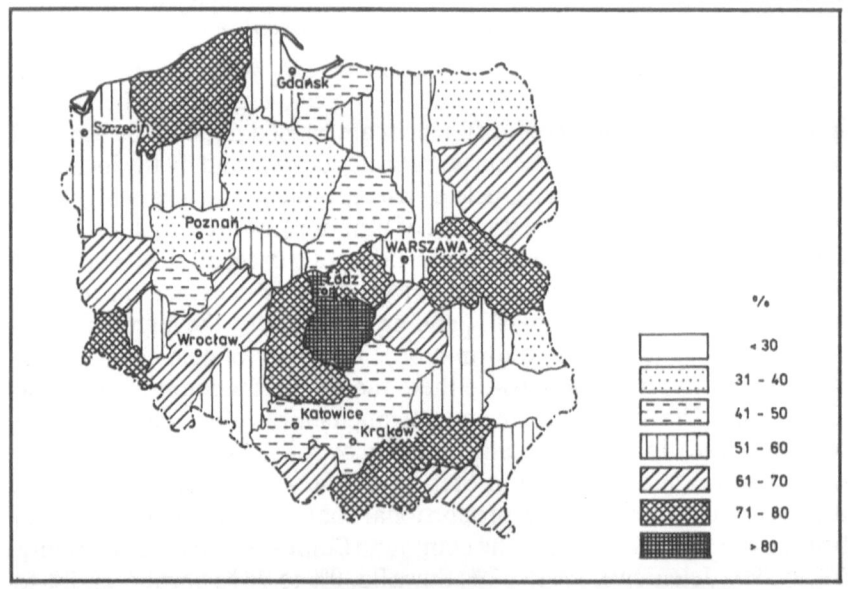

Figure 8–3. The proportional share of soils below 5.5 pH in 1n KCl

From 2912 soil samples taken in the five previously mentioned communes, 1691 samples had a reaction above pH 6.2, 913 samples in the range of 6.1 to 5.1, 306 samples in the range of 5.0 to 4.2, and 2 samples had a reaction in the range of pH 3.0 to 4.1. The 306 soil samples with a reaction in the range of pH 4.2 to 5.0 prove the fact that in this region the process of soil acidification caused by acidic precipitation has begun. This fact can be confirmed by a considerable amount of sulphates in water wells that exceeds 100 mg L^{-1}.

A very high proportion of acidic soils can be observed in the western and southern parts of our country: the districts of Zielona Góra, Jelemia Góra, Wałbrzych, Bielske-Biała, Newy Sacz, and Krosne. Besides soil and climatic factors (mountain areas) special weight should be attributed to the import of pollution from E. Germany and Czechoslovakia. As calculations have shown, the stream of SO_2 which reaches the breeding ground has a value of 20 t km^{-2}. In the direction of Gorzów-Legnica-Karpacz values of even 15 t km^{-2} occur. Similar values on the southern border range from 5 to 10 t km^{-2} (Juda et al., 1978).

The districts of Katowice and Kraków are an exception because the share of acidic soils is smaller in spite of the highest concentration of SO_2 in the air (see Table 8–4). Besides soil factors a special role should be attributed to the considerable share of dust particles in air pollution (Katowice, 637 thousand t; Kraków, 140 thousand t). This causes neutralization of SO_2 by alkaline dusts coming from metallurgical processes, cement plants, and power stations. The share of those dusts in the Katowice region is as follows: dusts from power stations, 428.4 thousand t; from cement plants, 69.5 thousand t; from iron works 71.5 thousand t. As was already mentioned, the rainwater rarely shows an acidic reaction in the Katowice district, particularly in its central part. That reaction can be noticed on the borders of the district. The later investigations on soils have proven that the problem of their acidification is more serious than was thought at the beginning, because acidification enters not only into the arable layer (0–20–30 cm), but even deeper into the lower layers to a depth of 150 cm (Kern, 1985. Table 8–5). The data of the table show that in the layer 0 to 50 the reaction below pH 5.5 (in 1nKCl) appears nearly in 61% of cultivated soils in Poland, in the layer 50 to 100 cm nearly in 39%, and in the layer 100 to 150 cm nearly in 27% of the soils. Thus the acidification proceeds down systematically though the lime doses used for deacidification of soils that have been increased in recent years.

The use of lime fertilizers in kg ha^{-1} on cultivated soils in Poland is as follows:

	1975–1976	1976–1977	1977–1978	1978–1979	1979–1980
CaO	118.6	124.5	144.2	133.1	159.7

	1980–1981	1981–1982	1982–1983	1983–1984
	133.0	126.9	143.4	154.3

Table 8–4. The stream of SO_2 that reaches the subsoil, dosage of agricultural limestone, and the proportional share of acidic soils in some districts (Strzyszcz, 1985).

District	SO_2 t km^2 a	CaO kg ha^{-1} 1980	Soil with pH < 5.5 %
Katowickie	169.8	166	49
Krakowskie	75.4	298	48
Konińskie	74.0	83	60
Łódzkie	55.8	152	81
Legnickie	52.9	206	57
Wałbrzyskie	52.1	151	65
Warszawskie	35.2	154	55
Płockie	30.0	168	42
Tarnobrzeskie	22.0	123	60
Szczecińskie	19.0	225	54
Radomskie	17.9	77	70
Opolskie	15.8	330	53
Bielskie	13.9	370	68
Gdańskie	12.0	232	59
Wrocławskie	11.3	210	65
Ostrołęckie	11.3	90	60
Toruńskie	10.0	183	34
Bydgoskie	9.3	214	38
Rzeszowskie	6.3	110	78
Poznańskie	5.3	205	40
Chełmskie	4.4	81	38
Nowosądeckie	3.7	77	76
Zamojskie	2.7	74	28
Zielonogórskie	2.3	405	66
Krośnieńskie	2.1	67	67
Koszalińskie	1.9	366	76
Leszczyńskie	1.9	255	46
Słupskie	1.7	279	80
Suwalskie	1.3	127	36
Pilskie	1.1	230	58

With regard to the whole country, areas of very acidic and acidic soils with occasional occurrence of $CaCO_3$ in the whole soil profile (150 cm) take up about 54% of the cultivated areas, soils with a light acidic and acidic reaction with frequent occurrence of $CaCO_3$ in the matrix take up an area of 16%, and soils with a neutral and basic reaction and with very frequent occurrence of $CaCO_3$ in soil profiles take up about 30% of cultivated areas (Figure 8–4). Even assuming that geological factors (the ex-

Table 8-5. Proportional share of soils with a specified reaction in different depth in relation to areas of arable land.

Depth (cm)	pH 1n KCl				
	<4.5	4.6–5.5	5.6–6.5	6.6–7.2	>7.2
0–50	21.4	39.5	26.6	8.6	12.6
50–100	12.6	26.3	24.9	24.9	11.3
100–150	10.4	16.5	13.0	19.1	41.0
Sum	44.4	82.3	64.5	52.6	64.9

tent of respective glaciations—the Baltic and the middle Polish ones) and also the granular-metric composition and farming culture have a special influence on acidification of cultivable soils, it is not possible to explain the high degree of acidification of cultivated soils to such a large depth. In districts with the highest percentage of farming culture (Poznań, Leszno, Bydgoszcz, and Opole) lime doses from 205 to 330 kg ha^{-1} CaO yearly have been used in the last 10 years and nevertheless the share of acidic soils is as follows: Poznań to 40%, Leszno to 46%, Bydgoszcz to 38%, Opole to 53% of cultivable areas (Table 8–4).

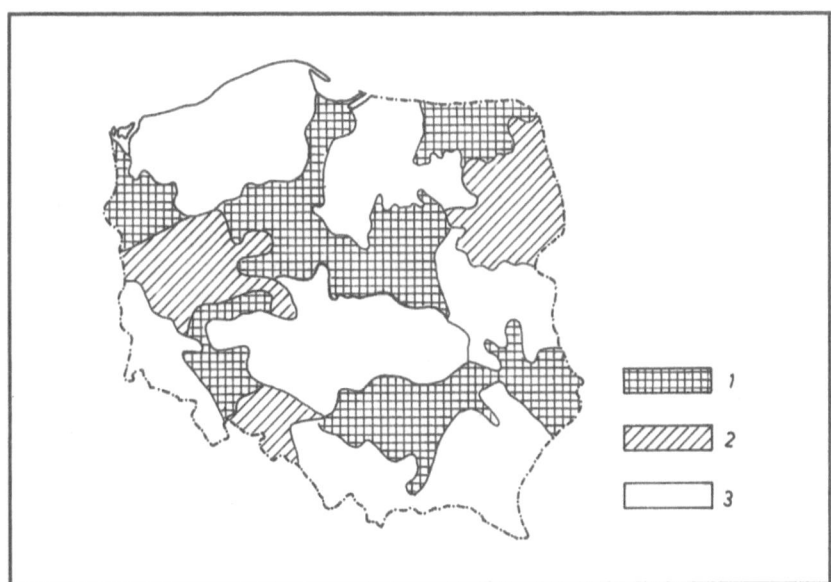

Figure 8-4. Regions with soils of a specified reaction (Kern, 1985). Regions with soils of a neutral or alkaline reaction with frequent occurrence of CaCO$_3$ (1); of a light acidic and acidic reaction with frequent occurrence of CaCO$_3$ (2); of very acidic and acidic reaction with sporadic occurrence of CaCO$_3$ (3).

Considerably fewer investigations connected with the impact of precipitation on forest soils have been carried out. The reaction of forest and cultivable soils was also investigated during research on the chemical composition and reaction of rainfall in the Katowice district (Czyż, Godzik, Greszta, and Olszowski, 1968). The investigations have shown that the reaction of forest soils measured in 1 nKCl in the layer 0 to 15 cm ranged from 2.7 to 4.5 pH whereas in cultivated soils at the same time the pH ranged from 4.5 to 7.9. This comparison shows that the reaction of forest soils in the years 1960 to 1965 was much lower than that of cultivated soils. The forest soils deeper than 100 cm had a reaction ranging from 2.7 to 4.5 pH in 1 nKCl when the average pH was 3.6. Similar values of pH were found in forest soils of the Niepołomice Woods in south Poland, where in a depth of below 100 cm a pH lower than 4.5 was found (Adamczyk et al., 1983).

In the northern part of Poland (Koszalin, Toruń) investigations connected with the buffer capacity of various types of forest humus were carried out. At the same time the reaction in various layers of the humus level was measured and it ranged from 3.5 to 4.4 pH, but the measurements were carried out in H_2O (Pokojska, 1985). In the years 1986 to 1990 complex investigations on standard experimental areas using Glatzel's method (Glatzel, 1983), adapted to Polish conditions, are expected within the research program connected with the measurement of the cycle of mineral ingredients in woods as a component part of the monitoring system of forest ecosystems. The standard areas are differentiated from 200 to 500 ha and even above 500 ha (Prusnikiewicz, Kowalkowski, and Królikowski, 1983). Altogether 19 testing areas were scheduled with more than 500 ha and more than 100 testing areas of 200 to 500 ha on a forest land of nearly 9 million ha.

Investigations dealing with the direct impact of acid precipitation on cultivated plants as well as woods have not been carried out in Poland. However, investigations referring to the influence of SO_2 on cultivated plants were conducted with the aim of assessing losses in their yields caused by various SO_2 concentrations in the air (Warteresiewicz, 1979). The global impact of air pollution (dry and wet) on woods in Poland by fixing the total sulphur content in needles of pines has been investigated by Molski and others (Molski, Dmuchowski, and Chmielewski, 1986, Figure 8–5).

As it follows from data shown earlier, in Poland there is no one central research program devoted to studying the impact of acidic precipitation on particular components of the ecosystems, though there is an urgent need of one (Białobok, 1986). The reason for this is less the existing conditions of organization but more the shortage of suitable apparatus (and hard currency for its purchase) which would allow such investigations to be carried out and the achievement of results comparable with data of other countries.

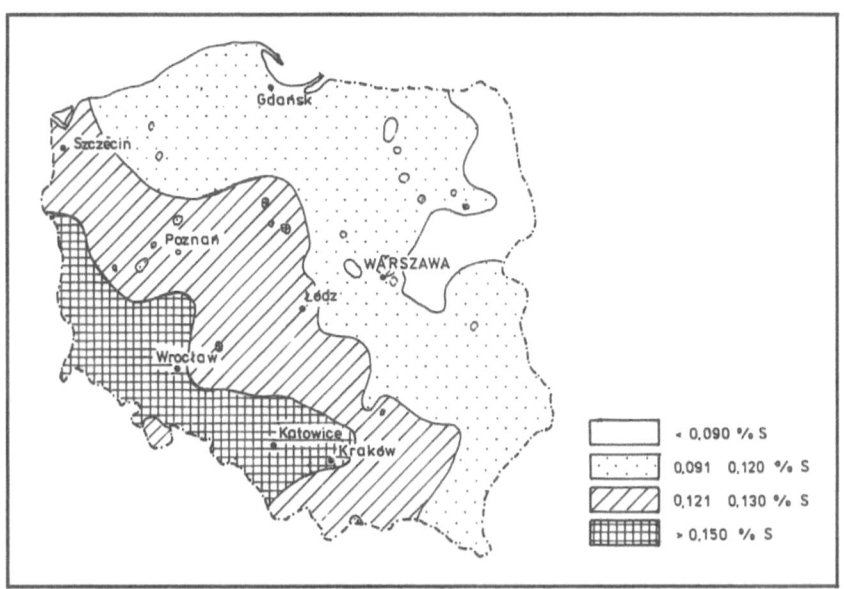

Figure 8–5. The total sulphur content in second year of growth pine needles in different parts of Poland (Molski et al, 1986).

References

Adamczyk, B., K. Oleksynowa, J. Niemyska-Łukaszuk, M. Drożdż-Hara, A. Miechówka, E. Kozłowska, and A. Fajto. 1983. *Roczniki Gleboznawcze* 4:81–92.

Białobok, S. 1986. *Kosmos* 1:43–63.

Chudzyńska, J. 1985. *Przegląd Techniczny* 4:12–13.

Chojnacki, A. 1967. *Pamiętnik Puławski* 29:171—184.

Czyż, A., S. Godżik, J. Geszta, and J. Olszowski. 1968. *Ochrona Przyrody* 33:309–338.

Dobiech, 1978. *Wpływ zanieczyszczeń powietrza na jakość wód atmosferycznych i powierzcniowych w rejonie prezemysłowym.* Część I i II. IPIS-PAN Zabrze/ manuscript.

Glatzel, G. 1983. *Die Messung der Deposition langzeitwirksamer Luftschadstoffe in Wäldern.* Osterreichischer Forstverein, Wien. 165 pp.

Juda, J., S. Chróściel, M. Nowicki, J. Jędrzejowski, W. Jaworski, K. Buddziński, A. Jagusiewicz, A. Warchałowski, and J. Ordega. 1978. Seminaria Instytutu Inżynierii Srodowiska Politechniki Warszawskiej, 96–109.

Kasina, S. 1980. *Atmos Environ* 14:1217–1221.

Kasina, S., J. Kwiek, and J. Lewińska. 1984. Concentration, transformation, and deposition of sulphur compounds. In Grodziński, et al. (eds)., *Forest Ecosystems in Industrial Regions,* 45–56. Berlin: Springer-Verlag.

Kern, H. 1985. *Odczyn i zawartość węglanu wapnia w glebach użytków rolnych Polski,* IUNG, Puławy. 97 pp.

Molski, B., W. Dmuchowski, and W. Chmielewski. 1986. *Biomonitoring for air pollution evaluation of countryside and urban areas.* World Commission on Environment and Development, 17th Meeting, Moscow. 14 pp.

Nürnberg, H.W., V.D. Nguyen, and P. Valenta. 1983. Jahresbericht 1982–1983 Kernforschungsanlage Jülich, 41–53.

Ochrona, S.I. and W. Gospodarka. 1986. GUS, Warszawa. 382 pp. Environmental Protection and Water Management. 1986.

OECD. 1977. The OECD Program on long-range transport of air pollutants. Paris.

Pokojska, U. 1985. Buffering properties and cation-exchange capacity of different forest humus types. In *Air Pollution and Stability of Coniferous Forest Ecosystems*, 121–130. University of Agriculture. Brno.

Prusinkiewicz, A., A. Kowalkowski, and L. Królikowski. 1983. *Roczniki Gleboznawcze* 3:185–201.

Stöhr, D. 1984. *Waldbodenversauerung in Österreich.* Österreichischer Forstverein, Wien. 165 pp.

Strzyszcz, Z. 1982. *Oddziaływanie przemysłu na srodowisko glebowe i możliwości jego rekultywacji.* Ossolineum, Wrocław. 91 pp.

Strzyszcz, Z. 1983. *Aquilo, Ser. Bot.* 19:71–79.

Strzyszcz, Z. 1985. *Archiwum Ochrony Srodowiska* 1/2:69–80.

Warteresiewicz, M. 1979. *Archiwum Ochrony Srodowiska* 1:95–166.

Acidic Precipitation Research in Italy

D. Camuffo*

Abstract

A concise scenario of the emissions and the main environmental problems in Italy are discussed together with the measures programmed by the National Energy Plan. The aims and characteristics of the main monitoring networks, past and present, are illustrated.

In addition to the emissions, meteorological and climatic factors are of considerable importance in determining the acidity of precipitation. A meteorological analysis has shown that, for every weather type, the observed acidity is a result of acidic emissions (mainly anthropogenic) and natural buffering factors originating on the local and mesoscale. The Sirocco and other Bora-type winds are associated with natural buffering agents. These reduce the acidic impact, especially on the eastern side of northern Italy and in southern Italy and Sicily.

In order to gain insight into the dynamics of acidic precipitation and its effects, real-time measurements of free acidity were carried out. Dry deposition contributes in two different ways: at times anthropogenic acidic particles with higher solubility increased the acidity at the beginning of rainfall; buffering particles with lower solubility tended, eventually, to decrease the acidity. For this reason deferred measurements of the bulk deposition often showed neutral or slightly alkaline rainfall, notwithstanding the actual effects on the environment.

The spring thaw of the Alpine snowpack does not result in increased acidity as it does in northern Europe or Canada, because during the winter some snow frequently melts, and the more mobile acidic ions are progressively removed by percolation.

The state of forests and vegetation is less worrying than in central and northern Europe. Here, the damage is more likely related to soil

*National Research Council (CNR-ICTR), Corso Stati Uniti 4, I-35020 Padova, Italy.

characteristics and climatology. Some arid periods and modification of the seasonal distribution of rainfall in past decades seem to be responsible for the main part of the damage.

The subalpine lakes are subjected to highly acidic precipitation. However, the rock is mainly calcareous and therefore provides for their natural liming. Several alpine lakes are characterized by granitic and metamorphic cristalline rocks, risking acidification. Extensive field tests have shown that only 3% of them had pH ranging between 5.5 and 6.0. Acidification is attenuated by some factors: (1) the low ratio between catchment area and water body volume; (2) the lower amount of precipitation and (3) the less acidic precipitation at the higher altitude; (4) the wind-borne Saharan dust or dolomitic particles.

The deterioration of monuments is a very serious problem in Italy, due to the high density of the monuments, their great artistic value, and observed weathering rate. This problem has been studied extensively both in terms of atmospheric pollutants and meteorological factors. For marble and limestone, the deterioration patterns (distinguishable in black, gray, and white areas) are related to the way the rainwater wets the surface. Dry deposition, and especially oil-fired carbonaceous particles, play a very important role.

Other activities undertaken and future research needs (especially for taking measurements in the homogeneous and heterogeneous phases) are also discussed.

I. Introduction

The anthropogenic sources of air pollution are not equally distributed over the Italian territory: Most of them are based in northern Italy in the Po Valley. This region, which covers an area of about 30% of Italy, is also the most industrialized and densely populated zone, having 45% of the total population and 45% of the vehicles. The local contribution of the densely populated towns is even more evident in winter, when emissions due to heating increase the washout and the free acidity of rainfall. Subalpine regions on the borders of the central Po Valley, where the emissions are greater, also experience acidic rain.

The natural sources of air pollution are located mainly in central and southern Italy. The volcano Vesuvius is now quiescent. However, at present three other volcanoes are active: Etna (recent eruptions in 1969, 1971, 1985); Stromboli (continued but moderate explosive activity); and Vulcano (since the eruption in 1888 to 1890 the "solfatara" emissions are moderate). Natural emissions occur from soil, for example, the "solfatara" at Pozzuoli (Naples), the "soffioni" at Larderello (Volterra), and Mount Amiata. The natural emission of SO_2 has been estimated as being only 10% of the anthropogenic sources in Italy (Dall'Aglio, 1984).

An analysis of the isotopic composition of sulphate in rainwater has shown that the main source would seem to be the oxidation of anthropogenic sulphur dioxide (Cortecci and Longinelli, 1970). The addition of seawater sulphate to rainwater is quite small even in Venice, despite its position facing the Adriatic Sea (Longinelli and Bartelloni, 1978).

Actual electrical energy is produced by means that may be mainly responsible for the increase in acidic deposition: petroleum and coal account for 59.8% of the production (i.e., petroleum 44.7% with 3% S content; coal 15.1% with 1% S); natural gas 15.8%; hydro-geo 22.6%; atomic power 0.1%; others 1.7%; imported 3% (ENEL, 1989). The goal is to reduce the proportion of oil power stations with a consequent increase of other sources, increasing substantially the energy produced from coal (e.g. coal was 9.7% in 1987) which has a lower sulphur content. The new power plants will be equipped with desulphurization units, as required by the European Parliament and Commission. The target is a 30% reduction by 1993 in comparison to 1980, to comply with the agreement signed at the Geneva Convention in 1985. At the same time, studies are in progress for a further reduction of these emissions (Pinchera, 1984a, 1984b; Project CERI by ENEA). A concise scenario of the emissions in the 1980s and of the planned measures to be taken was presented at the Amsterdam Conference on Acidification and Its Policy Implications by the Italian delegation (Italian Delegation, 1986).

The soil is sensitive to acidic rain especially in central Italy, Sardinia, in some parts of the Apennines, and in some limited areas of the central and western side of the Alpine chain, where several high mountain lakes are exposed to risk. However, the situation is not dramatic. The soil is much less sensitive in the rest of the territory, especially in northern Italy where it is calcareous, often with a pH value of 8 (CCS, 1966; Perelli and Franzin, 1984).

The state of lakes, forests, and vegetation is checked periodically and many studies are currently in progress. Fortunately, their state is of less concern than in central and northern Europe, and only in a few limited cases have the effects been ascribed to airborne pollutants. Climatic factors seem to be more dangerous in several cases.

Although the situation in terms of the natural environment is not as bad as in other countries, acidic deposition causes serious damage to monuments. For Italy they are an enormous "natural" resource, although tourism is not as well developed as it could be. The most precious monuments are principally in marble or limestone; several Greek and Roman temples and other buildings in southern Italy are of sandstone. All these materials are very sensitive to acidic impact: They may deteriorate and dissolve; thick crusts may form that are the results of the combined physicochemical weathering of the surface. Of course, these monuments constitute a heritage that cannot be renewed and are of inestimable cultural value for the entire world.

Extensive studies have been devoted to analyzing the effects of pollution and acidic deposition; local emissions and long-range transport have been monitored; a comparison of the problems of northern and central Europe and the alpine regions is in progress. The Mediterranean climate, where the rainfall regime and the production of photochemical pollutants play an important role, has also been studied. In the 1980s, a new impetus was provided by the National Research Council (CNR) in cooperation with the National Committee for Research and Development of Nuclear Energy and Alternative Sources (ENEA); the Commission of the European Communities (CEC); and finally the Italian National Electricity Board (ENEL), which finance several of the studies we discuss here.

II. Main Monitoring Networks

Many studies have been carried out in Italy in the past on the fundamental processes involved in acidic depositions and on their effects. Although there has been a great deal of research in this field, the actual monitoring networks are a relatively recent development and do not, as yet, cover the entire national territory homogeneously. The siting of the principal operating networks and the extent of territory that they cover are shown in Figure 9–1a to d. The first network was installed during the International Geophysical Year (1957), and it was composed of nine stations that operated for over four years (Picotti, 1963).

The first network, of those that are still active, was installed in 1975 by the Italian Air Force (IAF), which participated in the international Background Atmospheric Pollution Monitoring Network (BAPMoN) program, under the auspices of the World Meteorological Organization. This network is aimed at monitoring the wet background deposition in Italy at five stations: Verona Villafranca, Mount Cimone, Viterbo, S. Maria di Leuca, and Trapani. Verona was substituted for Parma in 1979. All the stations, except Mount Cimone (isolated, 2165 m a.s.l.) and Viterbo (suburban, 330 m a.s.l.), are situated on a plain and are suburban. The measurements are taken with wet-only automatic collectors or bulk collectors which are cleaned daily. The data resulting from the samples are expressed as average monthly events. In addition to the chemical data, the networks also supply useful information on atmospheric turbidity (Ciattaglia, 1975, 1979, 1981; Ciattaglia and Cruciani, 1977; Ciattaglia and Fiore, 1981).

Italy is taking part in the European Monitoring and Evaluation Program (EMEP), promoted by the CEC and coordinated by the Istituto Superiore di Sanità (Italian Health Ministry) in cooperation with CNR, the IAF, and the Ministry of Agriculture and Forestry (MAF). Data is being taken from sampling stations, belonging to different organizations:

three IAF stations, three MAF, one CNR (Institute for Air Pollution), one Avalanche Protection Center of the Regione Veneto. These are found at Plateau Rosà (isolated, 3480 m a.s.1.); Parco dello Stelvio (isolated, 1415 m a.s.1.); Arabba (isolated, 2094 m a.s.1.); Mantova Bosco Fontana (rural, 20 m a.s.1.); Mount Cimone, Vallombrosa (isolated, 840 m a.s.1.); Montelibretti (suburban, 48 m a.s.1.); and S. Maria di Leuca (suburban, 100 m a.s.1.). The deposition resulting from wet and dry automatic collectors and the concentration of both sulphur dioxide and airborne particles are analyzed.

Additional regional networks have been planned by local authorities, some of which are still active; others will be in the near future.

The most widespread network belongs to ENEL that, in association with MAF, has about 60 stations with wet and dry automatic collectors. This network was set up in 1984 and produces data which are representative of the distribution pattern of acidic deposition in Italy, including both background and local effects. This network was especially designed after having carefully examined the results of a special group formed to study the main features of precipitation in northern Italy (NI Group).

The NI Group was instituted in 1982 and monitored precipitation continuously for three years. It was formed by several CNR research institutes, universities, ENEL, local authorities, and industrial firms situated in northern Italy, that is, in the Po Valley and in the surrounding mountains: the Apennines and the Alpine chain. It also included the Ticino Canton (Switzerland) and the Tyrol region of Austria. The network was composed of 40 monitoring stations (isolated, rural, suburban, and urban) with bulk collectors operating on a weekly schedule of sample taking. This network gave a clear picture of the precipitation in this region (Ioannilli and Bacci, 1986; NI Group, 1985, Novo, 1986) with the result that the spatial patterns of the various chemical parameters could be seen and a distinction could be made between the contribution of local and distant sources.

All these networks are based on different operating criteria, and therefore it is not always possible to make an accurate comparison of the data. Some stations are located in remote areas and found high above sea level; others register the anthropogenic effects due to the presence of industry or power plants; others are situated near, or in big cities. However, a comparison of the resulting data has raised many important points that would suggest that ad hoc studies would prove useful in determining the real effects of buffering agents, which seem to be particularly active on the eastern side of the Po Valley and in southern Italy.

An interesting pioneering measurement was made at Venice in the 1950s (Sordelli and Zilio Grandi, 1972), which consisted of 13 years of taking monthly bulk deposition samples at Venice. The pH levels were virtually neutral or slightly alkaline throughout the whole period.

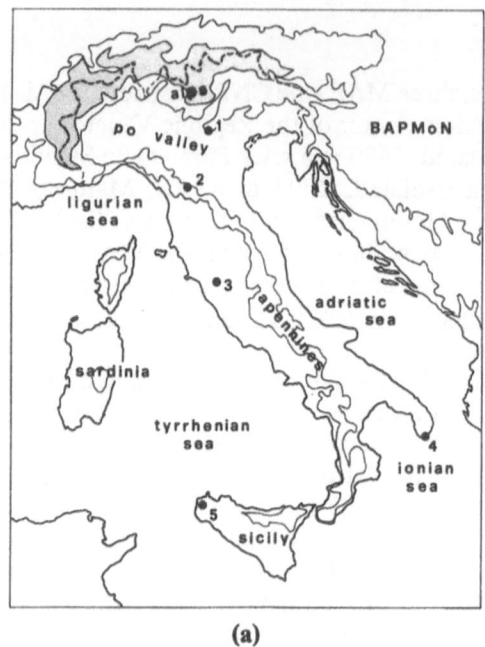

(a)

(b)

Figure 9–1. Main monitoring networks in Italy. (a) BAPMoN: (1) Verona Villafranca, (2) Mount Cimone, (3) Viterbo, (4) S. Maria di Leuca, (5) Trapani. (b) EMEP: (1) Plateau Rosa, (2) Stelvio, (3) Arabba, (4) Mantova, (5) Mount Cimone, (6) Vallombrosa, (7) Montelibretti, (8) S. Maria di Leuca.

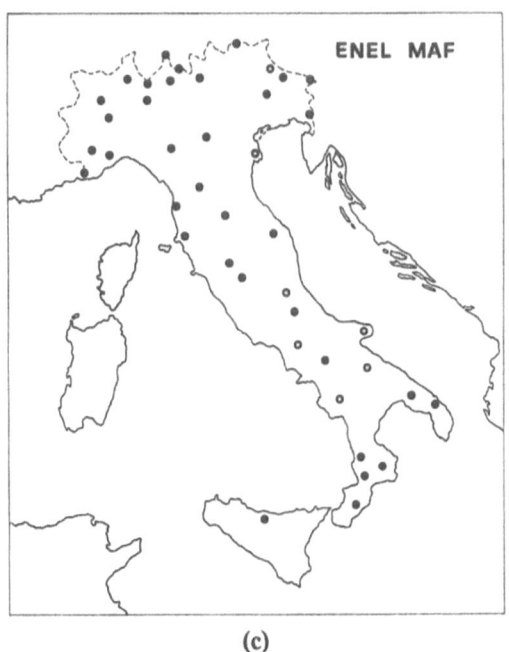

(c)

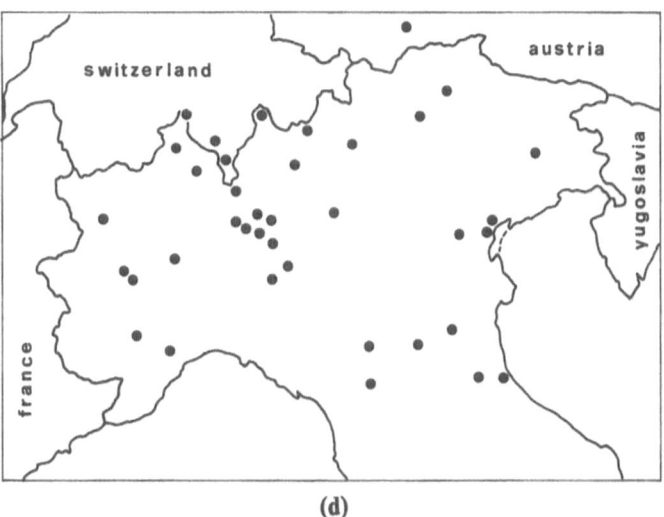

(d)

Figure 9–1 *(continued).* Main monitoring networks in Italy. (c) ENEL-MAF (solid dots: operative stations; circles: planned stations). (d) NI Group, sites active in northern Italy.

The values resulting from this long series of data seemed to be inconsistent with the high dissolution rate of the stone monuments in Venice. This observation suggested the existence of buffering agents (Camuffo, Del Monte, and Ongaro, 1984, Camuffo, Bernardi, and Zanetti, 1988), which could be shown by simply adding some distilled water to the collected rainwater. It was suspected that these buffering agents were possibly responsible for some departures from the norm when an analysis of the bulk deposition data measured on the eastern side of the Po Valley was made. They were also considered useful tracers in the interpretation of the dynamics of many precipitation events, resulting from some typical meteorological situations. The role of meteorology is fundamental in determining precipitation events, and it has also been recognized as being very important in determining the free acidity in rainfall. This role is different in the various regions, as can be seen in the next section.

III. Rainfall Climatology and Chemical Studies

A. Typical Meteorological Situations

The complex topography of Italy and its geographical position means that the territory can be divided into eight different climatic regions (Cantù, 1977). However, the precipitation regime is less complicated and is determined by the interaction of local features and large-scale phenomena (Gazzola, 1978). The orography and the rising air masses along the mountain slopes are very important: At first, it would seem that the rainfall distribution is closely related to the contour lines (Mennella, 1967).

The precipitation regime is well known from the very good network of rain gauges that, in the past, covered the territory and that are still active even though somewhat reduced. The secular trend of rainfall is also known, in that several secular series of precipitation exist, for example, Padova (since 1725), Milan (1778), Rome (1782), Udine (1803), Bologna (1813), Venice (1836), and so on (Camuffo, 1984a; Colacino and Purini, 1986; Eredia, 1911; Melicchia, 1939; Mennella, 1956; Santomauro, 1957).

In northern Italy, and over the Po Valley in particular, the distribution of the yearly precipitation is bimodal. This is due to the passage of fronts, mainly generated by Atlantic depressions and, in the cold season, possibly increased by cyclogenesis over the Mediterranean or on the lee of the Alps. The passage of fronts occurs in spring and autumn. In the summer the Azorre's Anticyclone dominates western Europe and the Mediterranean Basin; in mid-winter the Azorre's Anticyclone is joined by the Siberian High, so that the Atlantic depressions cannot reach Italy. The summer minimum of precipitation is attenuated by showers and thunderstorms generated by convective activity (Camuffo, 1984a), due to the soil being greatly heated and the low cloud condensation level caused by the high specific humidity (Camuffo, Bernardi, and Bacci, 1982).

On the Italian side of the Alps (mean height 2500 m) two precipitation regimes are dominant. In the first, precipitation is mainly due to upslope winds and convective activity with the seasonal maximum in summer Figure 9–2, see *a*). This distribution is similar to the continental one, but has a different origin. In the second, the seasonal distribution is characterized by two maxima in late spring and autumn (Figure 9–2, see *b*). The Po Valley is largely subjected to this regime, which is related to the penetration of Atlantic perturbations. The prevalence of one of these two regimes is due to geographic and topographic factors and whether these provide a shield or not for the area against synoptic influences. On the open lee side, the air masses rise along the upslope making precipitation more intense than on the plain.

In southern Italy, precipitation is typically Mediterranean, as tracks of depressions cannot usually reach this region. Precipitation occurs only in the cold season and is mainly due to instability resulting from the influence of the sea on the cold air flowing over it. The relatively warm water releases heat and moisture, which may reinforce the Atlantic perturbations or generate the deep depressions that are responsible for heavy precipitation and thunderstorms. No precipitation occurs in summer, when

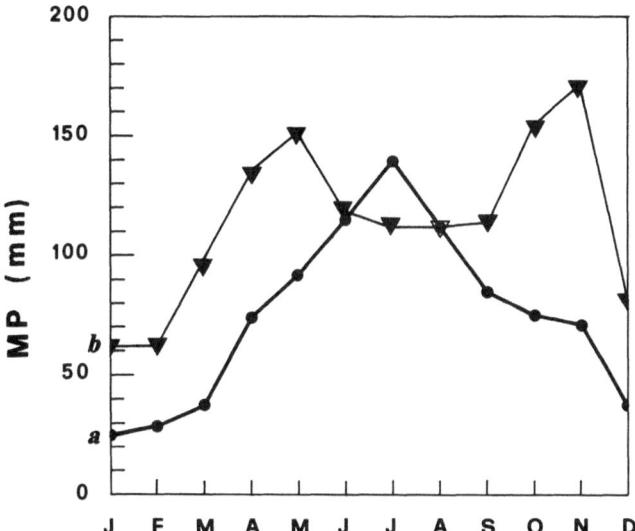

Figures 9–2. The two main precipitation regimes on the Italian side of the Alps. Monthly precipitation (MP): *a*. unimodal distribution with the maximum in summer (Dobbiaco, dots); *b*. bimodal distribution with the maxima in spring and autumn (Cencenighe, triangles). Period: 1921 to 1950.

the relatively cold water of the Mediterranean Sea stabilizes the air masses flowing over it, so that very modest gains in moisture are made and the cloud condensation level cannot be reached by the convective cells rising from the hot ground.

In central Italy and over the Apennines (mean height 800 m) the rainfall distribution passes gradually from the northern regime, characteristic of the Po Valley, to the southern regime, that is, the summer instability and the passage of fronts being progressively reduced, and winter precipitation increasing. This balance favors an autumn maximum rather than a spring one. In the cold season, the strong Mistral wind, blowing over the Tyrrhenian Sea from the northwest through the gap between the Alps and the Pyrenees, increases the intensity of rainfall on the western side of central Italy.

The Sirocco, a strong wind blowing from the Sahara, plays a special role. It may be generated when an extended Atlantic depression spreads over the Mediterranean basin, or when a trough extends south from northern Europe to the Mediterranean west of Italy, or when a deep depression lies over the central Mediterranean. The warm air masses with high moisture content, when flowing from northern Africa toward higher latitudes, are cooled, and may thus reach saturation and precipitate. Precipitation is particularly intense when the Sirocco rises over the mountain chains along its trajectory, that is, the Apennines and the Alps. In the latter case, several varieties of Bora-type winds are induced, which play a very important role in causing precipitation on the eastern side of the Po Valley (Camuffo, 1981; Camuffo, Del Monte, and Ongaro, 1984; Camuffo, Del Monte, Sabbioni, and Moresco, 1984; Defant, 1951; Meteorological Office, 1962; Reiter, 1971, 1975). As this situation is important and rather complex, the different Bora-types are explained.

When the Sirocco rises over the Alps, the adiabatic rise of the air masses is noticeable. As the Sirocco proceeds north, it generates a warm front that advances northward until it reaches the Alpine chain. The local, colder air remains entrapped below this front, which becomes stationary, and the warm Sirocco flows upward (Figure 9–3). The locally induced pressure pattern generates a cold northeastern wind of the Bora-type, coming from Yugoslavia, which increases the contrast between the two air masses. In addition to cumuli due to the saturation of the rising air masses, the frontal zone is also characterized by the presence of stratus clouds. When the depression lies over the central Mediterranean (Figure 9–4a), the cloud cover is very thick, and intense precipitation may last for a few days. Due to this characteristic cloud cover that reduces the visibility, this variety of Bora is named the Dark Bora (locally called "Bora Scura").

The Cyclonic Bora occurs when a depression lies west of the Alps (Figure 9–4b) and may be responsible for rainfall, but not so severe as in the case of the Dark Bora.

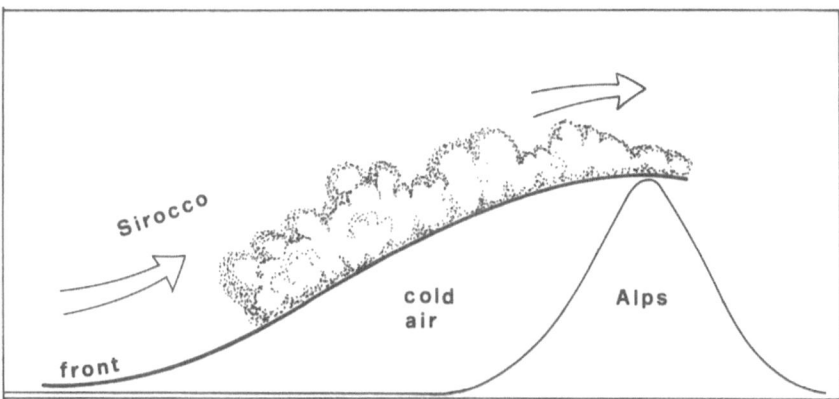

Figure 9–3. The stationary front formed when the warm Sirocco reaches the Alpine chain. The warm air masses rise along the cushion of the local cold air entrapped between the Sirocco and the Alps. Idealized section with different horizontal and vertical scales.

Another variety of Bora that may be responsible for precipitation, but having a synoptic situation symmetrical with the Cyclonic Bora, is the Anticyclonic Bora (Figure 9–4c). When a northern flow descends toward the Mediterranean basin in the south, it outflanks the Alps and blows in violently from the northeast through the gate in the Dynaric Alps at Trieste, spreading over the northern Adriatic Sea and causing a decrease in pressure locally. The low pressure on the lee side may be further deepened by cyclogenesis over the warm waters of the Adriatic in the cold season. This wind is not associated (except for reversed trajectories, see Prodi and Fea, 1978, 1979) with Saharan dust that, in the other varieties of Bora, is carried by the Sirocco.

In winter there is another variety of Bora, that is, the White Bora (locally "Bora Chiara," Figure 9–4d), which is a violent flow of continental polar air generated by the Siberian High or high pressure over central Europe. The air is still cold on arrival although it is warmed by katabatic compression and change of latitude. The warming makes it very dry (relative humidity 20% to 30%), and it is always associated with fine weather.

B. Wind-Borne Buffering Agents

The Sirocco and the first three Bora-type winds play a very important role in determining the chemical characteristics of rainfall over northern Italy, not only because they cause precipitation, but also because they may affect the pH. Both winds contain buffering agents, but these do not act with the same degree of efficiency. The Sirocco transports Saharan dust, which has proved to be a very effective buffering agent (Camuffo,

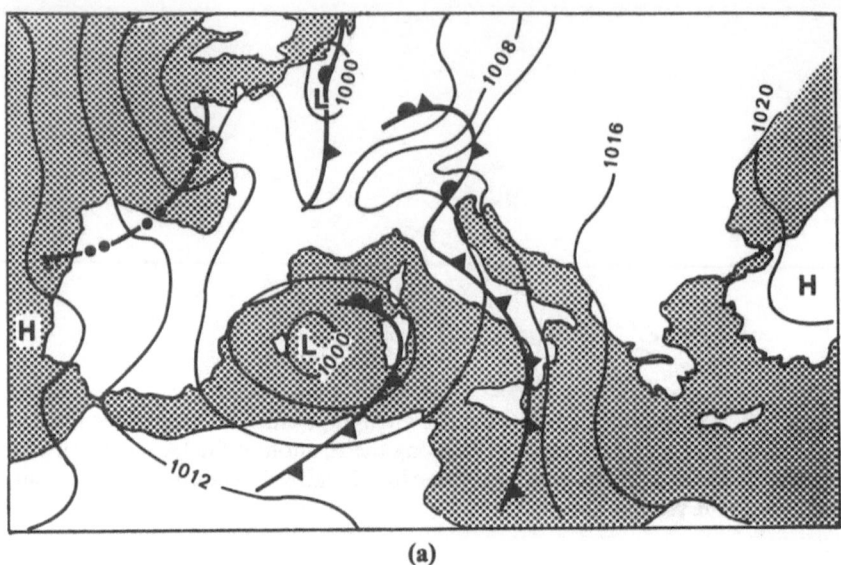

(a)

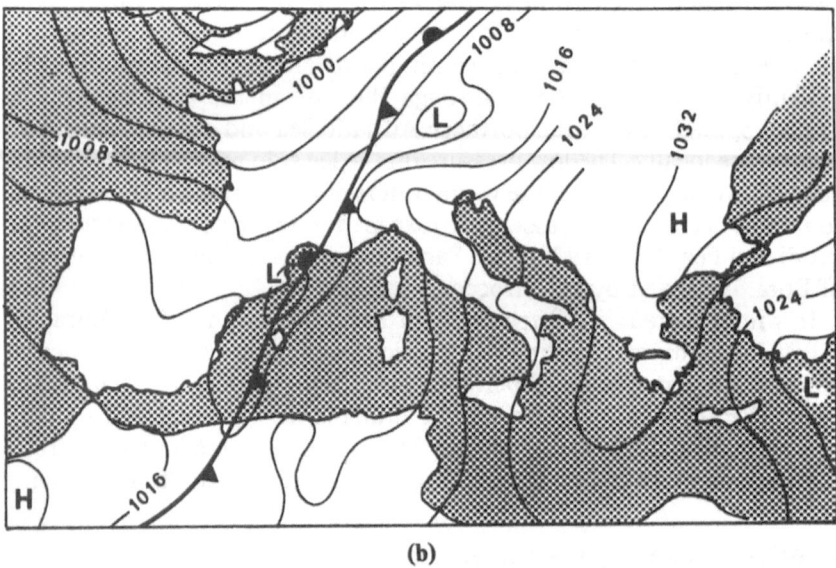

(b)

Figure 9–4. The four Bora-type winds. (a) Dark Bora, when a depression lies over the central Mediterranean. Synoptic situation, 00.00 GMT, October 23, 1982. (b) Cyclonic Bora, when a cyclonic area (or low pressure trough) dominates west of the Alps. Synoptic situation, 00.00 GMT, November 9, 1982.

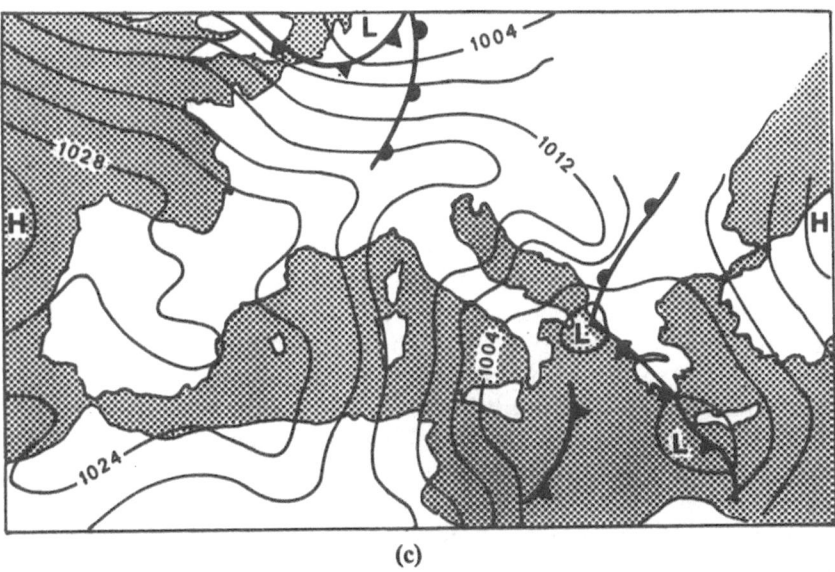

(c)

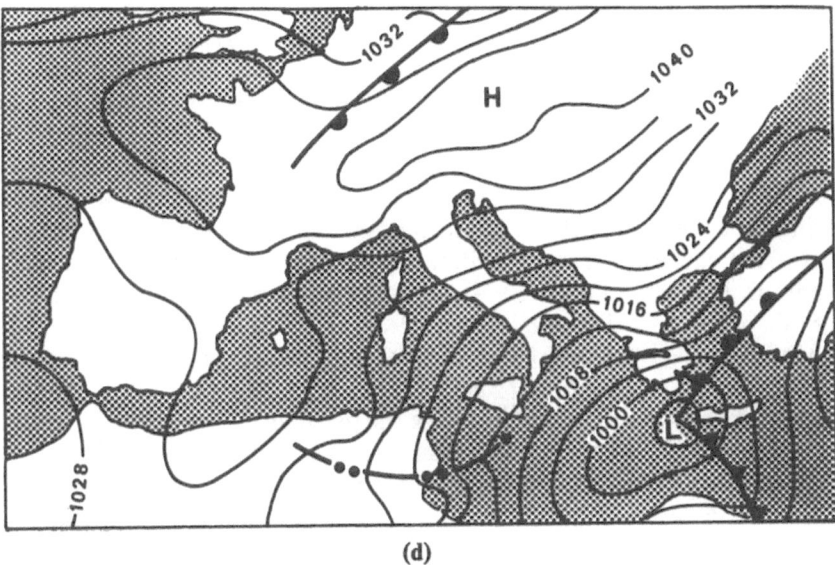

(d)

Figure 9–4 *(continued).* The four Bora-type winds. (c) Anticyclonic Bora, when an anticyclonic area (or high pressure ridge) dominates west of the Alps. Synoptic situation, 00.00 GMT, November 18, 1982. (d) White Bora, when an anticyclone or high pressure ridge lies over central Europe or Siberia. Synoptic situation, 00.00 GMT, January 9, 1981. Synoptic maps courtesy of the Meteorological Service of the Italian Air Force.

Del Monte, and Ongaro, 1984; Camuffo, Bernardi, and Zanetti, 1988; Camuffo, Del Monte, Sabbioni, and Moresco, 1984; Ganor and Mamane, 1982; Georgii, 1982; Prodi and Fea, 1978, 1979; Sequeira, 1982; Winkler, 1976). The Bora transports karst soil particles (from the plain between the Alps and the northern Adriatic Sea) and sea spray, which are also buffering agents, although they are less both in terms of quantity and effectiveness (Delmas and Gravenhorst, 1983; Psenny, McIntyre, and Duce, 1982). Both the Dark Bora and the Cyclonic Bora are associated with the Sirocco, which precipitates Saharan dust, with the result that the rainfall contains a very effective buffering agent.

For this reason, northern Italy (with the exception of urban sites) is often subjected to alkaline precipitation. Such rainfall is much more frequent on the eastern side of the Po Valley than on the western, as southern winds are often associated with depressions centered west of Italy over the central Mediterranean. Under such circumstances, the western side of Italy is affected by marine air masses. Their trajectories have shorter radii of curvature and flow all around the center of depression. The eastern side, on the other hand, is affected by air masses which are further away from the center of the depression, which may have passed over the Sahara.

In addition to the greater frequency of trajectories from the Sahara, the amount of Saharan dust released per episode is also greater over the eastern part of the country, as the southern wind moving aloft is often associated with Bora-type winds at ground level, resulting in more intense precipitation.

The Sirocco causes a net decrease in acidity on the northern side of the Po Valley where a stationary front lies, formed by the warm air masses rising over the cold cushion of local air entrapped by the Alpine chain. This front causes a greater precipitation of Saharan dust, as can be seen from the data gathered by the NI Group network during some case studies (Viarengo, Mosello, and Tartari, 1984). This situation is often found in winter over the eastern part of the Po Valley where acid precipitation is less frequent.

Over the rest of the Po Valley and at the periphery orography, the mean pH value is lower during winter than during summer. This is because of the high concentration of pollutants due to (1) increased emissions resulting from domestic heating and (2) climatic conditions characterized by low wind speed, fog, and radiative inversions. The emissions resulting from domestic heating have a low sulphur content giving a maximum NO_x/SO_2 ratio in winter (Anfossi, Bacci, Natale, and Viarengo, 1986). This ratio decreases in the south and islands. This is especially due to lower anthropogenic and higher natural emissions, in particular volcanic activity (Ioannilli, Bettinelli, Franciotti, and Ziliani, 1987).

In southern Italy, southern Sardinia and Sicily in particular, alkaline rainfall is frequent due to the Saharan dust and local soil particles (con-

sisting essentially of calcite, dolomite, and quartz generated from limestone and dolomite rocks) transported by the Sirocco wind. The Sicilian
inland winds have a higher concentration of Ca, whereas the offshore
winds contain Na, Mg, and Cl, which are closely correlated to them and
originate from the sea spray (Badalamenti et al., 1984). The dry deposition is always alkaline (Ioannilli et al., 1987).

C. Free Acidity and Weather Types

Precipitation in Italy is closely linked to the meteorological situation and
this, naturally, is related to the different trajectories of the air masses,
thus influencing the degree of acidity. It seemed opportune to see if there
was any relationship between the degree of acidity of the rainfall and its
trend, and weather types as classified on the basis of different criteria.
This analysis was designed to forecast under some well-defined circumstances when acidic precipitation would occur, in order to carry out special field tests or to take appropriate steps to neutralize the effects of
acidity in certain ecosystems.

The weather types used in classifying the meteorological situations
were those proposed by (1) Borghi and Giuliacci (BG) of the Italian Meteorological Service (Borghi and Giuliacci, 1979) because their research
had been especially conceived to classify circulation in northern Italy in
relation to primary circulation over Europe; (2) the United Kingdom Meteorological Office (UK; see Meteorological Office, 1962), which is very
general and widely used; (3) Aerospace Science Division (ASD) (Reiter,
1975) as it contained 28 weather types and allowed for a more detailed
classification. The BG and UK criteria in comparison to other criteria
had already been used successfully when considering pollution episodes
in Venice (Bernardi, Camuffo, Del Turco, Gaidano, and Lavagnini,
1987).

The analysis of the real-time measurement of the pH of rainfall at
Padova over a two-year period (Camuffo, Bernardi, and Zanetti, 1988)
showed that some weather types are more often associated with acidic
or alkaline rain, whereas other types are not associated with either. Only
in some cases may forecasting be possible, as several weather types are
associated with both acidic and alkaline elements. This approach
showed that the complex synoptic pressure pattern that causes great
spatial variability in the wind field did not result in the precipitation re-
taining the same characteristics over a long period. It showed also that
the detailed distribution of the synoptic pressure pattern is much more
important than the general features of the weather type. Due to the time
evolution of the wind field, the same weather type may be associated
with different trajectories when passing over Italy and the rainfall may
have different degrees of acidity. Given this situation, a detailed reconstruction of the trajectories shows that although air masses may be

enriched with alkaline particles originating in distant places (e.g., the Sahara) the pH value may be modified or changed by the addition of particles from local sources such as upwind towns or power plants, thus resulting in acidic precipitation.

When the local washout makes a greater contribution than rainout, the pH trend at the initial stage of precipitation shows outstanding variations. A similar weather analysis for the episodes of precipitation initially more acidic or more alkaline or without any trend variation was useful only in a few ASD classifications. It was noted in particular that the initial stage is acidic when associated with those classes in which the wind transports airborne pollutants from the most industrialized or inhabited parts of the Po Valley. As the pH value and the substances in solution in the rainfall may vary over time, it is important to measure them in real time.

Again, this means that a year should not be considered "normal" or "exceptional" in terms of acidic precipitation solely on the basis of the seasonal totals of acidic rainfall, as several authors do, but rather, on the relative frequency of the weather types associated with acidic precipitation events.

Several findings agree with Benarie and Detrie (1978) for central and northern Europe, namely (1) the emission of sulphur compounds is not the only cause of the observed acidity in rainwater; (2) transport of sulphates is only a partial explanation of rain acidity; and (3) short-range or microclimatic factors may influence the acidity of rainfall, especially where it may be neutralized by basic components of local origin.

D. Dynamics of the Aggressivity of Precipitation

To study the aggressivity of precipitation, the pH value was measured in real time, during and after precipitation, in order to show whether the acidic or alkaline elements present in the raindrops were already dissolved or whether the dissolution of these elements continued after the raindrops had reached the soil. At this point two questions of great importance can be discussed: If the data show that rains are more often alkaline than acidic in several regions of Italy, then does the problem of acidic rain actually exist in Italy? Or, is the frequency of alkaline rainfall the result of the method of using bulk collectors or, using wet-only collectors, analyzing rainwater after some time?

In general, it is the deposition that occurs in the wet phase that causes the main damage. Extensive damage may also result from the dissolution of the deposition that has occurred in the dry phase preceding precipitation, as has been shown on different occasions. For example, a comparison between bulk sampling and wet-only sampling has shown that the former increases the concentration of all the ions so that the acidity of the bulk sampling, on a yearly average in the Po Valley, is less than the acidity of the wet-only sampling (Figure 9–5, Ioannilli and Bacci, 1986; Novo,

1987). Real-time measurements show that acidic precipitation is more frequent (although less than 15%) in comparison to bulk sampling, even though rainwater does tend to become alkaline some hours after precipitation.

Again, using a funnel without subsequent cleaning to collect the water precipitated and taking pH measurements in real time, it was observed that the acidic values at the beginning of precipitation were more frequent as the dry period between one precipitation and the next was of greater duration (Figure 9-6, Camuffo, Bernardi, and Zannetti, 1988). This was due to the fact that the dry deposition contained acidic particles

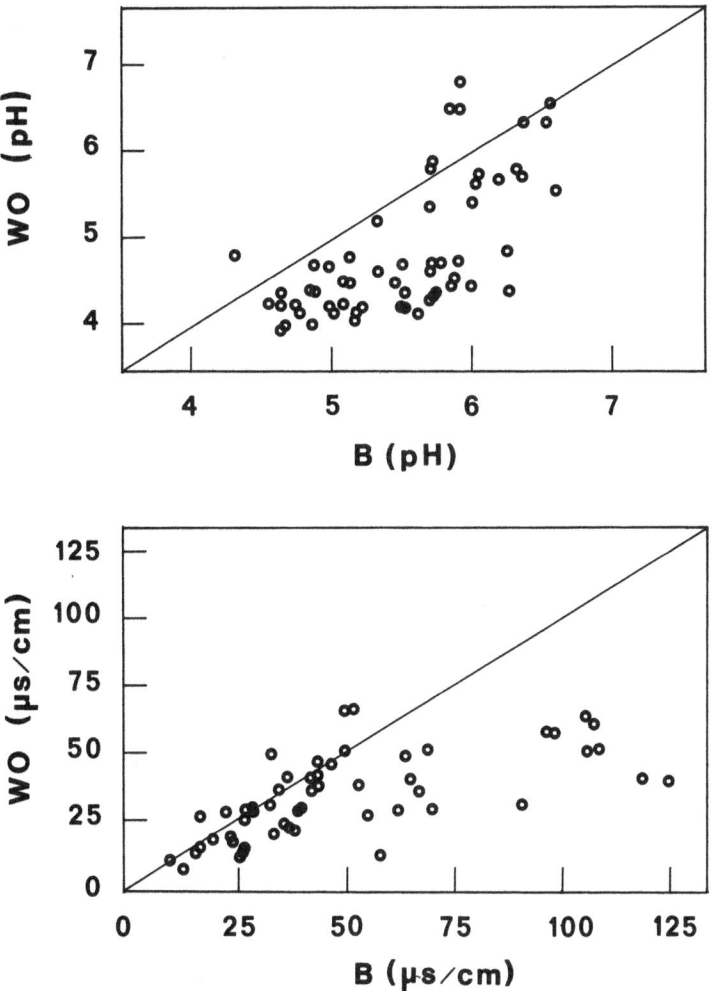

Figure 9-5. Comparison between bulk (B) and wet-only (WO) sampling, for pH and conductivity (μS cm^{-1}) at Piacenza. Period: 1983 to 1984. After the NI Group, 1987.

of an anthropogenic origin that have a more rapid kinetic rate of dissolution. The dry deposit also contained alkaline particles resulting from the soil or from the Saharan dust, but these have a slower kinetic rate of dissolution. Junge and Sheich (1969) and Delmas and Gravenhorst (1982) also found that the smallest particles gave an acidic reaction whereas the coarsest particles gave an alkaline one.

It would be appropriate here to remember that the rainout of the air masses carried by the Sirocco and uplifted on coming into contact with the Alps is very fast and occurs at low temperatures. Thus the rainwater that has already fallen continues to affect the dissolution of the condensation nuclei, which reach a more advanced chemical equilibrium in comparison to that found in the rainout stage. The chemical reactions become more important when the falling droplets are responsible for intense washout of the lower atmosphere.

Under these conditions, when the dry deposition is wetted, because the acidic component is more soluble than the alkaline, the rainwater immediately becomes acidic, thus increasing the aggressivity of the precipitation impact. The rainwater is successively buffered by the alkaline component, which has a slower kinetic rate, so that the rainwater appears alkaline when a deferred analysis is carried out. This means also that the harmful effects of the acidic rain are reduced in systems that are not sensitive to short fluctuations, such as large bodies of water or agricultural land. Monuments and foliage, however, are much more susceptible to fluctuations in the acidity of rainwater.

E. Atmospheric Scavenging

An important aspect of the complex cycle that links precursor emissions and acidic deposition is the chemical process that takes place when the pollutants come into contact with dispersed atmospheric liquid: cloud and fog droplets and precipitation elements. The final chemical composition of the droplets is determined by four main mechanisms: nucleation scavenging, scavenging of gas and particles during the evolution of the dispersed system, chemical reactions in the dispersed liquid phase, and microphysical evolution of the system. The overlapping of all these mechanisms makes the study of such systems highly complex. Several studies have been carried out on this subject at the CNR-FISBAT Institute (Caporaloni, Tampieri, Trombetti, and Vittori, 1975; Mandrioli, Puppi, Bagni, and Prodi, 1973; Prodi, 1976; Prodi and Nagamoto, 1971; Prodi, Prodi, and Fiore, 1970, Prodi, Santachiara, and Prodi, 1979; Prodi and Tampieri, 1982; Tomasi, Guzzi, and Vittori, 1975; Vittori, 1973, 1984; Vittori and Prodi, 1967; Vittori, Prodi, Morgan, and Cesari, 1969).

It has been proposed (Fuzzi, 1986a) that radiation fog would be a useful natural laboratory for the evaluation of the empirical rate of the physicochemical processes that occur in the dispersed atmospheric liquid

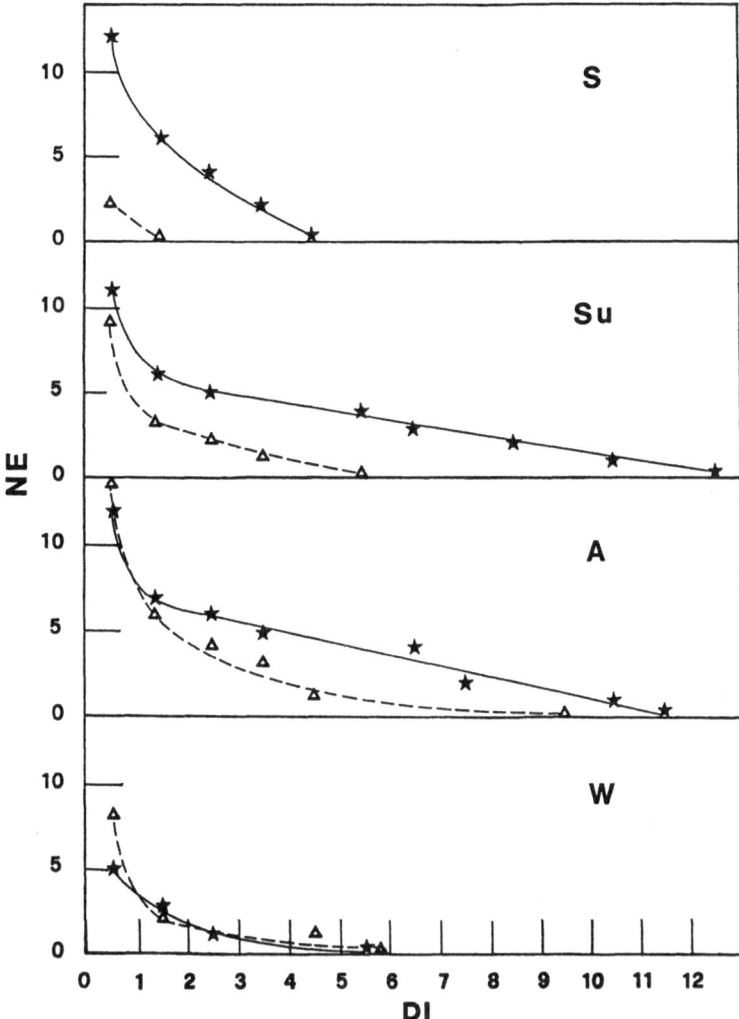

Figure 9–6. Cumulated number of episodes (NE) when the initial stage of precipitation contains more free acidity (star) or less (triangle), occurring with a dry interval (DI) shorter than the number of days in abscissa. Spring (S), summer (Su), autumn (A), winter (W). After Camuffo et al., 1987a.

phase. Because the atmosphere is very stable during radiation fog, a closed box approximation was assumed, as far as air mass motion is concerned, in order to perform a mass balance-based chemical study. An international project was established in 1984 at the CNR-FISBAT field station at S. Pietro Capofiume in the eastern Po Valley, where radiation fog may occur during 30% of total days in the cold season (Cantù, 1977). The first results of this project underline the preeminent role of nucleation scavenging and microphysical evolution of the droplets in determining the chemical composition of the fog water. Some early measurements indicated that the pH of the fog water was as low as 2.8, and sulphate and nitrate concentrations were in the order of several millimoles L^{-1} (Fuzzi, 1986b, 1987; Fuzzi et al., 1988; Fuzzi, Orsi, and Mariotti, 1983, 1985).

The results indicated that, in the Po Valley, the combined effect of high fog frequency and high ionic concentration of the solution droplets results in very dangerous deposition rates. In certain cases, for example, plant foliage, the damage is more closely related to the proton concentration rather than the total number of protons deposited. Therefore, it may be convenient to examine the effect of the deposition of acidic fog droplets in terms of single "intense" episodes and their frequency, rather than on the basis of seasonal estimates.

Another aspect that is under investigation is the role that organic compounds with a low molecular weight play in the chemistry of the dispersed atmospheric liquid phase. High concentrations of carbonyl compounds (Facchini, Chiavari, and Fuzzi, 1986) and organic acids (Winiwarter, et al. 1988) have been detected in the Po Valley fog water, and a study of their possible sources and sinks is underway.

IV. Other Activities Undertaken and Future Research Needs

A. Transboundary Transport of Pollutants

The problem of transboundary flows of airborne pollutants is very complex and needs international cooperation, especially as knowledge, in some areas, is inadequate. At present, there is an intensive research project, jointly undertaken by three CNR Institutes (i.e., Cosmogeofisica, FISBAT, ICTR) and ENEL-CRTN that involves the monitoring of the deposition and transport of airborne pollutants in two main Alpine passes: the Monginevro on the west and the Brenner on the east. The program is based on routine measurements of meteorological variables and chemical species, field tests, and reconstruction of the trajectories for case studies of special interest.

The first aim is to reconstruct, from the climatological point of view, the weather situations responsible for the import and export of pollutants and to attempt to find the balance of transboundary transport. A second aim is to distinguish between distant and local sources of pollution with the help of field tests and mathematical models. The Po Valley seems to be much more impervious to exchanges of pollutants than it would appear from popular belief. The episodes of transport through the Brenner pass were quite rare (i.e., some ten per year) and lasted for a few days with a concentration of pollutants very low. The analysis of the synoptic situation showed that the pollution was mainly transported from eastern and central Europe (Czechoslovakia and Germany) and only in a few cases from the Po Valley. In the former case, the ratio SO_2/No_2 was greater than 1; the opposite in the latter, due to the widespread use of methane in that region (Camuffo, Bernardi, Ongaro, Bacci, and Novo, 1988).

B. Improving Instruments and Methods

Several studies are presently devoted to improving instruments and methods of monitoring in urban and isolated locations. The research program has underlined many problems involved in collecting data from isolated sites situated at high altitude: meteorological measurements, teledetection of pollutants, and capture of droplets for rainout studies (Fuzzi, Gazzi, Pesci, and Vicentini, 1980; Fuzzi, Mariotti, and Orsi, 1982).

Another important, but still open problem, is how to measure the dry deposition (Mastino, Testa, Michetti, and Ghiara, 1987). Its rate is a function of atmospheric and surface variables, namely, the wind field near the surface and atmospheric turbulence, the interfacial exchange of heat and moisture, the concentration and nature of pollutants in the atmosphere, the diameter of particles, and the capture efficiency of the surface. A denuder to measure the concentration of pollutants in the atmosphere has been developed by CNR, Institute for Atmospheric Pollution (Allegrini, Buttini, Di Palo and Possanzini, 1987; Possanzini, Febo, and Liberti, 1983). This device gives accurate measurements but the problem of measuring the deposition velocities is still unresolved.

V. Brief Outline of Studies on the Effects of Acidic Deposition in Italy

A. Forest and Vegetation

Many case studies have investigated the effects of acidic rain in Italy. One of the earliest was carried out by Bottini (1939) who analyzed precipitation (with a pH less than 3) in the Naples area. He attributed the possible increase in acidity to the washout of emissions from the local volcano Vesuvius, with the result that crops growing on its slopes were damaged. After the eruption in 1944, Vesuvius is now in a quiescent stage.

In the 1960s and at the beginning of the 1970s, some studies were carried out to examine the factors that conditioned soil fertility, that is, the contribution of inorganic nitrogen, orthophosphates, and sulphates. The effects of the anthropogenic emissions, urban and industrial, as well as sea spray were also studied in various regions. These studies revealed a seasonal cycle and a spatial diffusion of these compounds in the emission zone (Gargano Imperato, 1963; Nucciotti and Rossi, 1968; Palmieri, 1964, 1965, 1973; Rossi and Nucciotti, 1968, 1974). A review of these experiments was made by Tartari and Mosello (1984).

In the case of forests and vegetation, direct damage due to SO_2 and oxidants is relatively well documented, but reliable exposure/effect relationships are lacking; various symptoms are often indiscriminately attributed to "acid rain." The real causes (e.g., increasing soil acidification, photooxidants, climate, pests, forestry management) are extremely complex and poorly understood (Ott, 1984).

The effects of atmospheric pollution on vegetation has been studied by many authors, in particular by Lorenzini (1983). The biomass of some agricultural products, such as potatoes, increases when subjected to acidic rain as simulated in the laboratory, while others, such as wheat, barley, oat, and triticale, decreases at a pH of 3.5. No foliar tissue damage was observed. An analysis of growth showed that early growth stages are more susceptible to the effects of acidic precipitation (Lorenzini and Materazzi, 1984). Long-term fumigation experiments with SO_2 have been carried out on sunflower plants, soybean, corn, and spinach. Visible symptoms have not been detected, other than a reduction in the chlorophyll content in soybean plants and corn (Lorenzini and Panattoni, 1986).

The effects of acidic smut emitted by oil-fired power plants were studied by Lorenzini, Panattoni, and Schenone (1988). The toxicity of smuts was largely dependent on their acidity and size, with a critical threshold diameter of 80 μm.

Twelve ecotypes and varieties of alfalfa were exposed to acute fumigation and chronic exposure to SO_2, NO_2, and O_3. The effects of acute fumigation showed no correlation to long-term exposures. Mixed treatments ($SO_2 + NO_2$) showed synergistic interactions, but the great variability in response makes it difficult to predict the impact on the environment (Lorenzini, Mimmack, and Ashmore, 1985). Field experiments carried out by ENEL with open-top chambers, aimed at estimating the actual effect of environment air pollution on several crops, started in 1987. The crops are continuously exposed to charcoal filtered or locally polluted air, and the growth parameters will be compared.

In 1987 ENEL set up a 24-station network for indirect measuring of O_3. The network utilizes indicator plants (Tobacco cv. Bel W3). Leaf damage for this plant is proportional to the O_3 dose above the 40 ppb threshold concentration (Schenone and Mignanego, 1988).

In reality, the precise mechanisms by which acidic depositions actually cause damage are still unknown, and as yet there is no universally accepted explanation of cause-effect in the determination of damage to vegetation. Although the growth of ecotypes may be partially controlled by suitable agricultural practice, the situation changes when forests are considered. It is not clear how much damage is due to acidic rain and how much is due to climatic, biotic, edaphic, or other factors, such as neglect and aging of forests. At the moment, the Italian Agricultural and Forestry Commission claims to have only scarce and fragmentary information on the effects on vegetation (Gisotti, 1984). In a CEC report (1983) on damage to forests, the relative importance of the challenges facing Italy was evaluated as follows: (1) very important: fires; (2) important: presence of the public, marine aerosols, damage resulting from the wildlife; (3) of little importance: unauthorized grazing, bad utilization, acidic rain, fluorine, heavy metals, insects, bacteria and funguses, temperature; (4) not important: other factors such as wind, snow, antifreeze salts, and unauthorized felling of trees.

It seems doubtful, therefore, that the damage to forests can be attributed to acidic rain and where it exists, it is very limited in comparison to central Europe (Sandroni, 1984). In practice, less than 1% of the forests have been seriously damaged by "obscure sickness," and 95% is defined as being in good condition (Alessandrini, 1987; Gisotti, 1984). On the contrary, the Laboratory of Agriculture and Forest Botany of Florence University has a pessimistic view of Tuscany, mainly ascribed to acidic rain in summer and other pollutants, especially ozone.

Serious symptoms of damage can be seen in some Italian localities (Figure 9–7), in particular: in Tuscany, the forest of Vallombrosa, on the

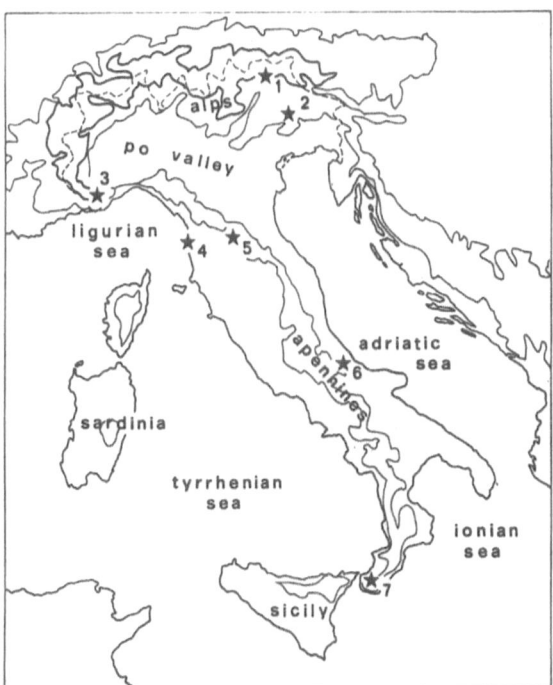

Figure 9–7. Location of the forests showing symptoms of damage: (1) Trentino Alto Adige; (2) Cansiglio (Veneto); (3) Mount Gouta (Liguria); (4) S. Rossore (Tuscany); (5) Vallombrosa (Tuscany); (6) Molise; (7) Aspromonte (Calabria).

side of the Apennines and on the coast; the forest of the Aspromonte in the Calabrian Apennines; in Molise; in Liguria and Trentino-Alto Adige in the Alps. The affected species are silver fir since 1970; Norway spruce, black pine, Douglas fir, and beech since 1982; chestnut, Turkey oak, Downy oak, European hop hornbeam, sycamore, Norway maple, Italian maple, lime, common yew, laburnum, willow, walnut, Aleppo pine, maritime pine, wild cherry, white beam, rowan, and ash since 1984. Several of these species have been affected by biotic factors: the silver fir and Norway spruce by parasitic fungi (i.e., *Armillaria mellea* and *Fomes*); the lime by aphids; the oak by oidium; the chestnut by others (Gisotti, 1984).

Among the worst hit is perhaps the silver fir forest at Vallombrosa. The soil is mostly limey sand, which is permeable, acidic, and lies on geological substratum of ground quartz. Rainfall is abundant for the Mediterranean climate (about 1400 mm yr^{-1}). A very important point is that the greatest damage has been observed during relatively dry summers on the silver fir and beech, which prefer an Atlantic-type climate where rainfall is evenly distributed throughout the year (Pagliari, 1986). Precipitation is often decidedly acidic, which, however, alternates with alkaline rain due to the wind-borne Saharan dust or by the marine aerosols that sometimes invade the area. During the day, the breeze in the valley sometimes transports urban emissions from Florence (25 km east); the forest is also affected by photochemical oxidants such as nitric oxides and ozone. In addition to these factors, there is no lack of criticism about the way the forest has been managed. The forest is mainly composed of high density monospecific stands; trees are of the same age and have suffered for lack of thinning. Several trees appear to suffer reduction in vitality due to scenescence (being more than 100 years old) rather than air pollution. To shed light on these complex factors, the zone is now being carefully studied by a combined MAF, ENEL, and Florence University project.

Accurate studies have been carried out in the Alto Adige forests (Minerbi, 1986) where the damage was mostly slight and more serious types of damage were not, to any great extent, observed. The mortality rate due to unknown reasons was low (i.e., less than 0.5%). There was no general tendency for the soil to become acidic. Leaf damage due to airborne pollutants was limited. An analysis of the climate has shown that the forests were subject to greater damage when there was little precipitation during the period of vegetation. This phenomenon has been associated with temporary changes in the seasonal pluviometric regime, in that the summer rains decreased while the spring and autumn rains increased, resulting from a different type of synoptic circulation. These climatic factors are believed to cause most damage to the vegetation.

The Cansiglio Forest (eastern Alps) is in a dramatic condition: Some 70% of the Norway spruce will have to be cut down in an area measuring 150 hectares. Also in this case the ultimate reason can be linked to climatic factors that influenced the host-parasite interaction of two hymenoptera, directly responsible for the enormous damage. The severe winter in 1985 (with minimum temperature −35°C in January) killed the *Trico-gramma cephalciae,* whose larva lives on the foliage and feeds on the eggs of the *Cephalcia arvensis Panzer.* The *Cephalcia,* at that period in the soil under the snow mantle, survived, and in spring devastated the foliage, without being controlled by its predator. When the spruces were defoli-ated by the *Cephalcia,* the *Xiloterus lineatus* attacked the trunks (Mezzalira, 1987).

Heavy rainfall—not acidic rain—is responsible for desertification, es-pecially in southern Italy, Sicily, and Sardinia, as a consequence of pro-gressive soil erosion. This is due to runoff on the relatively high relief with scarce vegetation (Aru, 1986; Rendell, 1986).

Damage to coastal vegetation, visible above all along the Ligurian-Tuscany coast, has been attributed to marine aerosols and to the presence of surface-active agents (Busotti et al., 1983; De Rossi and Gisotti, 1981; Gellini et al., 1985; Gellini, Pantani, Bussotti, and Raccanelli, 1981). There was a good correlation between the accumulation of chlorines in the plant and the severity of the damage, but there seemed to be no corre-lation between absorption by the leaves (or absorption by the roots) and damage. Damage to pines and oaks appeared, above all, to be due to the effects of the aerosols on the foliage (Guidi, Lorenzini, and Soldatini, 1986), especially in the cold season, when the strong *Libeccio* wind causes the main damage (Bottacci, et al., 1988).

B. Aquatic Environment

One of the main problems in Scandinavia is the acidification of lakes and streams. This leads to increasing concentrations of acids or toxic metals like aluminum leaching from the soil. In Sweden intense liming programs have been started to combat this acidification (Dikson, 1986).

In Italy, several rivers are heavily affected by urban and industrial wastes, and here, the problem of acidic rain is of secondary importance. At remote sites, fresh water is in general among the ecosystems least af-fected by acidification by atmospheric deposition. This is mainly due to the favorable lithological conditions of their watersheds, which are such as to buffer the actual acidity input. Natural subalpine and alpine lakes have been extensively studied by the CNR Institute of Hydrobiology (Mosello, 1984; Mosello and Tartari, 1982, 1983, 1984a, 1984b; Mosello, Pugnetti and Tartari, 1987).

The subalpine lakes, that is, Lake Maggiore, Lugano, Como, Iseo, and Garda, are large bodies of water located on the lee side of the Alps, some 300 m a.s.l., north of the central part of the Po Valley which is the most industrialized and populated part of Italy. For this reason, they are often subjected to highly acidic precipitation. However, the rock is mainly cal-careous and thus brings about the natural liming of the water bodies. The alkalinity increases from west to east, for example 0.8 meq L^{-1} for Lake Maggiore and 2.1 meq L^{-1} for Lake Garda.

The buffering properties of the Dolomites guarantee the survival of the small mountain lakes on the eastern part of the Alps. Other very small lakes exist at high altitudes in the central and western part of the Alps, consisting almost entirely of granitic and metamorphic cristalline rocks and characterized by small watersheds. Field tests carried out on 320 small alpine lakes have shown that 10 of these had a pH ranging between 5.5 and 6.0 and the total alkalinity is less than 0.2 meq L^{-1}, which means that they risk acidification. Only 5 had a pH lower than 5.6. Some are situated at the extreme reaches of the valleys that form part of the Alpine chain and lie along the same line as some of the larger subalpine lakes, thus being contaminated by the same pollutants originating in the Po Valley. The state of some small lakes in the central area of the Alpine chain near Lake Maggiore (i.e., Lakes Delio, Cavalli, Devero, and Toggia, between 1500 and 2160 m a.s.l.) is worrisome because of the relatively low pH value of precipitation associated with the disturbances which penetrate the zone. However, the conclusion was that the alpine lakes were not greatly altered by acidification processes.

It appears that for the majority of lakes the problem is less dramatic because of the following factors: (1) the low ratio between catchment area and water body volume; (2) the lower precipitation amount; (3) the less acidic precipitation at the higher altitudes; and (4) the wind-borne Saharan dust or other buffering particles from nearer rocks. For instance, at Arabba, in the eastern part of the Alps, precipitation during 1984 was always neutral or alkaline (Crespi and Monai, 1987), due to the buffering action of the dolomite and Saharan dust. An analysis of the meteorological situation during precipitation has shown that there, in 1984, 50% of the cases were associated with Saharan dust. On the western side of the Alps, due to the greater cyclonic curvature of the air masses, the frequency of trajectories from the Saharan regions is reduced by as much as 15%.

The effects of acidification on the biological life of these water bodies are under investigation (Setti, 1984). A comparison with the early measurements made by Tonolli in the 1940s (Tonolli, 1949; Tonolli and Tonolli, 1951) is in progress. However, the results obtained so far do not show any significant difference. The preliminary results of these studies, conducted by the biology department of the University of Milan, tend to exclude—both from a chemical and a biological point of view—the existence of any chronic acidification processes due to atmospheric inputs. The number of phytoplankton, zooplankton, and benthos species and their ecosystem are characteristic of oligotrophic, slightly acidic, cold water lakes without fish (Garibaldi, Mosello, Rossaro, and Setti, 1987). Further information is expected from a paleolimnologic survey of the lake sediments.

In order to obtain a complete scenario of the state of all the Italian lakes, including shallow lakes in the process of becoming turbid (with the consequent release of humic acids) and the volcanic lakes (where, as well as the rocks being poor buffering agents, there may also be endogenous acidic emissions) a joint ENEL-MAF program has been initiated (Pagliari, 1986).

C. The Impact of Melting Snow

During the winter, the Alps are covered by a thick mantle of snow and constitute the habitat of many rare species, some of which are on the way to extinction. It is well known that at high latitudes during the spring thaw the release of acids from the snowpack can result in acidic shock to the underlying soil and aquatic biota. The first meltwaters are often acidic because, during percolation, the acidic ions are released more quickly due to their greater mobility (Bjarnborg, 1983; Cadle Dasch, and Grossnickle, 1984a, 1984b; Oden, 1976; Tranter et al., 1986). However, the situation is not the same in the Alps, as the climate is different from that of the Scottish highlands, the Swedish mountains, or northern Michigan.

For two consecutive winters, a study of the snow mantle in the eastern Alps was made at S. Vito di Cadore (1000 m a.s.1.), at the altitude where the characteristic alpine forests of silver fir, Norway spruce, and larch grow. During the whole winter, in the early hours of the afternoon, the snow was subject to partial melting. Only in the depths of winter, and only for a very short period, did the air temperature remain below melting point. The frequent fusion of part of the snow mantle and some rainfall facilitated the migration of various types of ions, thus causing a progressive loss of acidic components by percolation. In addition, a natural liming results from the wind-borne transport of dolomitic particles or Saharan dust. Certainly these natural phenomena do not eliminate all the problems linked to the spring impact on the mountain ecosystem, but they do effectively reduce the adverse affects.

D. Monuments

In Europe, monuments have suffered more damage over the last 40 years than seen ever before. The question immediately arose as to whether this damage was in any way related to the high levels of pollution, although the exact mechanisms involved were not as easily determined. SO_2, for example, may be changed into sulphuric acid in the presence of catalysts and damage the marble, especially when its surface is wetted (Brimblecombe, 1978). This general hypothesis is correct; however, when applied to the monuments, it did not explain why some parts had deteriorated more than others even though the whole monument had been exposed to the same concentration levels of pollutants. The position of the black crusts is not even determined by the colder zones, where the deposition is greater and where the wetness time is longest. Laboratory tests have shown that pollution levels must be very high and associated with SO_2, NO_x, and high relative humidity to bring about the observed amount of damage. Moreover, no reliable quantitative relationship has yet been found between the concentration of pollutants and deterioration. In fact, in terms of physicochemical weathering, direct observation of monuments would seem to exclude condensation, the action of which is rather complex, as being primarily responsible, unlike rain (Camuffo, 1984c; 1986; 1988). Condensation may play a more important role in terms of biological life.

By reconstructing the number of foggy days and the SO_2 emissions in London from 1700, Brimblecombe (1987) showed that maximum pollution levels were reached at the end of 1800 and then declined continuously except for a fluctuation during World War II when oil began to be used in addition to coal. If deterioration and pollution levels had a linear relationship in London, this reconstruction would therefore (1) move the maximum rate of deterioration back in time some 50 to 60 years and (2) result in a present deterioration rate similar to the first half of the nineteenth century; however, both points are refuted by the evidence. The graph elaborated by Brimblecombe seems, on the contrary, to point to a relationship between recent deterioration and the new type of fuel, that is, oil. It could be deduced that there was a qualitative rather than quantitative factor involved in the new type of emissions which seem to exert a stronger catalytic and aggressive activity, mainly due to the black carbonatic particles (Bacci et al., 1983; Camuffo, Del Monte, and Sabbioni, 1983; Camuffo, Del Monte, Sabbioni, and Vittori, 1982; Del Monte, Braga Marcazzan, Sabbioni, and Ventura, 1984; Del Monte and Sabbioni, 1984; Del Monte, Sabbioni, Ventura, and Zappia, 1984; Del Monte, Sabbioni, and Vittori, 1981, 1984; Del Monte, Sabbioni, and Zappia, 1986).

The acidic rain hypothesis was proposed as an improvement on the increased pollution hypothesis (or substitution for the mechanism proposed by it). The main effect of airborne pollutants is, therefore, to acidify rainwater. The mechanism involved in the deterioration of marble monuments is thought to be the following: (1) atmospheric gases dissolved in rainwater result in a chemically active solution; (2) on the monument this forms a crust, transforming part of the calcite into gypsum; (3), the next rainfall removes this crust, which is more soluble (EPA, 1980).

Careful field observations have shown that, in the zones exposed to runoff, gypsum crystals were only rarely found, whereas a thin layer of spatic calcite crystals due to reprecipitation was common. On the other hand, maximum gypsum, which forms the black crusts, was found where runoff was prevented (see Figure 9–8). The acidic rain hypothesis was not sufficient either to explain the main phenomenon due to sulphation, that is, the black crusts, which are one of the main problems affecting marble monuments. By the way, the gypsum crusts due to simple sulphation by acidic rainwater should be whitish, not black.

The weathering of marble and limestone monuments was extensively studied by two CNR Institutes: ICTR and FISBAT. It has been determined (Bernardi, Camuffo, Del Monte, and Sabbioni, 1985a; Bernardi et al., 1985b; Camuffo, 1984b, 1986; Camuffo, Del Monte, and Sabbioni, 1983, 1987; Camuffo, Del Monte, Sabbioni, and Vittori, 1982; Del Monte and Vittori, 1985) that the visual features of the deterioration patterns could be related to the three main ways the water wetted the surface: (1) runoff was associated with "white areas" in the zones where the surface has been worn away and is covered with reprecipitated crystals of spatic calcite; (2) percolation and windborne droplets in zones where runoff was absent were associated with "black areas" where black crusts dominated; (3) absence of runoff, percolation, and windborne droplets, but possible condensation, was associated with "gray areas" where chemically the stone was not affected but covered instead by a layer of dust and particles.

Figure 9–8. Black crusts form in the zones prevented from runoff but wetted by rainfall. Runoff washes and dissolves the surface, thus resulting in the white area. This pattern is typically due to the geometry of the surface, so that it appears frequently on the faces of statues (Costantino Arch in Rome).

Carbonaceous particles (formed during the combustion of oil in particular) are very active in forming black crusts when they are wet. They contain sulphur compounds and catalysts, can absorb SO_2 from the atmosphere, and enucleate gypsum crystals when wet, so that they are embedded in a gypsum crust. The gypsum results partly from the transformation of the underlying rock, and is due partly to particle enucleation. All the black crusts examined, samples of which were taken from many European countries, contained carbonaceous particles, and it has been possible to reproduce this mechanism in the laboratory. The black crusts induce severe damage because they form an aggressive solution whenever they are wet. In comparison with this solution, wet deposition may be much less aggressive in many cases.

The white areas are subjected to dry deposition, but the runoff washes the monument surface and in general removes the main part of the attached pollutants. Their action is modest but is, however, continued by the falling rainwater. The result is a modest wearing of the surface that is only slightly dissolved. Calcite, which is less soluble than gypsum, may reprecipitate more abundantly on the surface, where the net flow of mass is negative.

The dry deposition that accumulates between two successive rainfalls cannot be ignored, especially in a typical Mediterranean climate, which may be subject to long dry periods. Under such conditions and in an urban environment, dry deposition is much greater than wet deposition. Modest drizzle can activate the dry deposition without removing it and can cause much more severe weathering than in the case of showers that wash the monuments. These findings (Camuffo, Del Monte, and Ongaro, 1984) are supported also by Brocco, Rotatori, and Tappa (1986) who compared the aggressivity of aerosols and rain and concluded that the acidic fraction in aerosols was generally much greater than that in rain of comparable electrical conductivity.

The dynamical regime of rainwater running over the monument is also an important factor. When the regime is laminar, the first layer in contact with the surface becomes saturated and is then chemically unreactive. When it is turbulent, the chemically reactive solution continually changes on the stone surface, which is subjected to heavy dissolution.

In conclusion, the pH of rainfall is not a representative parameter in the weathering of monuments. The problem seems to lie in the bulk deposition, both dry and wet, on the monument and how it is wetted. When the dry deposition is activated, the important factors are its time of residence on the monument and the way in which it is removed, that is, by a laminar or turbulent flow. When black carbonaceous particles are embedded in the monument, that is, the black crusts, they form a very aggressive solution whenever soaked with abundant rainwater, irrespective of the original pH of the rain.

VI. A Policy for Conservation of the Environment and Cultural Heritage

During the conference held in Amsterdam on acidification and what policy to follow, it appeared that many nations have similar problems, but very often special studies are needed to achieve general knowledge and resolve specific aspects. International cooperation is needed to take necessary experimental data and to interpret it. It was underlined that

> To enlarge the basis for the verification of nationally adopted measures to reduce emissions of relevant air pollutants and to broaden knowledge concerning their effects on the most endangered components of the environment, in particular forests, lakes and other ecosystems, materials, historical monuments and other cultural establishments, permanent integrated monitoring stations should be supported. To support the further elaboration of policy measures to reduce emissions, the interaction between scientists and policy makers should, for instance, be intensified (Final Report in Schneider, 1986, p. 340).

A very important point for both the cultural life and economy of Italy is the fate of its monuments. Carbonaceous particles are recognized as playing a very important role in the formation of black crusts. However, it has been also demonstrated that deposition alone is not an index of deterioration. The same monument zones exposed to greater deposition are less deteriorated than zones exposed to minor deposition, but exposed to the action of certain meteorological factors that are capable of triggering physicochemical processes (Bernardi, Camuffo, Del Monte and Sabbioni, 1985).

It is impossible to evaluate the damage to the monuments, as the damage is not proportional to their deterioration. The term *damage* includes an evaluation of several factors of a different nature and could be obtained by multiplying the deterioration by the not well-defined intrinsic value of the monument. For instance, the modest deterioration of the beautiful facade of the Orvieto Cathedral constitutes much more serious damage than the more intense deterioration found on the rough sides (Camuffo and Bernardi, 1988). However, if the deterioration of monuments is used as an invitation to people to look at the monument "for the last time," as in the case of Venice, then the actual damage is transformed into a present-day business, which then become a tremendous loss for future generations. These facts show how complex the problem is, and not only from the scientific point of view. It is hoped that people will arrive to visit our magnificent works of art in order to enjoy them for their actual appearance, and not just out of curiosity to look at them for the last time.

Reducing the emissions in order to reduce the negative impact of humans on their environment and on themselves should, clearly, be the prime target of a civilized society. Our only hope lies in a real concern for the future accompanied by a serious commitment to reduce the unfavorable conditions that humans themselves have created. It is unlikely that these conditions can be completely eliminated given that an inordinate amount of work would be necessary. However, by moving gradually and effectively in the political, technical, and scientific fields much could be accomplished. Science could help reduce the effects while solutions to the causes are being sought, especially in those areas where the causes are not yet clearly understood. The aim, therefore, of this brief review has been limited to underlining this objective.

Acknowledgments

Several of the studies mentioned in this review were supported by different organizations. The author acknowledges the Commission of the European Communities (CEC), DG XII, Environmental Research Program for Science Research and Development, "Conservation of Historic Buildings and Monuments" (contracts ENV 757/I/SB and EV4V-0051-I-A) and "Climatology and Natural Hazards" (contract EV4C-0082-I); the National Research Council of Italy (CNR), Research Finalized Project "Energetics," and Strategic Project "Climate and Environment"; the National Electricity Board—Thermal and Nuclear Research Center (ENEL-CRTN). Special thanks are due to Dr. D. Anfossi (CNR-Cosmogeofisica); Dr. P. Bacci (ENEL-CRTN); Dr. D. Brocco (CNR-IIA); Col. L. Ciattaglia (IAF); Dr. S. Fuzzi (CNR-FISBAT); Prof. R. Gellini (University of Florence); Prof. A. Longhetto (University of Torino); Prof. G. Lorenzini (University of Pisa); Dr. G. Schenone (ENEL CRTN); and Dr. F. Setti (University of Milan) for having kindly supplied updated information about their work. The author is also grateful to his colleague Dr. A. Bernardi for her suggestions. A publication with Dr. M. Zanetti (Bioprogram) on the acidic impact due to snowmelt is underway.

References

Allegrini, I., P. Buttini, V. DiPalo, and M. Possanzini. 1987. Preparation of Standard Atmospheres of Nitrogen Acid Compounds: the NO_2 Permeation Tube. In G. Angeletti and G. Restelli (eds.), *Physico-Chemical Behaviour of Atmospheric Pollutants,* 15–24. Dordrecht: Reidel.

Alessandrini, A. 1987. Foreste e Inquinamento. In Collegio degli Ingegneri di Milano (ed.). *Le precipitazioni acide e il controllo dell'atmosfera,* 95–99. Milan: Clup.

Anfossi, D., P. Bacci, P. Natale, and S. Viarengo. 1986. *Sci Total Environ* 55:329–338.

Aru, A. 1986. Aspects of Desertification in Sardinia-Italy. In R. Fantechi and N.S. Margaris (eds.), *Desertification in Europe*, 194–198. Dordrecht: Reidel.

Bacci, P., M. Del Monte, A. Longhetto, A. Piano, F. Prodi, P. Radaelli, C. Sabbioni, and A. Ventura. 1983. *J Aerosol Sci* 14:557–572.

Badalamenti, F., M. Carapezza, G. Dongarra, S. Hauser, A. Macaluso, and F. Parello. 1984. *Rendic Soc It Mineralogia Petrografia* 39:81–92.

Benarie, M., and P. Detrie. 1978. Assessment of an OECD Study on Long Range Transport of Air Pollutants. In M. Benarie (ed.), *Atmospheric Pollution, Studies in Environmental Science*, Vol. 1, 207–215. Amsterdam: Elsevier.

Bernardi, A., D. Camuffo, M. Del Monte, and C. Sabbioni. 1985a. *Sci Total Environ* 46:243:260.

Bernardi, A., D. Camuffo, M. Del Monte, C. Sabbioni, S. Vincenzi, and G. Zappia. 1985b. Weathering of Bronze and Stone Monuments as a Result of Both Atmospheric Pollution and Precipitation. In Commission European Communities (CEC) (ed.), *The effects of air pollution on historic buildings and monuments*, III 1–16. CEC XII/ENV/3/86; Bruxelles.

Bernardi, A., D. Camuffo, A. Del Turco, D. Gaidano, and I. Lavagnini. 1987. *Sci Total Environ* 63:259–270.

Bjarnborg, B. 1983. *Hydrobiologia* 101:19–26.

Borghi, S. and M. Giuliacci. 1979. *Circolazione atmosferica nella Valpadana Centro Occidentale e suo impatto nel trasporto delle particelle*. Milan: Osservatorio Meteorologico di Brera.

Bottacci, A., L. Brogi, F. Bussotti, E. Cenni, F. Clauser, M. Ferretti, R. Gellini, P. Grossoni, S. Schiff. 1988. *Inquinamento ambientale e deperimento del bosco in Toscana*. Florence: Regione Toscana. 134 pp.

Bottini, O. 1939. *Ann Chim Applic* 29:425–433.

Brimblecombe, P. 1978. *Tellus* 30:151–157.

Brimblecombe, P. 1987. *The big smoke*. London: Methuen. 185 pp.

Brocco, D., M. Rotatori, R. Tappa. 1986. *Sci Total Environ* 54:261–273.

Bussotti, F., C. Rinallo, P. Grossoni, R. Gellini, F. Pantani, and S. Del Panta. 1983. *La Provincia Pisana* 4:46–52.

Cadle, S.H., J.M. Dasch, and N.E. Grossnickle. 1984a. *Water Air Soil Pollut* 22: 303–319.

Cadle, S.H., J.M. Dasch, and N.E. Grossnickle. 1984b. *Atmos Environ* 18:807–816.

Camuffo, D. 1981. *Atmos Environ* 15:1543:1551.

Camuffo, D. 1984a. *Climatic Change* 6:57–77.

Camuffo, D. 1984b. *Atmos Environ* 18:2273–2275.

Camuffo, D. 1984c. *Water Air Soil Pollut* 21:151–159.

Camuffo, D. 1986. Deterioration Processes of Historical Monuments. In T. Schneider (ed.), *Acidification and Its Policy Implications*, 189–221. Amsterdam: Elsevier.

Camuffo, D. 1988. *European Cultural Heritage Newsletter* 2, (5):6–10.

Camuffo, D., and A. Bernardi. 1988. *Sci Total Environ* 68:1–10.

Camuffo, D., A. Bernardi, and P. Bacci. 1982. *Boundary Layer Meteorol* 22:503–510.

Camuffo, D., A. Bernardi, A. Ongaro, P. Bacci, and A. Novo, 1988. In *Transporto transfrontaliero di inquinanti atmosferici e stato dell'ambiente in zona alpina*. (Symposium in Brixen), 1–32. University of Padova, Padova.

Camuffo, D., A. Bernardi, and M. Zanetti. 1988. *Sci Total Environ* 71:187–200.

Camuffo, D., M. Del Monte, and A. Ongaro. 1984. *Sci Total Environ* 40:125–140.
Camuffo, D., M. Del Monte, and C. Sabbioni. 1983. *Water Air Soil Pollut* 19:351–359.
Camuffo, D., M. Del Monte, C. Sabbioni. 1987. *Bollettino d'Arte*, special issue "Materiali Lapidei" 15–36.
Camuffo, D., M. Del Monte, C. Sabbioni, and A. Moresco. 1984. In *Sep Pollution* 84. (Workshop in Padova), 227–248. Fiere di Padova, Padova.
Camuffo, D., M. Del Monte C. Sabbioni, and O. Vittori. 1982. *Atmos Environ* 16:581–587.
Cantù, V. 1977. The Climate of Italy. In H.E. Landsberg (ed.), *Climates of Central and Southern Europe*, 127–173. Amsterdam: Elsevier.
Caporaloni, M., F. Tampieri, F. Trombetti, and O. Vittori. 1975. *J Atmos Sci* 32:565–568.
Ciattaglia, L. 1975. *Riv Met Aer* 35:236–239.
Ciattaglia, L. 1979. *Riv Met Aer* 39:175–177.
Ciattaglia, L. 1981. *Riv Met Aer* 41:49–57.
Ciattaglia, L., and L. Cruciani. 1977. *Riv Met Aer* 37:323–325.
Ciattaglia, L., and G. Fiore. 1981. *Riv Met Aer* 41: 25–31.
Colacino M., and R. Purini. 1986. *Theor Appl Climatol* 37, 90–96.
Comitato per la Carta dei Suoli (CCS). 1966. *La carta dei suoli d'Italia.* Florence: Coppini.
Commission of the European Communities (CEC). 1983. *Degradations causées aux forets allemandes par les impuretés de l'air.* Bruxelles.
Cortecci, G., and A. Longinelli. 1970. *Earth Planetary Sci Letters* 8:36–40.
Crespi, M., and M. Monai. 1987. In *Contributo meteorologico alla problematica delle piogge acide sulle Alpi Venete.* Dip For QR.9, Regione Veneto, Venice. 47 pp.
Dall'Aglio, M. 1984. In *Precipitazioni acide in Italia,* 245–252. CNR PFE SC-12, Rome.
Defant, F. 1951. Local Winds. In American Met Soc (ed.), *Compendium of Meteorology,* 655–671. Boston: American Meteorological Society.
Delmas, R.J., and G. Gravenhorst. 1983. Background Precipitation Acidity In S. Beilke and A.J. Elshout (ed.), *Acid Deposition,* 84–109, Dordrecht: Reidel.
Del Monte, M., G.M. Braga Marcazzan, C. Sabbioni, and A. Ventura. 1984. *J Aerosol Sci* 15:323–327.
Del Monte, M., and C. Sabbioni. 1984. *Arch Met Geoph Biocl B* 35:93–104.
Del Monte, M., C. Sabbioni, and O. Vittori. 1981. *Atmos Environ* 15:645–652.
Del Monte, M., C. Sabbioni, and O. Vittori. 1984. *Sci Total Environ* 36:369–376.
Del Monte, M., C. Sabbioni, A. Ventura, and G. Zappia. 1984. *Sci Total Environ* 36:247–254.
Del Monte, M., C. Sabbioni, and G. Zappia. 1986. *Sci Total Environ* 50:147–163.
Del Monte, M., and O. Vittori. 1985. *Endeavour* 9:117–122.
De Rossi, C., and G. Gisotti. 1981. *Il Dottore in Scienze Agrarie e Forestali,* 31:17–25.
Dickson, W. 1986. Acidification Effects in the Aquatic Environment. In T. Schneider (ed.), *Acidification and Its Policy Implications,* 19–28. Amsterdam: Elsevier.
ENEL. 1989. *1988 Produzione e consumo di energia elettrica in Italia.* Rome: ENEL Report.

Environmental Protection Agency (EPA). 1980. *Acid rain.* Report EPA-600/9-79-036. US-EPA, Washington, DC 36 pp.

Eredia, F. 1911. *Il clima di Roma.* Roma: Bertero. 101 pp.

Facchini, M.C., G. Chiavari, and S. Fuzzi. 1986. *Chemosphere* 15:667–674.

Fuzzi, S. 1986a. In W. Jaeschke (ed.), *Chemistry of Multiphase Atmospheric Systems* (NATO AISI Vol G6) 213–226. Berlin: Springer Verlag.

Fuzzi, S. (ed.). 1986b. *Heterogeneous atmospheric chemistry project.* CNR-FISBAT HACP/1, Bologna.

Fuzzi, S. (ed.). 1987. *Heterogeneous atmospheric chemistry project.* CNR-FISBAT HACP/2, Bologna.

Fuzzi, S., M. Gazzi, C. Pesci, and V. Vicentini. 1980. *Atmos Environ* 14:797–801.

Fuzzi, S., M. Mariotti, and G. Orsi. 1982. *Sci Total Environ* 23:361–368.

Fuzzi, S., G. Orsi, and M. Mariotti. 1983. *J Aerosol Sci* 14:135–138.

Fuzzi, S., G. Orsi, and M. Mariotti. 1985. *J Atmos Chem* 3:289–296.

Fuzzi, S., G. Orsi, G. Nardini, M.C. Facchini, S. McLaren, E. McLaren, and M. Mariotti. 1988. *J Geophys Res* 93:11141–11151.

Ganor, E., and Y. Mamane. 1982. *Atmos Environ* 16:581–587.

Gargano, Imperato E. 1963. *Ann Fac Sci Agr Univ Napoli,* ser 3, 29:369–378.

Garibaldi, L., R. Mosello, B. Rossaro, and F. Setti. 1987. *Docum Ist Ital Idrobiol* 14:165–180.

Gazzola, A. 1978. *Distribuzione ed evoluzione delle temperature e delle precipitazioni in Italia in relazione alle situazioni meteorologiche.* Rome: CNR-IFA report SP13, 155 pp.

Gellini, R., F. Pantani, F. Bussotti, and E. Raccanelli. 1981. *Inquinamento* 10:27–30.

Gellini, R., F. Pantani, P. Grossoni, F. Bussotti, E. Barbolani, and C. Rinallo. 1985. *Eur J For Path* 15:145–157.

Georgii, H.W. 1982. Global Distribution of the Acidity in Precipitation. In H.W. Georgii and J. Pankrat (eds.), *Deposition of Atmospheric Pollutants,* 55–56. Dordrecht: Reidel.

Gisotti, G. 1984. In *Precipitazioni Acide in Italia,* 215–244. CNR PFE SC-12, Rome.

Guidi, L., G. Lorenzini, and G.F. Soldatini. 1986. *Agricoltura Ital* 5/6:55–65.

Ioannilli, E., and P. Bacci. 1986. In 79th Annual Meeting of the Air Pollution Control Association (APCA). Paper 34.2. Minneapolis, MN.

Ioannilli, E., M. Bettinelli, A. Franciotti, and G.L. Ziliani. 1987. A Study of Atmospheric Deposition Chemistry in Peninsular and Insular Italy. In G. Angeletti and G. Restelli (ed.), *Physico-chemical Behaviour of Atmospheric Pollutants,* 166–175. Dordrecht: Reidel.

Italian Delegation. 1986. Presentation by the Italian Delegation. In T. Schneider (ed.), *Acidification and Its Policy Implications,* 431–433. Amsterdam: Elsevier.

Junge, C.H., and G. Scheich. 1969. *Atmos Environ* 3:423–441.

Longinelli, A., and M. Bartelloni. 1978. *Water Air Soil Pollut* 10:335–341.

Lorenzini, G. 1983. *Le piante e l'inquinamento dell'aria.* Bologna: Edagricole. 359 pp.

Lorenzini, G., and A. Materazzi. 1984. *Riv Patologia Vegetale* 20:108–122.

Lorenzini, G., A. Mimmack, and M.R. Ashmore. 1985. *Riv Patologia Vegetale* 21:13–26.

Lorenzini, G., and A. Panattoni. 1986. *Riv Ortoflorofrutt It* 70:215–229.

Lorenzini, G., A. Panattoni, and G. Schenone 1988. *Water Air Soil Pollut* 42:47–56.

Mandrioli, P., G.L. Puppi, N. Bagni, and F. Prodi. 1973. *Nature* 246:416–417.

Mastino, G., L. Testa, I. Michetti, and E. Ghiara. 1987. *Acqua Aria,* 1:17–33.

Melicchia, A. 1939. *Variazioni climatiche nella pianura Padana e loro rapporti col regime del Po.* Bologna: Zanichelli. 142 pp.

Mennella, C. 1956. *L'andamento annuo della pioggia in Italia nelle osservazioni ultrasecolari.* Bologna: Mareggiani. 248 pp.

Mennella, C. 1967. *Il clima d'Italia,* vol 1. Napoli: Conte. 718 pp.

Meteorological Office. 1962. Weather in the Mediterranean, vol. 1, *General Meteorology.* MO 391/b, HMSO, London. 362 pp.

Mezzalira, G. 1987. *Airone* 78:134–135.

Minerbi, S. 1986. *L'influenza del clima recente sulla vegetazione in Alto Adige.* Provincia Autonoma di Bolzano, Alto Adige. 71 pp.

Mosello, R. 1984. *Schweitz Z Hydrol* 46:86–99.

Mosello, R., and G. Tartari. 1982. *Mem Ist Ital Idrobiol* 40:163–180.

Mosello, R., and G. Tartari. 1983. *Wat Qual Bull* 8:96–100.

Mosello, R., and G. Tartari. 1984a. Precipitazioni atmosferiche, caratteristiche chimiche e conseguenze sulla qualitàdelle acque fluviali e lacustri. In M. D'Aversa, C. Monguzzi, S. Remi, and G. Schulze (eds.), *Le piogge acide,* 81–88. Milan: Angeli.

Mosello, R., and G. Tartari, 1984b. In *Precipitazioni acide in Italia,* 215–244. CNR PFE SC-12, Rome.

Mosello, R., A. Pugnetti, and G. A. Tartari. 1987. Chemistry of Atmospheric Deposition and Lake Acidification in Northern Italy, with Emphasis to the Role of Ammonia. In H. Barth (ed.), *Reversibility of Acidification,* 85–94. London: Elsevier.

NI Group. 1985. *Acqua Aria* 8:721–735.

Novo, A. 1987. *Ingegneria Sanitaria* 6:5–63.

Nucciotti, F. and N. Rossi. 1968. *Agrochimica* 12:540–548.

Oden, S. 1976. *Water Air Soil Pollut* 6:137–166.

Ott, H. 1984. In *Precipitazioni Acide in Italia,* 29–37. CNR PFE SC-12, Rome.

Pagliari, M. 1986. In *Sep Pollution* 86 (Workshop in Padova), 207–218. Padova: Fiere di Padova.

Palmieri, F. 1964. In *Ann Fac Sci Agr Univ Napoli,* ser 3, 30:3–18.

Palmieri, F. 1965. In *Boll Soc Nat Napoli* 74:3–26.

Palmieri, F. 1973. In *Ann Fac Sci Agr Univ Napoli,* ser 4, 7:3–16.

Perelli, M., and F. Franzin. 1984. In *Sep Pollution* 84 (Workshop in Padova), 307–326. Padova: Fiere di Padova.

Picotti, M. 1963. In *Atti del XIII Convegno dell'Associazione Geofisica Italiana* (Conference in Rome), 69–80, AGI, Rome.

Pinchera, G. 1984a. La Conferenza di Stoccolma e le possibili strategie di intervento. In M. D'Aversa, C. Monguzzi, S. Remi, and G. Schultze (eds.), *Le piogge acide,* 143–165. Milan: Angeli.

Pinchera, G. 1984b. In *Precipitazioni Acide in Italia,* 279–295. CNR PFE SC-12, Rome.

Possanzini, M., A. Febo, and A. Liberti. 1983. *Atmos Environ* 17:2605–2610.

Prodi, F. 1976. In International Conference on Cloud Physics (Conference in Boulder), 70–75. Boulder, CO: American Meteorological Society.

Prodi, F., and G. Fea. 1978. In *Alpine Meteorologia* (Proc 15 Intern Tagung) 179–182. Grindelwald, Switzerland: Swiss Meteorological Institute.

Prodi, F., and G. Fea. 1979. *J Geophys Res* 84:6951–6960.

Prodi, F., and C.T. Nagamoto. 1971. *J Glaciology* 10:299–308.

Prodi, F., V. Prodi, and G. Fiore. 1970. *J Appl Meteorol* 9:283–288.

Prodi, F., G. Santachiara, and V. Prodi. 1979. *J Aerosol Sci* 10:421–425.

Prodi, F., and F. Tampieri. (1982). *Pageoph* 120:286–325.

Pszenny, A.A.P., McIntyre, and R.A. Duce. 1982. *Geophys Res Lettres* 9:751–754.

Reiter, E. 1971. *Digest of selected weather problems of the Mediterranean.* NAVWEARSCHFAC TP9/71, Norfolk, VA.

Reiter, E. 1975. *Handbook for forecasters in the Mediterranean.* NAVWEARSCHFAC TP 5/75, Norfolk, VA.

Rendell, H.M. 1986. In R. Fantechi and N.S. Margaris (eds.), *Desertification in Europe,* 184–193. Dordrecht: Reidel.

Rossi, N., and F. Nucciotti. 1968. In *Agrochimica* 12:240–250.

Rossi, N., and F. Nucciotti. 1974. In *Agrochimica* 18:454–462.

Sandroni, S. 1984. In *Sep Pollution* 84 (Workshop in Padova), 103–114. Padova: Fiere di Padova.

Santomauro, L. 1957. *Lineamenti climatici di Milano.* Comune di Milano. 251 pp.

Schenone, G., and L. Mignanego. 1988. *Acqua Aria* 9:1085–1090.

Schneider, T. (ed.). 1986. *Acidification and Its Policy Implications.* Amsterdam: Elsevier.

Sequeira, R. 1982. *J APCA* 32:241–245.

Setti, F. 1984. In *Precipitazioni Acide in Italia,* 303–325. CNR PFE SC-12, Rome.

Sordelli, D., and F. Zilio Grandi. 1972. In *Antinquinamento* 72 (Congress in Milan), Milan.

Tartari, G., and R. Mosello. 1984. Caratteristiche chimiche delle acque di pioggia. In M. D'Aversa, C. Monguzzi, S. Remi, and G. Schultze (eds.), *Le piogge acide,* 45–59. Milan: Angeli.

Tomasi, C., R. Guzzi, and O. Vittori. 1975. *J Atmos Sci* 32:1580–1586.

Tomasi, C., F. Prodi, and F. Tampieri. 1979. *Atmos Physics* 52:215–228.

Tonolli, V. 1949. *Mem Ist Ital Idrobiol* 5:39–93.

Tonolli, V., and L. Tonolli. 1951. *Mem Ist Ital Idrobiol* 6:53–163.

Tranter, M., P. Brimblecombe, T.D. Davies, C.E. Vincent, P.W. Abrahams, and I. Blackwood. 1986. *Atmos Environ* 20:515–525.

Viarengo, S., R. Mosello, and G. Tartari, 1984. In *Sep Pollution* 84 (Workshop in Padova), 211–226. Padova: Fiere di Padova.

Vittori, O. 1973. *J Atmos Sci* 30:321–324.

Vittori, O. 1984. *Nuovo Cimento* 7C:254–269.

Vittori, O., F. Prodi, G. Morgan, and G. Cesari. 1969. *J Atmos Sci* 26:148–152.

Vittori, O., and V. Prodi. 1967. *J Atmos Sci* 24:533–538.

Winiwarter, W., H. Puxbaum, S. Fuzzi, M.C. Facchini, G. Orsi, N. Beltz, K. Enderle, W. Jaeschke. 1988. *Tellus* 40B:348–357.

Winkler, E.M. 1976. *Water Air Soil Pollut* 30:873–878.

Development of Forest Damage and Air Pollution in Switzerland and Initiation of a Relevant National Research Program

H. Turner*

Abstract

The general situation and development of forest damage and air pollution in Switzerland is outlined, and a survey is given of a multidisciplinary research program which critically examines possible causal relationships between ambient conditions (air and soil pollution in particular, as well as natural ecological factors) and the present large-scale forest decline. Some 20 research groups from 9 universities and federal research institutes are participating in an experimental program which started in 1985 and concerns three experimental plots in the central Alps (Davos, 1640 m a.s.l.), in the Pre-Alps (Alptal, 1180 m), and in the northern foreland of the Alps (Lägeren, 680 m). Initial results of some projects are presented.

I. Introduction

In Switzerland the influence of pollutants on the physiology and growth of forest trees has been under investigation since 1964 (T. Keller, 1976, 1982). However, in the early days only local emittents and their detrimental effects on nearby forests were taken into consideration, although "latent" (nonvisible) damage by emissions of low concentrations was carefully investigated as well (e.g., T. Keller, 1977, 1978).

In the meantime, especially since 1983, widespread forest decline in Europe and North America seems to have reached portions of the northern hemisphere (Bucher, 1987). This "new" (large-scale) phenomenon was attributed by researchers primarily to anthropogenic air pollution. Nevertheless, at present the widespread degradation of forests is still in-

*Swiss Federal Institute for Forest, Snow and Landscape Research (FSL), CH-8903 Birmensdorf, Switzerland.

sufficiently explained, and a multiplicity of causes seems to be responsible. In any case, forest decline exhibits different symptoms in various regions, and only in a restricted number of regions can the damage be attributed satisfactorily to air pollutants (Cowling, 1985; Rehfuess, 1987). Other detrimental influences, for example, climatic, edaphic, or biotic factors, must also be taken into consideration and seem to play a major part in some regions.

Sometimes it is suspected that electromagnetic waves of high frequency might bring about forest damage. Investigations into the foliage density and radial growth of four tree species in the neighborhood of ultra-short wave transmitters in areas of high and low electromagnetic field strength could not detect such detrimental effects (Joos, Masumy, Schweingruber, and Staeger, 1988).

Forest decay has also a psychologic component (Bucher, 1987). For example, when in the spring of 1987 a widespread reddening of the conifers in an altitudinal belt between 800 and 1100 m a.s.l. occurred (Figure 10–1), this damage was instantly interpreted by many people as a "new form of forest decay due to air pollution," an interpretation that was mainly promoted by the mass media. Investigations by a multidisciplinary group of researchers revealed, however, that the damage was created by frequent and rapid thawing and freezing in the upper layer of an extraordinary temperature inversion zone, which occurred after an inflow of arctic/Siberian air masses (with temperatures of -20 to $-30°C$) and a subsequent development of a warm foehn current above the lake of cold air (Turner, 1987, 1988).

II. Forest Damage Situation

Forest decline in Switzerland up to now has not been manifested by the diverse types of damage that have been described from Germany, for example, yellowing of needles at high altitudes of central Germany, reddening of needles in old stands at low altitudes in southern Germany, and so on (Rehfuess, 1987). In Switzerland a yellowing of the foliage occurs only on an insignificant fraction of the area (Bucher, 1987). The degree of crown sparseness was used as an indication of forest damage. The great problem is, however, to define which amount of foliage or degree of crown sparseness is normal with respect to site and climatic period. In any case, from 1984 until 1987 the percentage of "damaged" trees in Swiss forests increased from 34% to 56%, where a "damaged" tree is defined as having lost 15% or more of its foliage (Figure 10–2).

Some researchers suggest that an actual damage threshold may be as high as 40% or 50% crown transparency, claiming that trees below that threshold may recover rather easily. Applying a limit of 40% crown transparency, only some 7% of broad-leaved trees and 11% of the conifers are

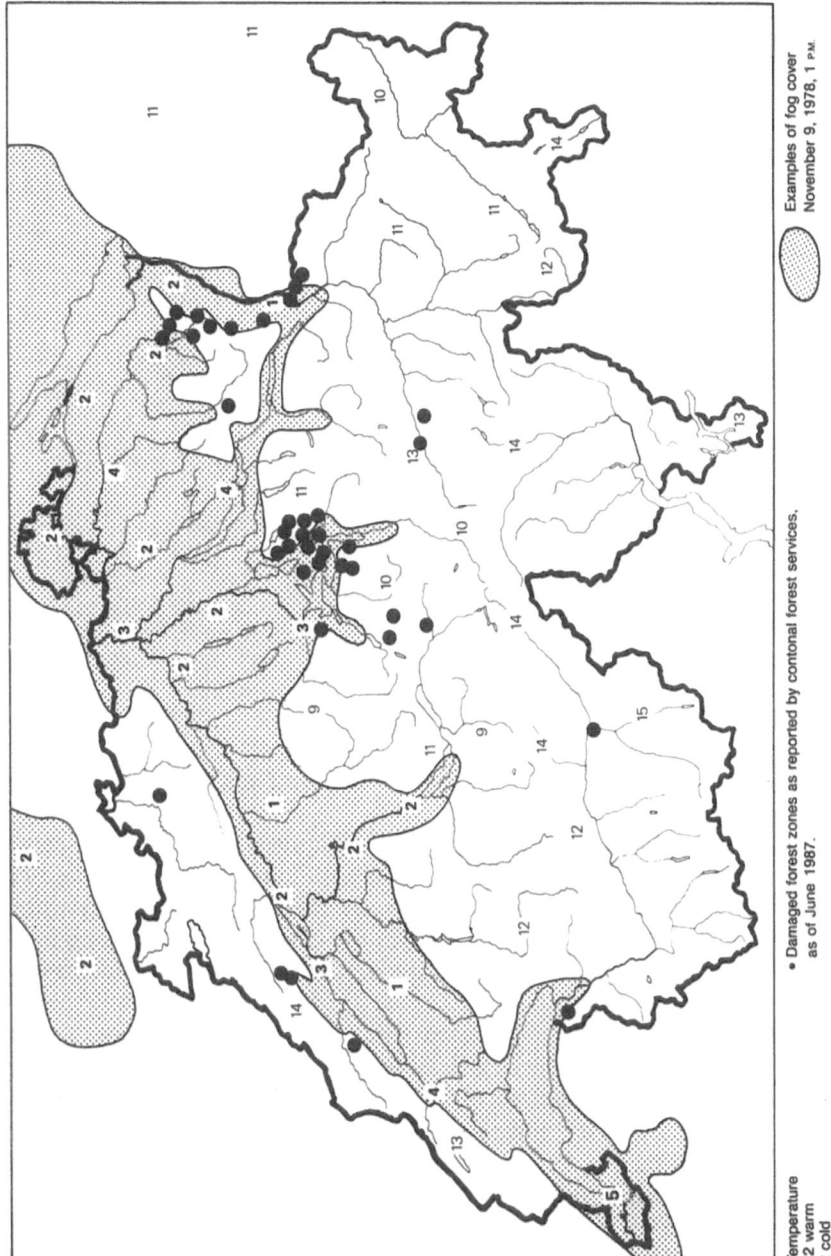

Figure 10–1. Winter frost damage of evergreen conifers. Reddening of needles was recorded most frequently in an altitudinal belt within the upper layers of fog and temperature inversion (800 to 1100 m a.s.l.) in combination with foehn. (From Turner, 1987.)

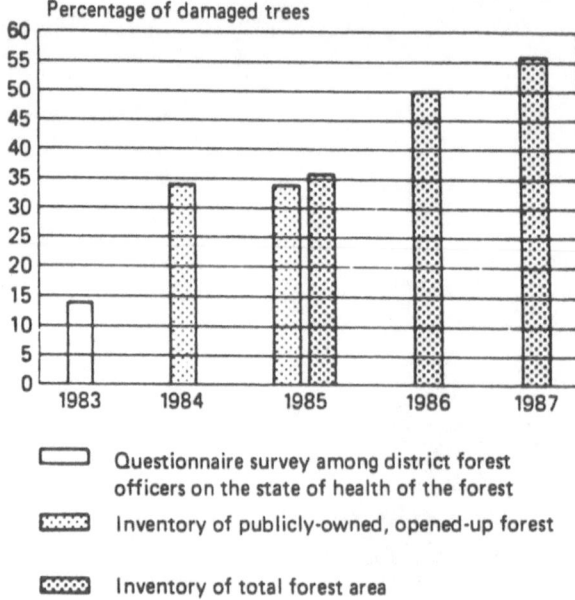

Percentage of damaged trees

Figure 10–2. Development of percentage of damaged trees (all tree species) according to the forest damage assessments, 1983 to 1987 (from Mahrer, 1987). Definition of a "damaged" tree: loss of foliage 15% or more.

classified as damaged; thus the extent of forest decline would appear rather small.

In the mountain areas the frequency of "damaged" trees is considerably higher (60%) than in the lowlands (48%) (Figure 10–3). The question arises whether the same assessment criteria may be applied over a whole country. There is some evidence that in the central Alps with a high degree of hygric continentality of climate, part of the needle loss has already existed for a long time.

Only recently could it be shown that the degree of foliage loss is closely correlated with the degree of radial growth reduction in all tree species, most pronounced, however, in conifers (Mahrer, 1987). For instance, conifers with a foliage loss of 15% to 25% show an average radial growth reduction of 15% (Norway spruce) and even 24% (fir) (Bräker, 1987). Furthermore, dendrochronological investigations have revealed that within the last 20 to 30 years, trees that exhibit at present a considerable foliage loss have suffered abrupt and long-lasting growth reductions as a consequence of climatic events (e.g., summer drought or severe late frost) (Bräker, 1987).

Recent studies have revealed that, in some cases, areas with high thinning have a distinctly higher foliage loss in dominant trees than areas with low thinning or selection management (Figure 10–4).

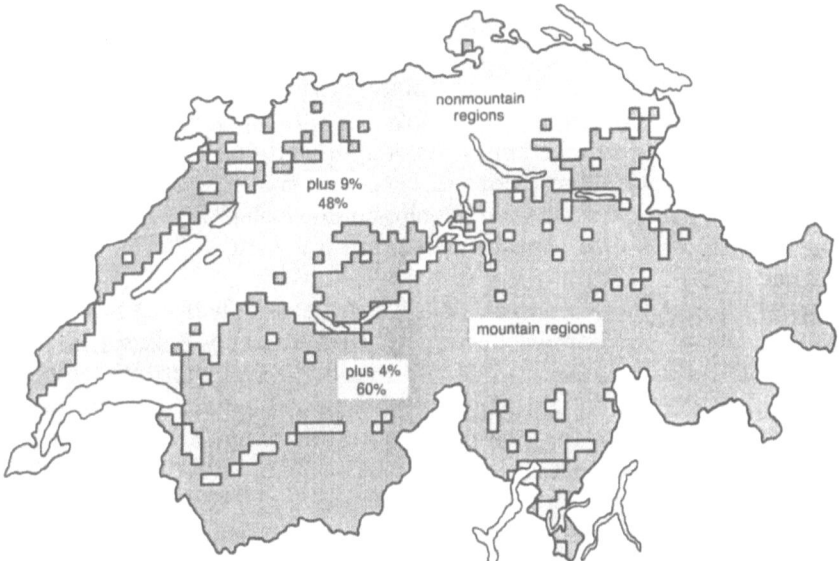

Figure 10–3. Development of forest damage from 1986 to 1987 and percentage of damaged trees as of 1987 in the mountain and nonmountain regions of Switzerland (from Mahrer, 1987). Definition of a "damaged" tree: loss of foliage 15% or more.

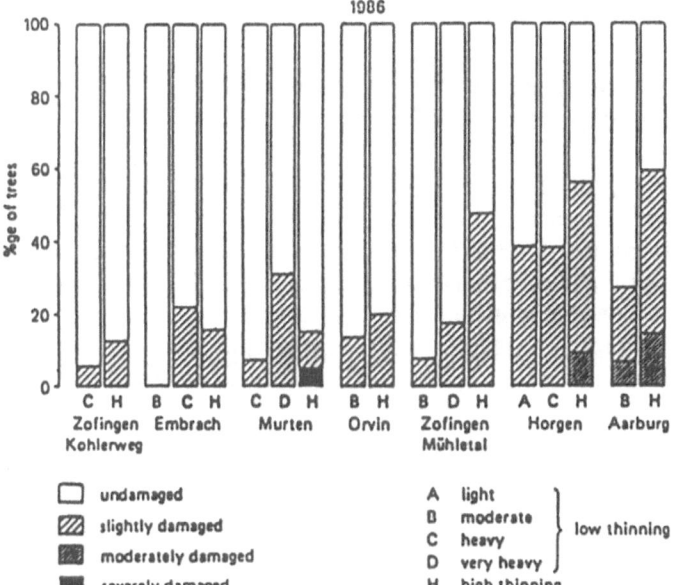

Figure 10–4. Distribution of damage classes in dominant beeches in relation to type of thinning. (Keller & Imhof, 1987.)

III. Air Pollution Situation

From 1950 to the present, the emissions of nitrogen oxides (NO_x) have increased from 30,000 to approximately 250,000 t per year. A similarly large increase concerns the emissions of hydrocarbons. The development of sulfur dioxide (SO_2) emissions reached a maximum of some 140,000 t per year in 1965 and since then has slowed down slightly and irregularly. The average emission concentrations are given in Table 10-1 (from Bucher, 1987).

In addition to these average values it may be demonstrated that maximum values of ozone emissions are problematical not only for forests, as the official upper limit value (Immissions-Grenzwert der schweiz. Luftreinhalteverordnung, 100 μg m^{-3}) is surpassed again and again during the major part of the year at every station (Figure 10-5).

Anthropogenic stress is clearly reflected by the distribution maps of element contents in conifer needles. In particular, the geographical distribution of the main indicator elements S, Cl, Pb, Cd, and Fe show most severe stress in the lowlands, due to whole groups of emission sources (Landolt, Bucher, and Kaufmann, 1984). The distribution map for lead contents of spruce needles (Figure 10-6) is indicative for the phenomenon that in 1983 the areas with the greatest pollution stress had the greatest forest damage (Bucher, Kaufmann, and Landolt, 1984); however, the strong increase of the number of damaged trees in the mountains (Figure 10-3) cannot be satisfactorily explained by pollution data.

IV. A National Research Program: "Forest Damage and Air Pollution in Switzerland"

An experimental program aimed at investigating the relationships between forest damage and air pollution was established in 1985. Some 20 research groups from 5 different universities and 4 federal research institutions participated in this multidisciplinary study, which ran until 1989, pending a possible continuation.

As the protection forests in the Alps are of particular importance in Switzerland (protection against avalanches and erosion), the study is not

Table 10-1. Typical values of emission concentrations in Swiss conurbations, rural areas, and at high altitude. Values in μg m^{-3} yr $^{-1}$.

	Conurbations	Rural areas	High altitude
SO_2	30–40	8–12	2–3
NO_2	30–50	20–30	2–3
HC ($-CH_4$)	50–200	5–10	5
O_3	30–50	40–70	60–90

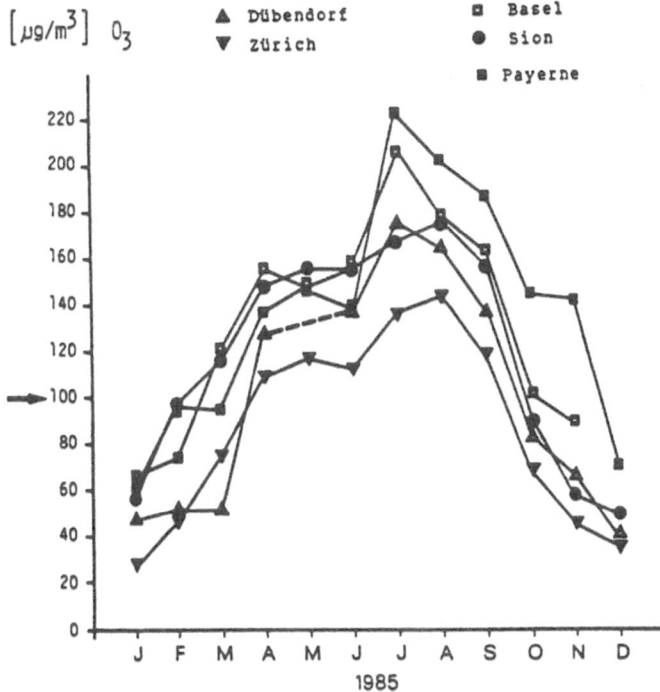

Figure 10-5. Annual course of maximum ozone concentrations. 98% values of half hourly means per month. Official upper limit value: 100 μg m⁻³ (98% value).

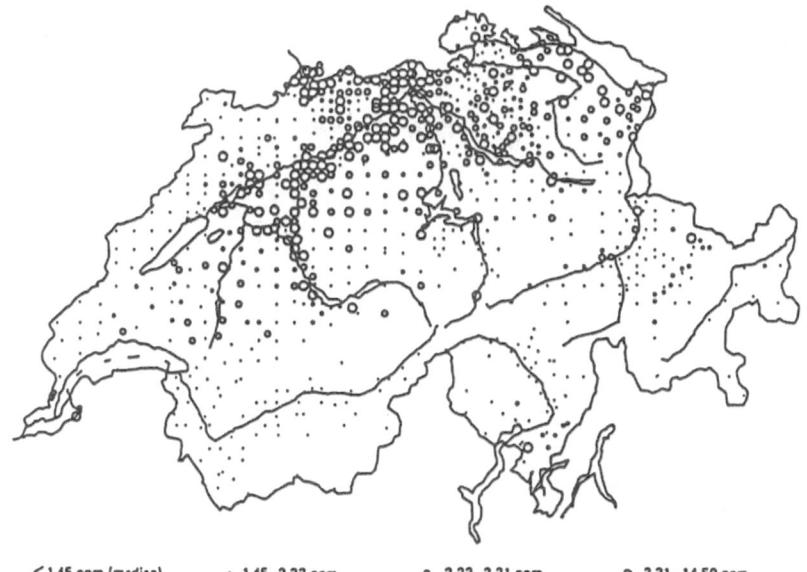

<1.45 ppm (median)   • 1.45–2.23 ppm   ∘ 2.23–3.31 ppm   ○ 3.31–14.50 ppm

Figure 10-6. Lead concentration of spruce needles in 1983. (Landolt et al., 1984.)

limited to the commercial forests in the lowlands, but also includes sites in the Pre-Alps (Alptal) and in the Central Alps (Davos).

At each station the following experimental design is employed. As far as possible, measurements are made at four levels: 10 m above the tree crowns, at the level of the highest treetops, within the crown zone, and on the ground. This permits determination of what occurs above the forest in the way of pollutants, what is filtered out by the trees, and what is eventually deposited on the ground.

Also of primary importance for the study are the weather conditions. The most important parameters, such as solar and atmospheric radiation (short- and long-wave), air and soil temperature, air humidity, and wind velocity and wind direction, are being measured by the Swiss Meteorological Institute and by the group Bioclimatology of the Swiss Federal Institute of Forestry Research (SFIFR), which is also studying the heat and water balance of the test sites. Stand precipitation and dry deposition are being investigated by the hydrology group of SFIFR in terms of volume and chemical composition. Attention is also being paid to chemical changes during the passage through the crowns, and the spatial distribution of the precipitation reaching the ground. Particular emphasis is given to the study of mist and fog (bioclimatology group of SFIFR). Both volume and chemical composition of the fog are investigated, and it is planned to supplement the fog studies with measurements of the liquid water content of the atmosphere (Figure 10–7).

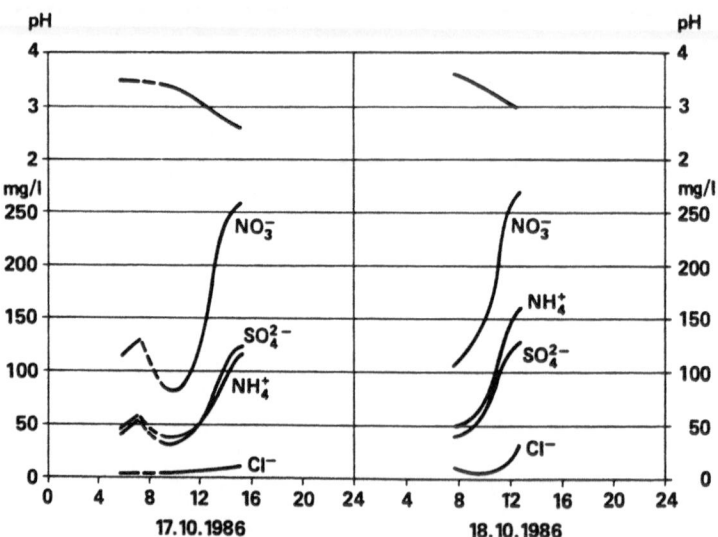

Figure 10–7. Diurnal course of pH acidity and of sulfate, nitrate, chloride, and ammonium concentrations in fog water during a stable high pressure weather situation with temperature inversion above the submontane broad-leaved/conifer mixed forest of Lägeren (680 m a.s.l.).

Gaseous pollutants are being studied quantitatively and qualitatively by the Swiss Federal Laboratories for Materials Testing. By means of gas chromatography, not only the classic harmful substances, such as sulfur dioxide, nitrogen oxides (NO_x), and ozone, but also PAN (peroxyacetyl nitrate) and some low-order hydrocarbons (ethylene, acetylene, isopentane, ethane) are determined. For some of these measurements it has not yet proved possible to achieve the degree of accuracy needed to distinguish definitely between differences at the various levels (Figure 10–8).

How do the trees react to harmful substances? The gas exchange in the light and shade crown of a spruce is being measured at two different levels (ecophysiology group of SFIFR) by means of thermoelectric climatized cuvettes (system Koch/Walz). In addition to net photosynthesis and nighttime respiration, transpiration is being continuously measured. To obtain supplementary information for the interpretation of the tree's behavior, leaf conductance and the concentration of carbon dioxide in the

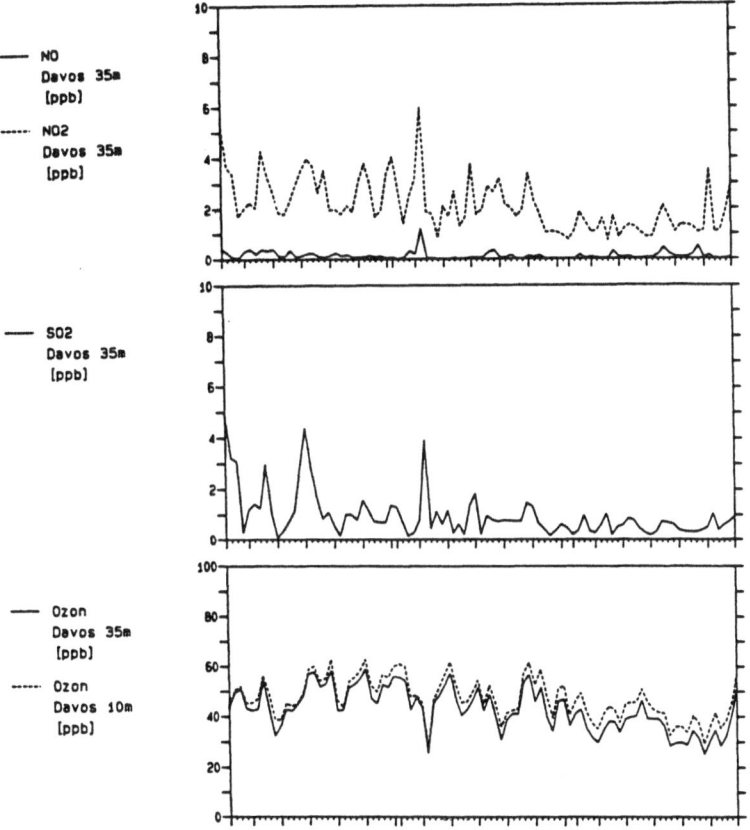

Figure 10–8. Daily mean values of NO, NO_2, SO_2, and ozone during the months April to June in Davos (1640 m a.s.l.). Values in ppb.

needles are also being determined. The ultimate aim is to confirm direct reactions of the tree to harmful substances in the case of peak exposure. As a further parameter of the water relationships, the flow rate in the xylem is being measured (Figure 10–9).

Test plants are being fumigated in open-top chambers under conditions close to those in open country with filtered air (low in pollutants) and unfiltered air (containing pollutants) (biochemistry group of SFIFR). Because of the experimental design, any differences in plant development must be attributed to the quality of the air in each case.

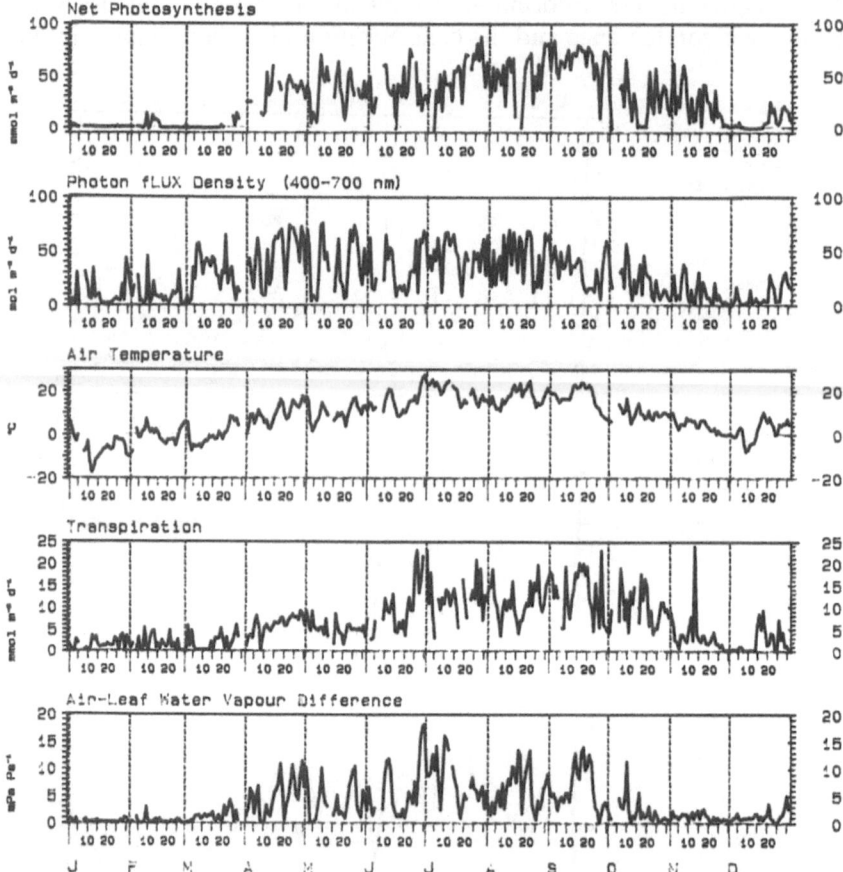

Figure 10–9a. Daily means of net photosynthesis, photon flux density, air temperature, transpiration, and leaf-air water-vapor difference for Norway spruce (*Picea abies*) from January through December 1987 in Läêgeren (680 m a.s.l.). Unpublished data by R. Häsler (FSL).

Both visible injury (chlorosis and necrosis) and invisible injury (physiological and biochemical) are regarded as symptoms. In each chamber six young spruces are fumigated throughout each entire vegetation period as long-term indicators. In addition, typical sensible indicators such as Red Clover (*Trifolium pratense*) are exposed to short-term fumigation.

Further investigations concern soil chemistry and physics, for example, episodic peak stresses by pollutants in soil water, concentration of heavy metals, and soil matrix water potential. Root fungi (mycorrhizae) are being studied by two groups which use morphological, anatomical,

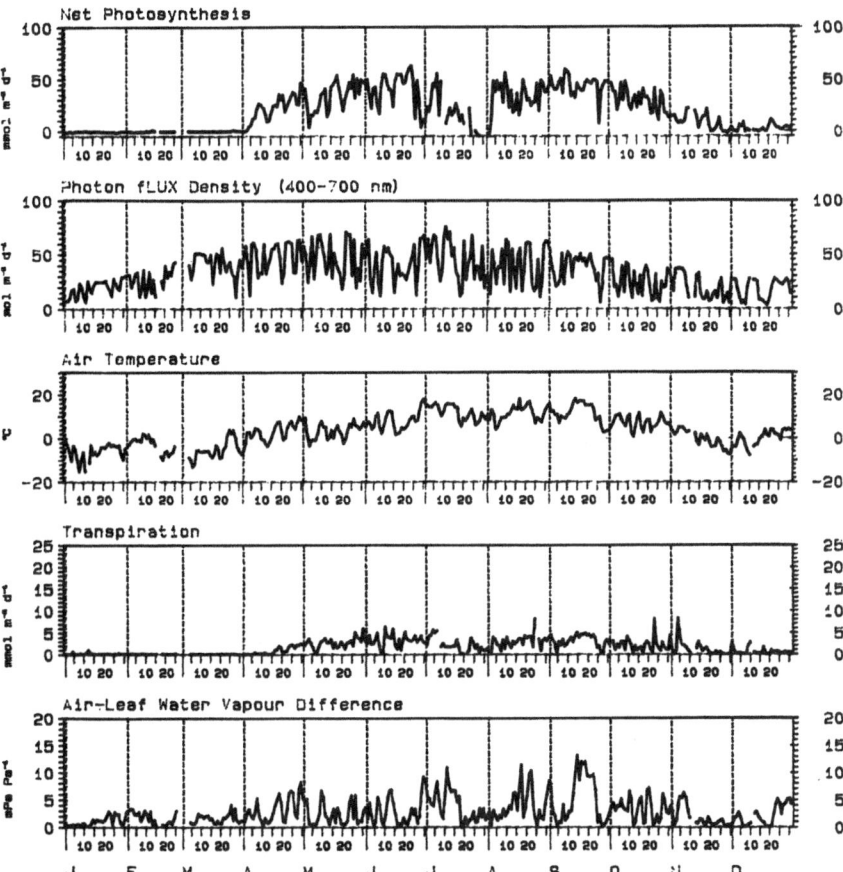

Figure 10–9b. Daily means of net photosynthesis, photon flux density, air temperature, transpiration, and leaf-air water vapor difference for Norway spruce (*Picea abies*) from January through December 1987 in Davos (1640 m a.s.l.). Unpublished data by R. Häsler (FSL).

and physiological methods to study dynamics of mycorrhizae in relation to toxicological inputs or natural environmental events.

In addition the superficial layer of wax is studied with regard to morphology and chemical composition in relation to the influence of pollutants. Finally of the broad spectrum of microorganisms, parasitic fungi (endophytes) have been selected for investigation. Comparison of the situation in healthy and damaged trees provides information on the role of these endophytes.

The continuously measured data (weather parameters, pollutant levels, gas exchange data, etc.) are recorded every 10 minutes, stored in a computer, and transferred every hour by telephone lines to a central computer. From here the data from every station can be retrieved with only a short delay. This complex system is run by a group of the Federal Institute of Nuclear Research.

Publication of a synthesis report on the results of the program outlined here is expected in 1990.

References

Bräker, O.L. 1987. Jahrringanalysen. In *Sanasilva-Waldschadenbericht 1987*, 21–24. Bundesamt f. Forstwes. u. Landschaftsschutz/Eidg. Anst. forstl. Versuchswes.

Bucher, J.B. 1987. Forest decline in Switzerland—critical remarks after four years of observation. *Joint Seminar on Global Habitability, Preprint,* 8 pp., 9 figs. Geneva, Switzerland.

Bucher, J.B., E. Kaufmann, and W. Landolt. 1984. Waldschäden in der Schweiz 1983 (1. Teil). Interpretation der Sanasilva-Umfrage und der Fichtennadelanalyse aus der Sicht des forstlichen Immissionsschutzes. *Schweiz, Z. Forstwes. 135:* 271–287.

Cowling, E.B. 1985. Comparison of regional declines of forests in Europe and North America; the possible role of airborne chemicals. In *Effects of Air Pollutants on Forest Ecosystems,* 217–234. The Acid Rain Foundation, St. Paul, MN.

Joos, K., S. Masumy, F.H. Schweingruber, and C. Staeger. 1988. Untersuchung über mögliche Einflüsse hochfrequenter elektromagnetischer Wellen auf den Wald. *Technische Mitteilungen PTT 1/1988:* 2–11. Bern, Switzerland.

Keller, T. 1976. Luftverunreinigungen und Wald; EAFV-Publikationen 1964–1975 mit Kurzzusammenfassungen. *Eidg. Anst. forstl. Versuchswes., Ber. 155,* 14 pp.

Keller, T. 1977. Begriff und Bedeutung der "latenten Immissionsschädigung". *Allg, Forst- u. Jagdztg. 148:* 115–120.

Keller, T. 1978. Einfluss niedriger SO_2-Konzentrationen auf die CO_2-Aufnahme von Fichte und Tanne., *Photosynthetica 12:* 316–322.

Keller, T. 1982. Luftverunreinigungen und Wald (Teil 2); EAFV-Publikationen 1976–1981 mit Kurzzusammenfassungen. *Eidg. Anst. forstl. Versuchswes., Ber. 242,* 17 pp.

Keller, W., and P. Imhof. 1987. Zum Einfluss der Durchforstung auf die Waldschäden in Buchenbeständen. *Wald+Holz 68:* 561–567.

Landolt, W., J.B. Bucher, and E. Kaufmann. 1984. Waldschäden in der Schweiz 1983 (2. Teil). Interpretation der Sanasilva-Umfrage und der Fichtennadelanalysen aus der Sicht der forstlichen Ernährungslehre. *Schweiz, Z. Forstwes. 135:* 637–653.

Mahrer, F. 1987. Die Ergebnisse der Waldschadeninventur 1987. In: *Sanasilva-Waldschadenbericht 1987:* 7–18. Bundesamt f. Forstwes. u. Landschaftsschutz/ Eidg. Anst. forstl. Versuchswes.

Rehfuess, K.E. 1987. Perceptions on forest diseases in central Europe. *Forestry* 69 (1): 1–11.

Turner, H. 1988. Frostschäden und Witterungsverlauf. In *Verrötungen immergrüner Nadelbäume im Winter 1986/87 in der Schweiz.* Eidg. Anst. forstl. Versuchswers. Ber. 307: 35–44.

Turner, H. 1987. Forest decline and air pollution in Switzerland; a national research programme. In T. Fujimori and M. Kimura (eds.), *Human Impacts and Management of Mountain Forests,* Proc. IUFRO Pl.07-00 Workshop Sept. 5–13, 1987, Susono, Japan. PP. 107–121. Forestry For Prod. Res. Inst. Ibaraki, Japan.

Acidic Precipitation Research in Finland

P. Kauppi*,
P. Anttila*, and K. Kenttämies*

Abstract

Because forests and lakes are such essential elements of the Finnish land-scape, it is important to study the ecological effects of acidic precipitation and to control air pollution accordingly so that all major damages can be prevented. A special interagency research program was initiated on acidic precipitation in Finland in 1985.

Finland escapes the highest air pollution levels because the country is rather remotely located on the outskirts of Europe. However, in the winter as high daily averages of sulphur dioxide concentration as 20 to 40 $\mu gX \ m^{-3}$ are observed in forested "background" areas in southern Finland. Total sulphur deposition ranges from 1.5 to 2.0 $gX \ m^{-2}X \ yr^{-1}$ in southeastern Finland to 0.3 to 0.5 $gX \ m^{-2}X \ yr^{-1}$ in northwestern Finland. In this way there is a declining gradient of sulphur levels from southeast to northwest. Interestingly, the rain pH is rather high (on the average 5.0) in southeastern Finland due to the high level of alkaline dust deposition in that area. The highest concentrations and depositions of nitrogen compounds are measured in southwestern Finland where, for example, the deposition of oxidized nitrogen is estimated at 0.2 to 0.4 $gX \ m^{-2}X \ yr^{-1}$. Nitrogen deposition estimates are, however, more uncertain than those of sulphur deposition.

A declining trend of alkalinity has been observed over the past decades in small lakes as well as in rivers. This has been documented also in areas where land use has not changed and cannot contribute to the trend. Fish and crayfish populations are at risk in small forested lakes. Unlike southern Norway, for example, major declines have thus far not occurred. The Norwegian watersheds have extremely thin soil layers. Finnish soils are also fairly thin but the ecosystems are, nevertheless, somewhat less sensi-

*Finnish Research Project on Acidification (HAPRO), Ministry of the Environment, E. Esplanadi 18 A, P.O. Box 399, SF–00121 Helsinki, Finland.

tive to acidification than those in Norway. Yet, many precursor effects to a forthcoming fish population decline have been documented in Finnish aquatic biota.

Forest growth on the average has substantially increased in Finland throughout this century. A particularly strong increase of forest growth has been observed over the past 20 years in southeastern Finland. A major cause is forest management. Cuttings have been less than the forest growth, implying an increase of forest inventory. Additional growth has then been formed upon this extra capital. However, air pollutants may contribute to the increase of forest growth. Nitrogen deposition acts as a forest fertilizer, sulphur deposition mobilizes calcium and potassium reserves, and the increase of atmospheric carbon concentration stimulates photosynthesis.

An increase of forest growth due to air pollutants is viewed as a warning signal. It indicates a rapid change in forest ecosystems, which is not based on sustainable growing conditions, and possibly precedes forest decline. Recent forest damage surveys indicated marked canopy defoliation in southeastern Finland, that is, in the area of high sulphur and nitrogen levels, and a strong increase of forest growth. According to statistical analyses, unfavorable weather conditions or bad forest management are not the only causes of this canopy defoliation. An additional warning signal has been a strong decline of epiphytic lichens observed in southern Finland since the 1950s and obviously associated with sulphur pollution.

Sulphur deposition levels are still regarded as too high in Finland. Gradually nitrogen deposition is also causing nutrient imbalances in ecosystems. Recommendations of actions to control and prevent air pollution damage include development of surface water liming and forest vitality fertilization, improvement of silvicultural practices, and, in particular, implementation of further emission reduction.

I. Introduction

Only 13.4% of the 305,000 km² land area of Finland is used for purposes other than forestry or natural protection. Agricultural fields cover less than 9% of the land area whereas the coverage of "forest land" and "scrub land" is as much as 76.3%. Yet the fraction of forests is still increasing as agricultural land is being converted to forest in order to overcome the current agricultural overproduction.

No mountains exist in the southern half of the country. The highest hills rarely reach elevations above 250 m. The bedrock is of old Precambrian origin, largely granite, but the soils are young and are still undergoing a process of change that started 10,000 years ago after the latest glaciation. Some major ridges were then formed containing gravel layers as deep as 10 to 30 m. Apart from those ridges, however, the layers of mineral soil are rarely deeper than 0.5 to 10 m. Organic peat covers as much as one-third of the forest land area. In some regions agriculture is also being practiced on peat soils. The thickness of the peat layer varies between 0.2 and 5 m. The largest peatlands are located in the central and northern parts of the country where thick and acidic *Sphagnum* peat layers cover more than half of the land surface area in some regions.

Finland stretches 1000 km north of the latitude 60° across the main fraction of the northern boreal biome. The dominating tree species in Finland are conifers as elsewhere within this biome. Just two tree species contribute 81.8% of the standing timber volume of Finnish forests: Scots pine (44.9%) and Norway spruce (36.9%). Spruce is the dominating species in southernmost Finland and on fertile and moist sites. Total annual growth of pine and spruce accounted for 40.1% and 36.6%, respectively, of the total forest growth in Finland as a countrywide average from 1977 to 1984 (Yearbook of forest statistics, 1987).

Broad-leaved trees contribute 18.2% of the standing volume and 23.2% of the forest growth. The main broad-leaved tree species are birch (*Betula pendula* and *B. pubescens*), aspen (*Populus tremula*), and alder (*Alnus incana* and *A. glutinosa*). A small number of oak trees (*Qurcus robur)* grow in the very southernmost part of the country but, for example, beech (*Facus sylvatica*), which is common in central Europe, does not tolerate the Finnish climate and is absent.

Finland is known as a land of thousands of lakes. The total number of lakes larger than 1 ha is about 55,000 and the number of very small lakes (or rather ponds) of the size 1.0 to 0.05 ha is as high as 132,000. Big lakes contribute the largest fraction of the total lake area, however. Lakes larger than 10 km^2 make up 64% of the total lake area of 32,000 km^2. Lakes in Finland are relatively shallow, however. The mean depth is only 7 m and the total lake volume about 230 km^3. The theoretical retention time is only 2.3 years as the yearly runoff from Finnish watercourses is about 100 km^{3}X yr^{-1}.

The total length of streams in Finland is not known. However, it is clear that the length of small streams and brooks, which are especially sensitive to acidification, is tens of thousands of kilometers.

Groundwater resources in Finland are not very plentiful because soils are thin and thus aquifers are shallow. Of the total groundwater basins, 700 km^3, only 25 km^3 is technically easily available. In addition, the low permeability of typical morainic and clay soils does not favor groundwater formation. Groundwater is, however, the source for household water for one-third of the Finnish population, that is, for about 1.5 million inhabitants.

Internationally air pollution damage to ecosystems was observed on a regional scale first in southern Norway and then in southern Sweden in the 1950s and 1960s. As Finnish air pollution levels are similar to those in neighboring countries where large-scale damages have already been observed, there is much concern that such damage will expand into the Finnish territory. Research on this issue started in the late 1960s and grew gradually over the following 15 years. Interest on this matter increased considerably in the early 1980s when air pollutants were suspected as being at least partially responsible for the forest damage in central Europe. In Finland, a substantial intensification of research took place in 1985, when a national HAPRO program was launched to strengthen research and assessments on the acidic precipitation issue. This program is a fixed-term effort scheduled to provide its main results from 1989 to 1991 (See Appendix for a list of research projects). Some results were already available to the authors when preparing this chapter. The main research document of the program is in press (Kauppi, Anttila, and Kenttämies, 1990).

The aim of this chapter is to review the results of Finnish air quality research and research on air pollution damage to ecosystems. The scope is restricted to damage on forests and to surface and groundwater systems.

II. Emissions

Sulphur dioxide emissions in Finland have their principal origins in energy production and in the paper and pulp industry. Between 1950 and 1970 the emissions of sulphur dioxide increased annually on the average by 6% but since the 1970s a declining trend has prevailed due to a decrease of the sulphur content in oil, to the substitution of oil by coal and nuclear power, and to the structural change in the process industry. Since the year 1970 the total emissions of SO_2 declined by 40% to about 350,000 t in the 1980s. The largest emission sources are located in southern Finland and on the coast of the Gulf of Bothnia in western Finland.

The sources of nitrogen oxides emissions differ slightly from those of SO_2, as half of the NO_x emissions come from traffic. While SO_2 emissions have declined over the past 15 years, NO_x emissions continue to increase (about 240,000 t in 1985). The regional distribution of NO_x emissions is very similar to the distribution of SO_2, that is, emissions are concentrated in southern Finland. However, when comparing the emissions of sulphur and nitrogen oxides one must keep in mind that nitrogen oxides emission inventories are less accurate than those of SO_2. Methods of calculating NO_x emission factors for Finland are being improved at the Technical Research Center of Finland and at Åbo Akademi.

Ammonia contributes to the total nitrogen deposition to ecosystems and has an acidifying net effect on forest soils. Ammonia is produced in large quantities from biogenic sources. Ammonia emissions evaporating from agricultural fertilization have been inventoried in Finland (Keränen and Niskanen, 1987). A major part of the NH_3 emissions is from livestock manure. Emissions of ammonia nitrogen from fertilizers is estimated at 38,000 t annually, which is about half of the amount emitted as NO_x–N from industrial sources.

Alkaline particles that contain elements like Ca and Mg are of importance in neutralizing acidity and some of them are also biologically active nutrients. Airborne alkaline dust comes not only from industrial sources but also from wind erosion, cultivation, building works, and traffic on unpaved roads. Those mechanical processes, however, produce such large-size particles they are likely to influence the atmospheric chemistry only locally, near the emission source.

Preliminary estimates have been made of the industrial emissions of alkaline particles in Finland (Anttila, 1987). Energy production was estimated as a minor source. Biggest industrial sources were the building material industry, iron and steel industry, and chemical forest industry (waste pulp mill liquor). Total industrial emissions of Ca, K, and Mg, for example, were each only a few tonnes per year according to these preliminary estimates.

III. Atmospheric Processes

Compounds emitted to the atmosphere are subject to a variety of physical and chemical processes which depend on seasonal and meteorological conditions. These processes are examined in several studies of the Finnish Research Project on Acidification.

The formation of acidic aerosols by gas-to-particle conversion is being studied at the University of Helsinki, Department of Physics in cooperation with the Finnish Meteorological Institute. In the theoretical part of this study special attention is paid to the thermodynamic properties and growth of gaseous solution droplets. The experimental work includes both laboratory and field experiments. Preliminary laboratory experiments were carried out to study heteromolecular nucleation between acidic molecules and water vapor. Results show that sulphuric acid and water molecules can form new particles in atmospheric conditions while nitric acid molecules can only dissolve into previous water droplets (Viisanen, Kulmala, Hillamo, Hatakka, and Keronen, 1987).

The characteristics of the long-range transported airborne pollutants were studied in an intensive research expedition over the Baltic Sea in the spring of 1983. The expedition was carried out in cooperation with the Institute of Applied Geophysics, Moscow, and the Finnish Meteorological Institute. Gaseous and particulate compounds (SO_2, SO_4^{2-}, NO_3^{-}, HNO_3, NH_3, and NH_4^{+}) were analyzed and interpreted with the help of concomitant meteorological measurements. Results of this work demonstrate that simple estimates of pollutant origin, prepared on the basis of wind direction classes, are not very reliable for predicting the strength of pollutant episodes. Moreover, the commonly used 850 mb trajectories were shown often to describe badly the transport within the boundary layer, especially over the sea (Salmi and Joffre, 1987). To make quantitative estimates of the dry deposition velocity and transformation rates of pollutants over the Baltic Sea, individual transport flow patterns are being investigated with the addition of air quality data from coastal stations.

Dry deposition of aerosols and gases to forests is being studied at the University of Helsinki, Department of Physics. Estimates based on the average dry deposition velocities and existing background concentrations of pollutants and surface areas of foliage suggest that deposition data achieved from traditional wet/dry collectors cannot be used to evaluate deposition to forests (Anttila, Kulmala, and Raunemaa, 1987). Models are being developed, therefore, to estimate forest deposition using air concentration measurements and forest surface characteristics as input data.

Detailed case studies on the transport, transformation rates, and deposition velocities are necessary for improving the model formulations describing the vertical and horizontal transport and dispersion of acidic substances and their precursors. The Finnish Meteorological Institute has developed a mesoscale dispersion model for estimating the regional deposition patterns of anthropogenic sulphur compounds (Nordlund and Tuovinen, 1988). In this model the so-called K theory of atmospheric diffusion is applied in calculating the vertical distribution of concentrations. For the horizontal concentrations common Gaussian plume assumptions are employed. The transformation of sulphur dioxide into sulphate and the effects of dry deposition and scavenging are taken into account.

The sulphur transport model of the Finnish Meteorological Institute is best applied to calculating deposition within 300 km from the source. It has been used in Finland to estimate the deposition due to the indigenous sulphur emissions. The contribution to deposition of the emissions in other European countries has been calculated with the ECE/EMEP model. The deposition estimates thus obtained have been added up to get the best estimate for the regional pattern of sulphur deposition in Finland. (Fig. 11–1).

A similar dispersion model for nitrogen oxide emissions is being developed. The modeling of nitrogen oxides is more complex due to nonlinearities of the key reactions. The physical and chemical processes need to be described jointly. These problems emphasize the importance of reliable basic information.

IV. Air Quality and Deposition Monitoring

A. Sulphur Dioxide and Sulfate Concentrations

Annual average air concentrations of sulphur dioxide in Finnish background areas in the 1970s and 1980s were around $10\,\mu gX\,m^{-3}$ in southern Finland and about half of that in central and northern Finland. In the winter months of January and February, sulphur dioxide concentrations are at their highest, and daily averages of 20 to 40 µgX m⁻³ of sulphur dioxide are not uncommon (Kulmala, Estlander, Leinonen, Ruoho-Airola, and Säynätkari, 1982).

Annual average air concentrations of sulfate sulphur are at around 2 to 3 μgX m⁻³ in southern Finland and below 1 μgX m⁻³ in central and northern Finland. Eighty percent of daily samples in Ähtäri contain less than 1 μgX m⁻³ SO_4^{2-}–S. In the winter months concentrations of 10 μgX m⁻³ SO_4^{2-}–S have been measured frequently.

A large part of samples are hence at the detection limit. This makes it difficult to study temporal and spatial gradients of concentrations, for example. A new method based on chemosorption and better suitable for measurements in low concentrations is being developed within the Finnish Meteorological Institute.

B. Ozone and Nitrogen Dioxide Concentrations

Measurements of ozone started in July 1985 at Utö, which is a rocky, treeless island on the Baltic Sea about 80 km southwest of mainland Finland. Another station was established in the summer of 1986 at Ähtäri in central Finland which is an inland forested region. Measurements of NO_2 were also started at both stations.

Annual average ozone concentrations in Utö and Ähtäri were 72 µgX m⁻³ and 54 µgX m⁻³, respectively, during the first year of the study. Utö data indicate a typical maritime environment with small diurnal amplitude and high average concentrations.

Correlations between ozone and NO_2 concentrations have indicated that high ozone concentrations (over 100 µgX m⁻³) tend to occur when NO_2 concentrations are 7 to 9 µgX m⁻³ (Hakola, Taalas, Lättilä, Joffre, and Plathan, 1987). Thus episodic high ozone concentrations are connected to increased NO_2 concentrations.

High ozone concentrations are produced either by long-range transport, by photochemical processes, or by stratospheric intrusions. Further investigation is needed to determine the relative importance of the different processes. Maritime environment has a marked influence on ozone concentrations in coastal regions. More detailed measurements from central Finland are needed to evaluate the ozone concentration levels in mainland Finland.

C. Deposition

The Water and Environment Research Institute carries out bulk deposition measurements on a monthly basis at 40 stations, and the Finnish Meteorological Institute runs two networks: one with 3 stations where samples are taken on a daily basis and one with 10 stations with monthly sampling. Networks were established in the beginning of the 1970s.

Annual bulk deposition of S, NO_3^--N, and Ca^{2+} show a clear geographical gradient with deposition decreasing northward. For strong acids and NH_4^+-N the gradient is less clear and, for example, for K there appears to be no geographical gradient (Järvinen, 1986).

The correlation between measurements and calculated deposition of S is quite reasonable, the annual S deposition being about 1000 mgX m^{-2} in southern Finland and about 300 mgX m^{-2} in northern Finland (see Figure 11-1). The sulphur load in southern Finland is of the same order of magnitude as in southern Norway and Sweden but that of strong acids is only half the value recorded there. Annual bulk deposition of NO_3^--N is about 300 mgX m^{-2} in southern and about 100 mgX m^{-2} in northern Finland, and that of NH_4^+-N is about 400 mgX m^{-2} and 100 mgX m^{-2}, respectively (Järvinen, 1986).

Annual bulk deposition of Ca^{2+} in southern Finland is about 700 mg m^{-2} and in northern Finland about 100 mg m^{-2}. The high deposition of calcium in southeastern Finland is interesting. Correspondingly the rain pH in that area is not very low in spite of the relatively high S deposition values. These may be due to the extensive mining and use of oil shale in the electric power stations in Estonia, the USSR. In 1985 about 30 million t of oil shale is being produced in Estonia from seven mines and four open pits. The flying ash from shale-fired power plants is documented as a serious problem in that area (Aarna, 1985).

The Finnish Meteorological Institute has carried out a detailed meteorological analysis of six years' data of daily bulk deposition observations at three Finnish stations (Utö, Virolahti, Ähtäri) (Laurila and Joffe, 1987). The variability of the ion concentration of rainwater (SO_4^{2-}, NO_3^-, NH_4^+, Mg^{2+}, and H^+) was studied together with meteorological parameters. The main emphasis of the work was on seasonal averages and episodic aspects of deposition, on studies on the ratios of ions inprecipitation, and on the relative importance of SO_4^{2-} and NO_3^- in contribution to the rainwater acidity.

Sulphate deposition has its maximum in summer, and NO_3^- deposition is almost constant having only a small peak in winter. Ammonium deposition events are concentrated in the spring and, on the southern coast, also in the autumn. The different behavior of sulfate and nitrate is also observed in the sulfate to nitrate ratios of the individual precipitation samples. The SO_4^{2-} to NO_3^- ratio (in equivalents) is about 3 for summer rain, but slightly above 1 for winter snow.

This material also confirms the fact that pollution events are associated with southerly winds. In addition to the general pattern of decreasing deposition toward the north, an east-west gradient of SO_4^{2-} and NO_3^- is observed. The NO_3^- deposition decreases eastward while the SO_4^{2-} deposition increases in this direction. This is connected to the relative large sulphur emissions on the Finnish south coast and also in the Leningrad region and in Estonia and that emissions of NO_x are more concentrated

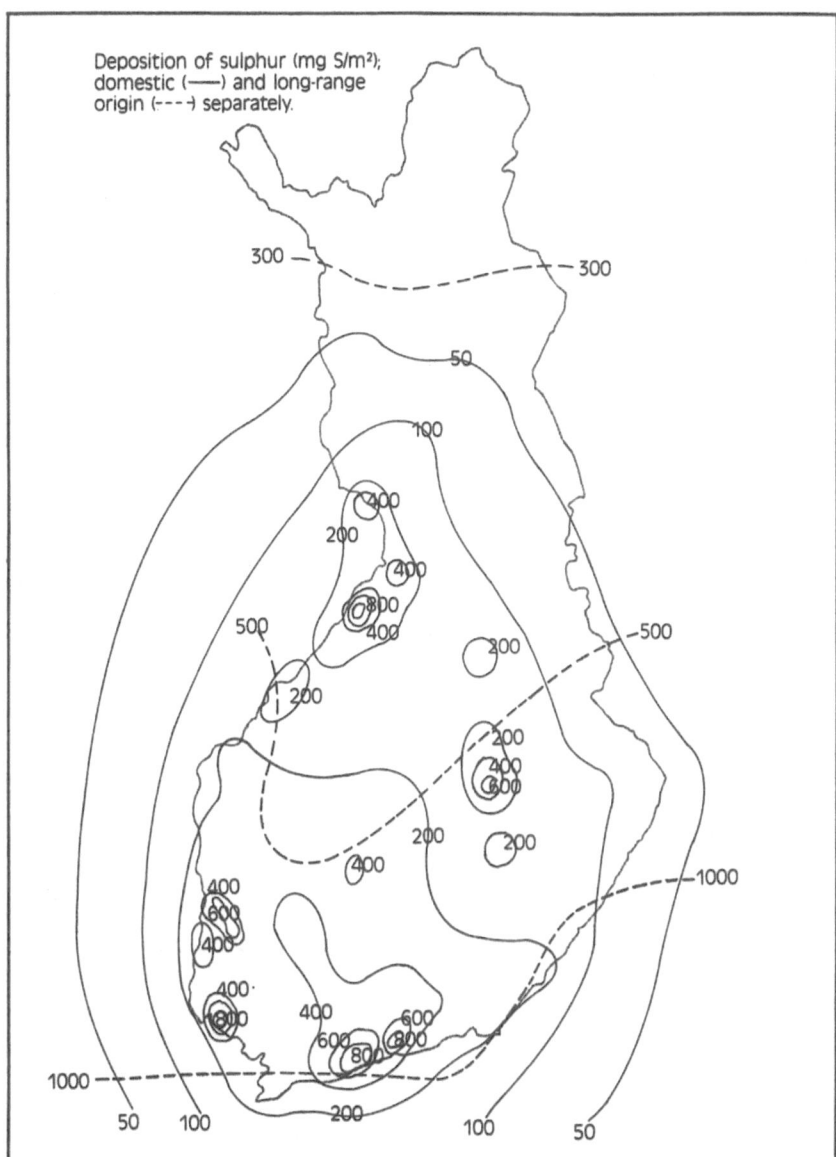

Figure 11–1. Annual deposition of sulphur in Finland (mgS m⁻²). Based on estimated emissions for 1982 to 1983 as assessed by the Finnish Meteorological Institute (unbroken lines). Proportion of foreign origin estimated on the basis of the
ECE/EMEP project (broken lines) (Nordlund and Tuovinen, 1988).

in Western Europe. The study of individual rain samples also confirms this conclusion.

Atmospheric heavy metal deposition in the Nordic countries was monitored in 1985 by means of moss technique (Rühling, Rasmussen, Pilegaard, Mäkinen, and Steinnes, 1987). Samples of the moss species *Hylocomius splendens* and *Pleurozium schreberi* were collected during the summer months of 1985 from 534 localities in Finland. The concentrations of As, Co, Cr, Cu, Fe, Pb, Ni, V, and Zn were analyzed. The regional background deposition pattern shows a decreasing gradient for all metals from relatively high values in the southern parts of Scandinavia to low values toward the north. The gradient is steep for As, Cd, Pb, and V, whereas the concentrations of Cr, Cu, Fe, Ni, and to some extent Zn, show weaker gradients. Important local enhancements of the concentrations in moss were found superimposed on the regional background pattern in Finland at the steel mill in Tornio and at the copper smelter in Harjavalta, and in northernmost Finland adjacent to the copper-nickel mining and smelting area in the USSR.

V. Aquatic Effects

A. The Sensitivity of Lakes and Streams to Acidification

Generally speaking, Finnish lakes and streams can be characterized as sensitive to acidification. Old, measured, or reconstructed alkalinity values, representing the preacidified phase, show very low alkalinity production of Finnish catchments (Alasaarela and Heinonen, 1984; Kämäri, 1985). In fact, the prevailing bedrock and soil types in Finland are very much like those in the acidified parts of southern Scandinavia. However, in practice, prominent differences in the sensitivity of lakes have been identified inside regions of homogenous, acidic bedrock types like granites. Weathering rate of rocks, cation exchange capacity, runoff, and the occurrence of surface flow events are perhaps the most prominent characteristics affecting alkalinity production. Regional nationwide scale maps of rock types, soil properties, runoff, and terrestrial reliefs were used by Kämäri (1986) to measure these four processes. Four (only three in weathering) classes were formed inside each of four factors and the prevailing classes were identified in each 10 x 10 km quadrant on the map. The weighted relative values of the classes were applied to produce a map of sensitivity which indicates the relative sensitivity of surface waters to acidic deposition. Lakes in which acidification has already been documented tended to be located on areas indicated as sensitive to acidification on this map.

B. Historical Evidence of Chemical Trends

Systematic monitoring of water quality in major Finnish watercourses has been carried out only since the beginning of the 1960s. However, an interesting view into the history of watercourses has been provided by Alasaarela and Heinonen (1984) who compared monitoring data acquired between 1911 and 1931 by the Hydrographical Bureau of Finland on 10 rivers discharging into the Gulf of Bothnia (Holmberg, 1935) with data compiled by the National Board of Waters in the same sites during the period from 1962 to 1979. The analyzing methods of alkalinity and chemical oxygen demand (COD) were so similar that comparisons between the two monitoring periods were possible. The mean annual alkalinity values were lower from 1962 to 1979 than from 1911 to 1931 in all sites.

C. Observations from the Finnish Water Monitoring Networks

The Finnish water monitoring networks have been examined in order to find out the possible effects of acidic precipitation, that is, trends in pH, alkalinity, conductivity, and total sulphur. Sampling sites of these networks are located in large lakes and rivers, which are subject to water quality changes due to waste waters and nonpoint sources. According to Laaksonen and Malin (1980) the growing trend of conductivity was very common. The decreasing trend of alkalinity was observed only in some sampling stations. Statistically significant trends of pH changes were very few.

The concentrations of cations and anions, with the only exception of hydrocarbonate, increased in monitored large lakes between the late 1960s and the early 1980s (Laaksonen and Malin, 1984). A shift toward lower alkalinity values was already found in statistical frequency distribution analyses of the same lake population by Laaksonen (1982).

The general monitoring programs have not necessarily been optimally designed to reveal the effects of atmospheric pollutants. In 1972 a study dealing only with lakes near which there is no industry or agriculture was made by the water authorities (Kenttämies, 1973). According to the study the pH values of surface water in winter had decreased by about 0.1 pH units from 1962–1964 to 1970–1972.

In 1977 another study on lake water acidification was made by Kenttämies (1979). The purpose was to compare the pH, alkalinity, and conductivity levels from 1975 to 1977 to those from 1970 to 1972 and from 1962 to 1965. The study was restricted to surface water (1 m) of lakes in an unpolluted, "natural" state. Alkalinity was lower and conductivity higher from 1975 to 1977 than from 1962 to 1964. There were no significant differences in pH values.

D. Acidification of Small Lakes

The general research program of small forest lakes was started in Finland in the late 1960s. In the mid-1980s the lakes that were first sampled at least 10 years earlier were resampled. Results of the program show that the alkalinity of small, less than 1 km^2 clear water lakes has in general decreased during the last decade both in summer and winter samples (Forsius, Kenttämies, and Kämäri, 1987).

The survey of 8,000 lakes (16 % of the total number of lakes larger than 1 ha) revealed about 150 clear water lakes in southern Finland with a minimum pH value of less than 5.0. These lakes are situated in areas of acidic bedrock and thin rocky soils. In fact, the number of acidic lakes is much larger, because acidic brown water lakes prevail in Finland. However, the effect of air pollutants on the acidity of humic lakes is negligible. Air pollution effects on lakes were observed mainly in southern Finland. In southwestern parts of Finland the proportion of lakes with no alkalinity was over 14%, elsewhere less than 10% (Forsius et al., 1987). Indirect methods for quantifying lake acidification suggest that lakes in southern Finland have lost their alkalinity on the average by more than 100 µeq L^{-1}. Because the lakes in this study were not chosen according to statistical principles, the results may not be considered representative for estimating the number and regional distribution of acidified lakes. Therefore a new survey of 1200 lakes was conducted in the autumn of 1987 in order to obtain statistically representative data.

E. Paleolimnological Evidence of Trends

Paleolimnological diatom analyses have been used commonly in Finland to study the acidification history of lakes (Simola, Kenttämies, and Sandman, 1985; Tolonen and Jaakkola, 1983; Tolonen, Liukkonen, Harjula, and Pätilä, 1986). The actual time of acidification is regarded as valuable, even if indirect, evidence of the recent airborne anthropogenic acidification of lakes. The main paleolimnological types of acidic lakes in Finland may be divided in the following groups:

1. Humic lakes that have acidified in the remote past due to paludification and podzolization in their catchments.
2. Groundwater seepage lakes in which the pH has remained stable for a long time.
3. Rainwater-predominated (surface flow) lakes which have recently become acidified due to the acidic deposition (Tolonen et al., 1986; Simola et al., 1985).

In southern Finland the majority of acidic lakes belong to group 3; in central and northern Finland the natural acidic lakes of groups 1 and 2 predominate. The present (<40 years old) acidification of lakes seems to concentrate in areas of the highest (0.8 to 1.8 gX m^{-2}X a^{-1} S) sulphur deposition in Finland.

F. Acidification of Groundwater

A network of 54 groundwater observation stations has been monitored since 1975. Time series of water quality measurements in groundwater in Finland are still too short to demonstrate acidification of groundwater (Soveri, 1985). However, early indications of acidification can be observed even from these data. The concentrations of sulfate, calcium and aluminum in groundwater have significantly increased from 1975–1976 to 1982–1983 (Soveri, 1985). Particularly marked change took place in the case of sulfate. However, the simultaneous increase of calcium shows that the infiltrating acidity has been buffered in the upper soil layers.

G. Heavy Metals in Lake Water and in Sediments

Tolonen and Jaakkola (1983) found a considerable increase in the contents and sedimentation rates of Pb, Zn, Ti, Cd, Cu, and Hg in the sediments of four small lakes in southern Finland. This increase began a few decades ago, except in the case of Pb where the growth in sediment concentrations had already started during the latter half of the nineteenth century. After correction for input due to erosion from the catchment area, the enrichment factors that indicate airborne anthropogenic influence were +28 for Pb, Cd +9.4, Ti +3.1, Zn +3.0, Hg +2.7, and Cu +1.6. New research from 16 Finnish small headwater lakes showed a general increase in several major cations (Na, K, Mg, Ti) at the surface of the cores only in southern and central Finland (Verta and Mannio, 1987). That indicated changes caused by soil acidification and/or erosion in the catchment area. The most prominent increase was observed again in Pb concentration, beginning as early as the seventeenth or eighteenth century in southern Finland. The biggest (up to fifty-fold) increase in Pb concentration and sedimentation was found in the twentieth century, however.

H. The Effects of Weak Organic Acids on the Acidity of Lakes and Streams

The content of natural organic matter in lakes and streams is fairly large in Finland. The sources of this acidic humic material are peatlands and waterlogged podzolic soils. The mean concentration of total organic carbon (TOC) in river waters running to the Baltic Sea is over 10 mgX L^{-1}, and in peat catchments the annual discharge of TOC exceeds 10 tX km^{-2}. When considering the acidification of surface waters it is important to distinguish between strong mineral anthropogenic acidity and natural organic acidity. In the studies of Kortelainen and Mannio (1986, 1987) three methods of separation have been applied.

1. Calculating organic anion concentrations as a difference between the total amount of cations (Ca^{2+} + Mg^{2+} + K^+ + Na^+ + Fe (III) + Al (III) + H^+) and inorganic anions (SO_4^{2-} + Cl^- + HCO_3^-),
2. Estimating organic anion concentrations by the empirical model of Oliver, Thurman, and Malcolm, 1983, and
3. Calculating organic anion concentrations from the measured (Gran's titration) weak acidic concentrations using the model of Oliver (1983). In humic lakes where organic matter is responsible for most of the weak acidic concentrations, the estimates from these three methods corresponded well with each other. The share of organic anions amounted to an average of one-third of the total amount of anions. In 60 slightly acidic humic lakes TOC explained the variance of pH better (r = 0.614, P <0.001) than nonmarine sulphate (r = 0.074).

I. Biota of Lakes and Streams

No lake has yet been found in Finland with all of its fish killed by airborne acidification. However, drastic effects on the species composition and densities of fish populations in some recently acidified lakes have occurred. The densities of perch, which is the most common and tolerant fish in small forest lakes, varied in the acidified lakes from 0 to 6 ind.X ha^{-1} to 150 to 200 ind.X ha^{-1} implying 5 to 500 times lower fish densities compared to the neutral reference lakes (Rask, 1987). Reproduction failures were the main reasons for decreased populations. The most sensitive phases of the fish life cycle occur soon after fertilization. The fry is also very sensitive whereas adult perch are much more tolerant (Rask, 1987). Delayed spawning of perch and whitefish was observed in field studies, and verified in laboratory tests (Tuunainen, Vuorinen, Rask, Järvenpää, and Vuorinen, 1987). For example, in a small lake with pH value 4.4 to 4.9 and total Al 275 to 979 µgX L^{-1} the perch spawned one month later than usual and all the eggs died. The ranking of the fish species according to the sensitivity of their newly hatched fry to low pH was as follows: roach > pikeperch > perch > whitefish > pike. At the lowest pH tested, 4.0, fertilization of pike eggs did occur, but the hatching was delayed and largely failed (Tuunainen et al., 1987).

Biological characteristics of small acidic lakes were studied as a part of the integrated research of watershed, water chemistry, and lake biota of 140 forest lakes. According to the preliminary results (Kenttämies, Haapaniemi, Hynynen, Joki-Heiskala, and Kämäri, 1985) the effects of pH level on macrophytes, benthic invertebrates, and zooplankton could clearly be observed. In humic waters, however, many organisms were found to tolerate a lower pH level than in clear water lakes. This fact makes it harder to develop a general biological indicator organism system for surveying the acidification status of lakes. The diatom analysis of surface sediments proved to be a very promising tool in the surveying and monitoring of acidic and acidic-sensitive lakes (Turkia and Huttunen, 1987).

VI. Forest Effects

A. Background

Forests are of exceptional importance in Finland both culturally and economically. A number of favorable preconditions form the basis for the economic success of forestry and the exporting forest industries. Firstly, the main natural tree species, Scots pine and Norway spruce, are among the most valuable species in the world as the raw material of sawlogs, pulp, and paper. Secondly, Finland is located close to the major importing countries, in particular the European Community. Thirdly, forestry operates on a sustainable basis, which promotes a stable and robust economic structure. The standing volume of all major tree species has slightly increased over the past decades in spite of the intensive use of forest resources for about 100 years. Historical records indicate that twice as much timber has been removed from Finnish forests this century as there are trees standing in the forests today. Unlike agricultural production, forestry and forest industries have been able to operate successfully without any major subsidies.

Forest policy in Finland has been rather conservative on importing exotic tree species. They contribute, therefore, less than 0.1% of the standing volume and forest growth. Forest plantations of indigenous species cover about 20% of the forest area, the remaining 80% of the stands being of natural origin. Local seed origin is used also for plantations. (Some failures were experienced in the 1960s when importing spruce seed from foreign sources.)

Finland would have very much to lose if air pollutants were to deteriorate Finnish forests. The concentrations and deposition of air pollutants are low in Finland compared to many other parts of the industrialized world. This does not guarantee, however, that forest effects will remain absent. The climate is severe, with the mean annual average temperature ranging from 4.5° C in southern Finland to −1.0° C in the northernmost forests. The soils are thin, containing only limited nutrient reserves, and there is a chance for unfavorable synergistic effects between air pollutants and the severe growing conditions. Even slight forest damage would have more significance in Finland than in most other countries of the world where the culture and economy are based less on forest resources.

B. Indicator Variables and Approach

Forest damage and forest decline are somewhat problematic concepts, as there have been no standard measurements that can describe them directly. The selection of indicator variables depends on the purpose of the investigation, on the type of forest, and on the purpose of forest on management. *Forest growth* and *forest morbidity* are of great importance in Finland because of the economic significance of the forest sector.

Ovaskainen (1987) argues that forest growth is an important variable mainly in the long term (> 50 years). Forest morbidity (expected average lifetime of forest stands) is more important in the short term (5 to 30 years). This is based on the reasoning that if existing forests suddenly start to deteriorate and die out on a large scale, the forestry infrastructure cannot cope with the pressure for sanitation fellings. This causes immediate perturbations in the timber market. If, instead, forest growth is affected but the trees stay alive, there will be fewer short-term economic effects. Fellings can exploit standing timber resources, forestry and the forest sector will have time to adapt, and only later on will there be a shortage of resources.

Even in Finland where the forest sector is responsible for about one-half of the national export earnings, one should not pay attention to direct economic effects only. People feel that forests should be protected even if they themselves do not earn their income from the forest. Much concern has been expressed about air pollution in natural reserves. Protected areas in Finland, in which no forest fellings are permitted, cover more than 8,000 km². The timber resources within these reserves, if delivered into the market, would be worth billions of Finnish marks (hundreds of millions of U.S. dollars). As the timber is *not* delivered into the market, one can estimate that these reserves, as people view them, are worth more than this. One can set geographical boundaries to protect these areas, forbid fellings and other forestry practices, but one cannot provide the areas with filtered air.

A clear choice has been made in recent acidic rain related research of describing *regional* instead of *temporal* variability. Although old time series of emissions, concentrations, deposition, soil chemistry, forest composition, and so on, are being collected and analyzed, no special effort has been made to initiate monitoring programs for obtaining entirely new time-series data. The time frame of the present investigations is from 1985 to 1990. From experience we know that air quality in rural areas can change very slowly so that the interannual variability obscures possible five-year trends in air quality parameters. A newly established monitoring station cannot reveal trends until after an operation period of 10 to 15 years.

The regional survey approach, however, is not an easy choice. New information will have to be collected using statistical predetermined sampling methods. Only from such data can the accuracy of the results be quantified. Statistical sampling of any forest variable over a whole country is a demanding exercise both theoretically and in practice. Special attention is needed to train the people who carry out the fieldwork. Regional surveys are relatively expensive to carry out.

Forest resources in Finland have steadily increased over the past decades. The primary reason for this is forest management, mainly the introduction of sustainable forest management and peatland drainage. Part of the trend, however, seems to be due to the increasing atmospheric CO_2 concentrations that have stimulated photosynthesis, and to the deposition of nitric acid and ammonium that act as nitrogen fertilizers (Arovaara, Hari, and Kuusela, 1984; Hari, Arovaara, Raunemaa, and Hautojärvi, 1984).

A statistically located systematic grid of 3,000 permanent plots was established in Finland in 1985–1986 for damage survey purposes. The grid is permanent in the sense that the plots can be reidentified and remeasured. The method of assessing forest damage on these plots was roughly the same as the one being used in many other European countries, for example, in the Federal Republic of Germany. It is based on an assessment of the tree crown with the main emphasis on defoliation. Damage is indicated on a percentage basis comparing each tree to a reference tree and estimating the amount of needle loss. The method is fast to apply in the field, and trained experts can reproduce the measurement in a fairly consistent fashion. The results do not refer to air pollution damage in particular, but instead general forest damage that can be due to any cause. The role of air pollutants can be estimated from these data with the assistance of a statistical data analysis that takes into account the amount of pollutants, site factors, and stand characteristics. A great deal of background data were measured from the same plots and thus it was possible to analyze the statistical relationships between damage variables and other forest variables.

C. Results

The statistical analysis of the data suggested that the high damage frequency observed in northern Finland is due to "natural" factors, that is, age-class distribution and climate. But in the other area of high damage frequency, that is, southernmost Finland, forests can, in fact, suffer from air pollutants. Additional analyses will be carried out to test this hypothesis. It will be examined whether exceptionally high stand densities or unfavorable weather conditions in 1984–1985 contributed to the observed pattern of crown damage in southernmost Finland.

Retrospective investigations have been carried out in order to detect changes in the chemical composition of soils, plant material and, for example, tree bark. Changes in the species composition of forests have also been studied. Raunemaa et al. (1982) conducted chemical analyses of Scots pine needles collected from 1959 to 1979. The samples were from three sites: one in the vicinity of the Helsinki metropolitan area, one in a rural area in southern Finland, and one in a remote forest region in Lapland (68° northern latitude). No statistically significant trends in the chemical composition of needles were observed in Lapland. In the rural site in southern Finland, P and Ca concentrations had increased from 1959 to 1979. The same development was observed on the suburban site where, in addition, increasing trends were observed in the concentrations of Si, S, Ti, and Fe. None of the eleven compounds investigated showed statistically significant declining trends on any of the three sites. Later, however, the same authors published results hinting that from 1975 to 1982 the K concentrations on the rural site in southern Finland may have declined (Raunemaa, Hari, Kukkonen, Kulmala, and Karhula, 1987). Kuusinen and Jukola-Sulonen (1987) studied the trends in the frequencies of different species of lichens in litter samples collected from 1962 to 1982. These data were from three Scots pine stands in rural areas. A clear declining trend in the frequency of beard lichen (*Usnea*) was observed in the two southernmost sites. This trend was not observed in Lapland.

Raunemaa, Hari, Kukkonen, Anttila, and Katainen, (1987) have emphasized the advantages of tree bark over tree rings as a retrospective source of information documenting time series of the chemical composition of trees and the environment. Tree bark, being merely a protective tissue, is assumed to act as a sink for potentially toxic compounds such as heavy metals. Having a low moisture content tree bark has little if any metabolism between bark cells. Bark tissue stores the chemical composition of cells intact over time.

A pioneer in the Finnish research on the effects of air pollutants on forests is Satu Huttunen. Her group at the University of Oulu started the work doing case studies on the effects of industrial and energy production units on surrounding forests (Huttunen, 1973, 1974). Thereafter, the group did research on the morphological symptoms of air pollution damage on tree needles (e.g., Huttunen and Soikkeli, 1984). Particular emphasis has been given to the interaction of pollution stress with other unfavorable components of the environment like severe winter climate (Huttunen, 1983).

In late winter Finnish trees rely on their internal water reserves because the soil water is still bound in frost. Yet on sunny days the irradiance is rather high, inducing high needle temperatures. This enhances transpiration thus bringing about a risk of desiccation. Trees in polluted areas are at a particular risk, because pollutants tend to cause erosion of the wax structures of tree needles (Huttunen and Laine, 1983). A study on the wintertime water economy of Scots pine shoots indicated that trees in polluted areas, indeed, tend to be more susceptible to late winter drought than trees in clean areas (Huttunen, Havas, and Laine, 1981).

In 1985 Finnish forests suffered from unusually severe frost damage. A spatial correlation existed between the intensity of damage and the level of air pollutants in such a way that trees with the most severe damage were observed in southernmost Finland (Kubin and Raitio, 1985). However, the causal link remained uncertain, because the natural tolerance of trees to frost also varies over time and space. The nutritional status of forest soils is being studied experimentally at the Finnish Forest Research Institute, at the University of Helsinki, and within a simulated acidic rain experiment at the subarctic research station of Kevo (University of Turku). The special emphasis at Kevo is on the interaction of acidic deposition and pests.

Derome, Kukkola, and Mälkönen (1986) investigated previously established 20- to 24-year-old experiments on the application of lime on forest ecosystems growing on mineral soils. The results indicated that, 20 years after the application, 65% to 75% of the lime was still present in the humus layer and the top 20 cm of the mineral soil. An increased level of cation exchange capacity and a decreased level of mobile aluminum was observed particularly within the humus layer. However, no enhancement of tree growth was observed associated with these seemingly positive changes in soil conditions.

The modeling approach has for the time being a strong effect on the Finnish research focusing on the impacts of air pollutants on forests. Models are used in order to obtain links between experimental and statistical research, to extract and solicit scattered scientific information, to generate scenarios, and to quantify and compare the different sources of scientific uncertainty. First promoters of the modeling approach were Hari and Raunemaa from the universities of Helsinki and Kuopio. A number of Finnish scientists have done research on the ecological effects of air pollutants at the International Institute for Applied Systems Analysis (IIASA) in Laxenburg, Austria (Alcamo et al., 1987).

Modeling has been done at a general level demonstrating the synergistic effects of different air pollutants (Hari, Raunemaa, and Hautojärvi, 1986). Process-oriented modeling has been done on acidification of forest soils (Kauppi, Kämäri, Posch, Kauppi, and Matzner, 1986). A sensitivity study indicated that the results of the soil model developed (and obviously of other similar models as well) are sometimes reliable and sometimes less reliable (Posch, Kauppi, and Kämäri, 1985). The better the data, in particular on the cation exchange capacity and the initial base saturation of soils, the more reliable are the results from such models. An attempt has been made in collaboration with Göran Ågren of Sweden to model the gradual saturation of forest ecosystems by nitrogen (Ågren and Kauppi, 1983).

Annikki Mäkelä has been the principal investigator in developing a model on the impact of sulphur dioxide on stand foliage (Mäkelä, Materna, and Schöpp, 1987). The model calculates a dose of "sulphur strain" as a function of sulphur dioxide concentration and time. The susceptibility of forest stands to this dose is expressed as a function of climate. The model uses a qualitative risk indicator "forest degradation."

All models require input data on the concentrations and fluxes of pollutants. Studies on the mass balances of forest ecosystems were initiated in the 1980s especially within the Finnish Forest Research Institute. Preliminary results indicated, for example, that pine canopies absorbed substantial fractions of the deposited fluxes of nitrate and ammonium. The topsoils contributed magnesium into deeper soil layers at a rate that exceeded the rate of magnesium deposition (Hyvärinen, 1987; Katainen and Mälkönen, 1987). The models are all consistent in suggesting that forest ecosystems in Finland are undergoing a slow process of change due to air pollutants.

VII. Summary and Conclusions for Policy Recommendations

The presence of sulphur and nitrogen compounds, transported over long distances partly from abroad, can clearly be observed in air quality measurements in Finnish rural areas. Especially the sulphur flux can further be traced in forest ecosystems and in lakes and rivers and also in groundwater. A large number of chemical and biological consequences can be observed in forest and lake ecosystems. The following observations are relevant regarding the theories on the impacts of sulphur and nitrogen pollution on ecosystems.

Alkalinity of large lakes has only slightly declined over the past 20 years. Instead the increase of Ca, Mg, Na, and K concentrations has been very general. An unusual measurement series taken between 1911 and 1931 and repeated from 1962 to 1971 indicates consistent alkalinity decline in large Ostrobothnian rivers including rivers without any major changes in land use. Another measurement series from the early 1960s until today indicates alkalinity decline in small clear water lakes both in summer and winter samples. Paleolimnological studies indicate that pH values in some lakes have been low for centuries but that a considerable decline has taken place over the last few decades in many lakes. Both SO_4^{2-} and Ca^{2+} concentrations in groundwater stations have increased from 1975–1976 to 1982–1983 indicating that sulphur penetrates through the terrestrial ecosystems but acidity is buffered in upper soil layers.

Forest growth has increased in Finland, particularly in the southeastern part of the country where nitrogen and sulphur deposition are at a high level. Mostly this is due to changes in forest management, but statistical analyses suggest also some air pollution effect. The increase of growth is viewed as a warning signal, because it is based on unsustainable growing conditions and may precede forest damage in terms of increased tree mortality. Forest damage, as assessed with the crown density method, is substantially higher in that same region compared to regions with 40% to 60% smaller air pollution loadings. An unusual data series on forest litter, collected without interruption in several forest stands between 1962 and 1982 indicates a clear decline of beard lichens (*Usnea*) on sites in southern Finland. This trend was absent on sites in northern Finland where sulphur air concentrations and deposition are substantially lower. Different experiments lend support to the view that air pollution damage to trees is a result of a combination of high air pollution levels, poor soil conditions, reduced stress resistance of the forest stand, and severe climatic conditions especially over the winter period.

Measures to control air pollution damage to ecosystems include development of surface water liming and forest fertilization programs, strengthening of sustainable forest management and, as the main target, implementation of air pollution abatement. Much emphasis is given on international joint agreements on reducing air pollution emissions. Finland, on a unilateral basis, would have had to reduce its own emissions by 85%, in order to achieve the same reduction of sulphur deposition to Finland that will be achieved by implementing the multilateral UN/ECE agreement to reduce emissions by 30%. Yet, even after this first agreement, both sulphur and nitrogen emissions are still regarded too high in Finland to allow ecosystems to function on a sustainable basis.

Appendix

FINNISH RESEARCH PROJECT ON ACIDIFICATION (HAPRO)
Research program

1. Forest Research

   Survey and Monitoring
   The effect of air pollution on forest ecosystems: the national monitoring program

   Situation of air pollution since the 1950s to the present time according to composition of litter materials

   Impaction Mechanisms
   The effect of acidic deposition on the fungi in forest humus

   Influence of air pollutants on the chemical contents of surface soil and on the decomposition of litter in polluted forests

   Air pollution and Finnish forest injury: mode and tempo in pest succession

   Effects of air pollution on the susceptibility of forest trees to diseases

Effects of air pollutants on peatland sites monitored by mass-balance studies and peat microbiology

Acidic deposition and wintering of forest plants

The transport of pollutants into the forest

Effects of changes in the atmospheric composition on the productivity of forests: Effects of changes in soil properties

The role of toxic metals in forest die-off

Identification of the causes of injuries to conifers in Finnish industrial environments

The effect of acidic deposition on the biological properties of mineral soil sites

Application of a forest model to evaluate the long-term effect of acidic deposition on forest growth

Microbial flora of coniferous trees

Acidification Control
Effects of liming on peat chemistry and forest growth on peatlands

Economic Effects
Pollution impacts on European forestry

2. Water Research

Regional Distribution and History of the Acidification of Surface and Groundwater

Inventory of lake acidification

Acidification history of small lakes in Finland

Effects of air pollutants on the quality of groundwater

National well water study

Diatoms of surface sediments and lake acidity

Relations Between Emission, Deposition, Soil and Bedrock Type, and the Quality of Water

Effects of acidic deposition on aquatic ecosystems at different sulphur emission scenarios

Roles of acidic deposition and humic substances in the acidification of Finnish lakes

Influence of hydrological factors on the acidity of watercourses

Deposition quality in Finland

Buffering capacity of soil and its dependence on geological factors

Proton budgets in small drainage basins

Effects of Acidification of Aquatic Ecosystems on Fish and Other Aquatic Organisms

Effects of acidic deposition on fish

Physiological regulation of fish in acidic environment: disturbances in acid-base and ionoregulation of eggs, embryos, fry, and adults

Airborne heavy metal load on lakes

Effects of airborne acidification on the biota of lakes

Acidification Control
Liming of acidified lakes, neutralization efficiency, and cost

The role of humic substances in the liming of water

3. Atmospheric Studies
NO$_x$ emissions in the 1980s in Finland

Ammonia emissions in the 1980s in Finland

Extreme values of sulphur compounds in background air

Occurrence of ozone in Finland

Formation of acidic aerosols in the Finnish atmosphere

Chemical transformation of organic pollutants in simulated atmospheres

Transformation and deposition of gaseous and particulate pollutants over the Baltic sea

Acidification episodes climatology

Deposition of nitrogen oxides in Finland

The atmospheric nitrogen compounds in Finland

Intercalibration of the analytical methods

Role of the hydrocarbon emissions of forests in the ozone formation

4. Effects on Soil and Materials
Impacts of Air Pollutants on Crops and Agricultural Soils;

Changes in pH and in the contents of extractable metals in soil during the last 13 years

Effects of acidifying air impurities (SO$_2$) on materials

The Geochemical Impacts of the Acidification in Finland

5. Emission Control
Treatment, placing, and utilization of desulphurization products from coal-fired power plants

Effective choice of SO$_x$ and NO$_x$ emissions control methods

6. General Studies
Development of an overall environment impact assessment model for Finnish conditions

References

Aarna, A. 1985. *Estonian Nature* 3:130–137 [In Estonian].

Ågren, G.I., and P. Kauppi. 1983. IIASA collaborative paper, 83–28.

Alasaarela, E., and P. Heinonen. 1984. Publications of the Water Research Institute, National Board of Waters, Finland 57:3–13.

Alcamo, J., M. Amann, J.-P. Hettelingh, M. Holmberg, L. Hordijk, J. Kämäri, L. Kauppi, P. Kauppi, G. Kornai, and A. Mäkelä. 1987. *Acidification in Europe: A simulation model for evaluating control strategies.* Ambio 16:232–245.

Anttila, P. 1987. In P. Anttila and P. Kauppi (eds.), *Abstracts of the Symposium of the Finnish Research Project on Acidification* (HAPRO), Abstracts Ministry of the Environment Environmental Protection Department, series A 64:15–16.

Anttila, P., M. Kulmala, and T. Raunemaa. 1987. *Aquilo Ser Bot* 25:20–28 [In Finnish].

Arovaara, H., P. Hari, and K. Kuusela. 1984. *Commun Inst For Fenn* 122:1–16.

Derome, J., M. Kukkola, and E. Mälkönen. 1986. National Swedish Environment Protection Board report 3084.

Forsius, M., K. Kenttämies, and J. Kämäri. 1987. Int. symp. on Acidification and Water Pathways, 4.-5. May 1987, Norway, Proc. vol. II:177–186.

Hakola, H., P. Taalas, H. Lättilä, S. Joffre, and P. Plathan. 1987. Finnish Meteorological Institute Reports 6 [In Finnish].

Hari, P., H. Arovaara, T. Raunemaa, and A. Hautojärvi. 1984. *Can J For Res* 14:437–440.

Hari, P., T. Raunemaa, and A. Hautojärvi. 1986. *Atmospheric Environment* 20:129–137.

Holmberg, L. 1935. Hydrografisen toimiston tiedonantoja (Finland) V:1–54.

Huttunen, S. 1973. *Aquilo Ser Bot* 12:1–11.

Huttunen, S. 1974. *Aquilo Ser Bot* 13:23–34.

Huttunen, S. 1983. In M. Treshow (ed.), *Air Pollution and Plant Life,* 321–356. New York: Wiley.

Huttunen, S., P. Havas, and K. Laine. 1981. *Holarct Ecol* 4:94–101.

Huttunen, S., and K. Laine. 1983. *Ann. Bot. Fennici* 20:76–86.

Huttunen, S., and S. Soikkeli. 1984. In A. Kozial and C. Whatley (eds.), *Gaseous Air Pollutants and Plant Metabolism,* 117–127. Butterworths. London.

Hyvärinen, A. 1987. In P. Anttila and P. Kauppi (ed.), *Abstracts of the Symposium of the Finnish Research Project on Acidification* (HAPRO), Abstracts Ministry of the Environment Environmental Protection Department, series A 64, p. 77.

Järvinen, O. 1986. Papers of the National Board of Waters 408 [In Finnish].

Joffre, S., H. Lättilä, H. Hakola, P. Taalas, and P. Plathan. 1988. In I. Isaksen (ed.), *Tropospheric Ozone NATO ASF Series,* 137–146. Dordrecht: Reidel.

Jukola-Sulonen, E-L., K. Mikkola, S. Nevalainen, and H. Yli-Kojola. 1987. The Finnish Forest Research Institute Research Bulletin 256.

Kämäri, J. 1985. *Aqua Fennica* 15(1):11–20.

Kämäri, J. 1986. *Aqua Fennica* 16(2):211–219.

Katainen, H.S., and E. Mälkönen. 1987. In P. Anttila and P. Kauppi (eds.), *Abstracts of the Symposium of the Finnish Research Project on Acidification* (HAPRO), Abstracts Ministry of the Environment Environmental Protection Department, series A 64, p. 70.

Kauppi, P., J. Kämäri, P. Posch, L. Kauppi, and E. Matzner. 1986. *Ecol. Modelling* 33:231–253.

Kauppi, P., P. Antilla, and K. Kenhämies (eds.), 1990. *Acidification in Finland*, Springer Verlag, Heidelberg.

Kenttämies, K. 1973. *Vesitalous* 3:22–23.

Kenttämies, K. 1979. Publications of the Water Research Institute, National Board of Waters, Finland 30:42–45.

Kenttämies, K., S. Haapaniemi, J. Hynynen, P. Joki-Heiskala, and J. Kämäri. 1985. *Aqua Fennica* 15(1):21–33.

Keränen, S., and R. Niskanen. 1987. Ministry of the Environment, Environmental Protection Department; series D 30 [In Finnish].

Kortelainen, P., and J. Mannio. 1986. *Aqua Fennica* 16(2):221–229.

Kortelainen, P., and J. Mannio. 1987. Int. Symp. on Acidification and Water Pathways, 4.-5. May 1987, Norway, Proc. vol. II:229–238.

Kubin, E., and H. Raitio. 1985. The Finnish Forest Research Institute Research Bulletin 198 [In Finnish].

Kulmala, A., A. Estlander, L. Leinonen, T. Ruoho-Airola, and T. Säynätkari. 1982. Commun. of the Finnish Meteorological Institute 36 [In Finnish].

Kuusinen, M., and E-L. Jukola-Sulonen. 1987. In P. Anttila and P. Kauppi (eds.), *Abstracts of the Symposium of the Finnish Research Project on Acidification* (HAPRO) Abstracts Ministry of the Environment Environmental Protection Department, series A 64, p. 49.

Laaksonen, R. 1982. *Hydrobiologia* 86:159–160.

Laaksonen, R., and V. Malin. 1980. Publications of the Water Research Institute, National Board of Waters, Finland 36. 70 pp.

Laaksonen, R., and V. Malin. 1984. Publications of the Water Research Institute, National Board of Waters, Finland 57:59–60.

Laurila, T., and S. Joffre. 1987. In R. Perry, R.M. Harrison, J.N.B. Bell, and J.N. Lester (eds.), *Acid Rain, Scientific and Technical Advances*, 167–174. Selper Publ., London.

Mäkelä, A., J. Materna, and W. Schöpp. 1987. IIASA working paper 87-57.

Ministry of the Environment. 1988. Environmental Protection Department Series A. 67. 1988.

Nordlund, G., and J.-P. Tuovinen. 1988. Modeling long-term averages of sulfur deposition on a regional scale. In: Proceedings of the WMO conference on air pollution modeling and its application (Vol. III), 19–24 May 1989, Leningrad. WMO/TD No. 187, pp. 53–61.

Oliver, B.G., E.M. Thurman, and R.L. Malcolm. 1983. *Geochim Cosmochim Acta* 47:2031–2035.

Ovaskainen, V. 1987. IIASA working paper 87-37.

Posch, M., L. Kauppi, and J. Kämäri. 1985. IIASA collaborative paper 85-45.

Rask, M. 1987. *On the ecology of perch*, Perca fluviatilis L., *in acid waters of southern Finland*. Dept. of Zool. and Lammi Biol. Station. University of Helsinki, 105 pp.

Raunemaa, T., P. Hari, J. Kukkonen, P. Anttila, and H-S. Katainen. 1987. *Can J For Res* 17:466–471.

Raunemaa, T., P. Hari, J. Kukkonen, M. Kulmala, and M. Karhula. 1987. *W.A.S.P.* 32:445–453.

Raunemaa, T., A. Hautojärvi, K. Kaisla, M. Gerlander, R. Erkinjuntti, T. Tuomi, P. Hari, S. Kellomäki, H-S. Katainen. 1982. *Can J For Res* 12:384–390.

Rühling, Å., L. Rasmussen, K. Pilegaard, A. Mäkinen, and E. Steinnes. 1987. The Nordic Council of Ministers NORD 1987:21.

Salmi, T., and S. Joffre. 1987. In P. Anttila and P. Kauppi (eds.), *Symposium of the Finnish Research Project on Acidification* (HAPRO), Abstracts Ministry of the Environment, Environmental Protection Department, series A 64, p. 21.

Simola, H., K. Kenttämies, and O. Sandman. 1985. *Aqua Fennica* 15(2):245–255.

Soveri, J. 1985. Publications of the Water Research Institute, National Board of Waters, Finland 63. 92 pp.

Tolonen, K., and T. Jaakkola. 1983. *Ann Bot Fennici* 20:57–78.

Tolonen, K., M. Liukkonen, R. Harjula, and A. Pätilä. 1986. *Developments in Hydrobiology* 29:169–199.

Turkia, J., and P. Huttunen. 1987. *Proceedings of Nordic diatomist meeting,* Stockholm 1987. University of Stockholm, Department of Quarternary Research (USDQR) Report 12 (1988), pp. 99–102.

Tuunainen, P., P.J. Vuorinen, M. Rask, T. Järvenpää, and M. Vuorinen. 1987. Rep. of Finnish Game and Fisheries Res. Inst. 67. 72 pp.

Verta, M., and J. Mannio. 1987. Int. Symp. on Acidification and Water Pathways, 4.-5. May 1987, Norway, Proc. vol. II:343–352.

Viisanen, Y., M. Kulmala, R. Hillamo, J. Hatakka, and P. Keronen. 1987. *J Aerosol Sci* 18:829–831.

Yearbook of forest statistics. 1987. Folia Forestalia 690.

Acidic Precipitation Research in France

M. Bonneau* and C. Elichegaray†

Abstract

Some cases of forest decline were known 10 or 20 years ago in certain regions of France, mainly in the vicinity of industrial cities, but also as a consequence of the severe drought of 1976. But, beginning in 1983, forest decline, with the same symptoms as observed in Germany, become visible in the Vosges and was noticed in the following years in the Jura, the Alps, and the Massif Central. At the end of 1986, 10% to 20% of the trees in these regions had lost more than 25% of their foliage.

A monitoring and research program, DEFORPA, was launched by the Ministry of Environment in the autumn of 1983 and today is administered by three ministeries: Environment, Agriculture, and Research. DEFORPA is divided into four main parts:

1. damage survey, symptomatic studies, and relationship with ecological conditions,
2. characterization of atmospheric pollution,
3. experiments on the direct action of atmospheric pollutants, and
4. study of the effects of air pollution through the soil.

The level of SO_2 in the Vosges is on the average rather low (15 μg m^{-3}) but there are several strong peaks in winter. The level of O_3 is more even, with peaks in summer and rather weak day-night variation.

The yellowing of Norway spruce and silver fir needles are related to nutrient deficiencies, mainly Mg and Ca, but sometimes K, according to the region.

Tree core analysis on silver fir in the Vosges and the first results of experiments in open-top chambers and in controlled conditions, as well as

---

*Institut National de la Recherche Agronomique, Centre de Recherches Forestières de Nancy, Champenoux, 54280 Seichamps, France.

†Agence pour la Qualité de l'Air, Tour GAN, Cédex 13, 92082 PARIS La Défense, France.

physiological characterization of declining trees in the forests, suggest that the current decline is the result of past drought periods, whose consequences might be worsened by acidic pollution and perhaps SO_2 peaks in winter.

I. Introduction

Since 1983, forest decline has become very obvious in France where the Ministry of Environment had already initiated a program some years before to measure and study acidic precipitation. The decline in the health of the forests naturally led to an acceleration in this program, which was augmented by another study into the causes of forest damage.

II. A Brief Historical Rundown of Research into Acidic Rain

A. The Rain Monitoring Network

The survey of the chemical composition of rain was, until 1987, undertaken by 10 monitoring stations that were set up within two different networks.

The *EMEP network* (European Monitoring and Evaluation Program) was created in 1977 and is composed of five stations (Figure 12–1): Cholet, La Crouzille, La Hague, Valduc, and Vert-le-Petit. They are managed by the Commissariat à l'Energie Atomique [1] or the Institut National de Recherches Chimiques Appliquées.[2]

The *BAPMON network* (Background Air Pollution Monitoring Network) was also created in 1977 and it, too, is composed of five monitoring stations (Figure 12–1): Abbeville, Carpentras, Gourdon, Phalsbourg, and Rostrenen. They are managed by the Météorologie Nationale.[3]

These 10 stations are equipped with rainfall gauges that open automatically (in order to collect the wet deposition only). Analysis is done weekly in the BAPMON network and daily in the EMEP network. The latter also has measuring appliances for aerosols and SO_2.

At the very beginning of "forest decline," the Ministry of Environment decided to increase the number of stations in the French network. Eight new stations were added to the original 10: Mt. Aigoual, Bonnevaux, Le Casset, Chartreuse, Donon, Iraty, Morvan, and Revin (Figure 12–1). These new monitoring stations were chosen to measure rural zone pollu-

---

[1]French Atomic Commission.
[2]National Institute of Applied Chemical Research.
[3]French Meteorological Office.

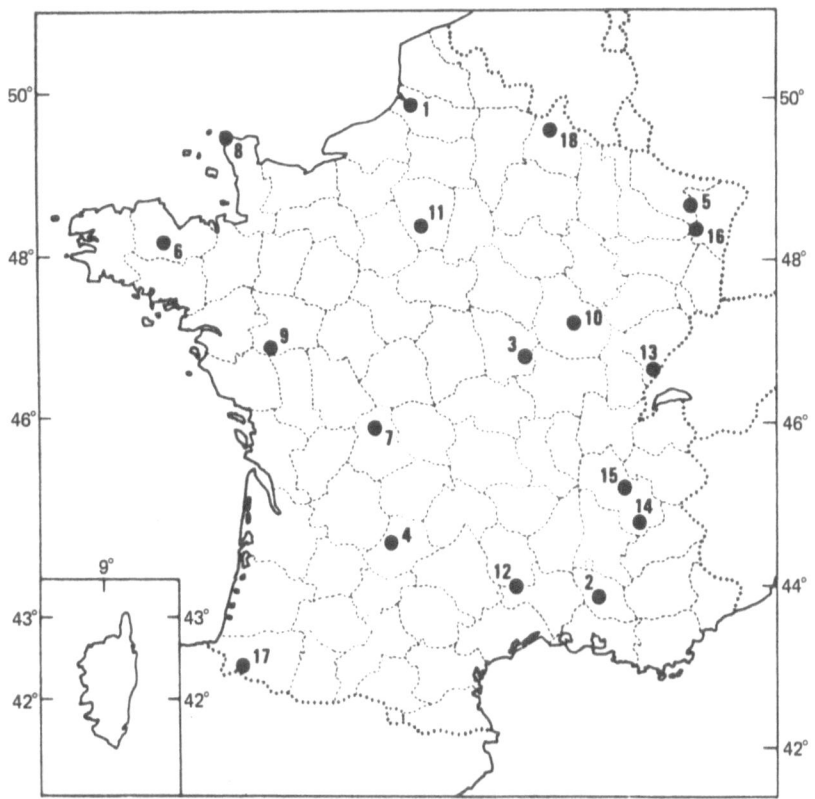

Figure 12-1. Stations of the RENAMERA rain monitoring network: (1) Abbeville; (2) Carpentras; (3) Morvan; (4) Gourdon; (5) Phalsbourg; (6) Rostrenen; (7) La Crouzille; (8) La Hague; (9) Cholet; (10) Valduc; (11) Vert-le-Petit; (12) Mont Aigoual; (13) Bonnevaux; (14) Le Casset; (15) Chartreuse; (16) Donon; (17) Iraty; (18) Revin.

tion. As with the EMEP stations, they are composed of rain gauges and other apparatuses for measuring aerosols and SO_2. They are managed by the Associations Locales de Surveillance de la Pollution de l'Air,[4] the Directions Régionales de l'Industrie et de la Recherche, [5] and the Administration des Parcs Naturels.[6]

The set of these 18 stations is called "Réseau MERA" (Réseau de Mesure des Retombées Acides).[7]

---

[4]Local Air Pollution Survey Associations.
[5]Regional Services of Industry and Research.
[6]The National Parks Administration.
[7]National Acidic Deposition Monitoring Network.

B. The Beginning of Forest Decline

Cases of forest decline have been known for some time in certain French regions. Since 1960, for example, the foresters in the Luchon region in the Pyrenees have noticed a yellowing at the tops of the fir trees in some hundreds of hectares, which was followed in certain cases by the death of the trees. No provable cause could be found, but flurohydric acidic emissions from an aluminum factory, a few tens of kilometers away, were suspected.

At roughly the same time an important mortality rate in coniferous and also broad-leaved trees was noted in Maurienne in the Alps, and the cause was clearly due to fluorine emissions from aluminum factories. It was also proved, some years later, that numerous Scotch pine deaths in the outskirts of Rouen were due to SO_2 and fluorine emissions from the chemical industries in the city. In these two cases, measures were taken to reduce the emissions, and the forests in these regions subsequently recovered.

After the drought of 1976, an important mortality rate in the pedunculate oak was noticed in the center of France and in the Pays Basque (southwest). At about the same time, silver fir stands in the Laigle forest (Normandy) underwent an important health loss characterized by a loss of needles and in some cases death. It is quite certain that the forest decline in these three regions was due to the drought of 1976, and also others of preceding years.

For a long time certain stands at low altitude on the Alsatian hills of the Vosges have been known for their low production and short life span. This has been attributed to the insufficient rainfall in the region.

Apart from the preceding examples, the French forests stayed healthy until the end of 1982, whereas decline in Germany had been recognized since the 1970s.

Beginning in the spring of 1983, however, the same symptoms of decline as in Germany became visible in the Vosges; abnormal needle loss was already noticed but discolored foliage was still very rare. During the summer of 1984, there was a clear worsening of the situation, with an increase in needle loss and a higher number of spruces affected by the yellowing process, which continued again in 1985. In 1984 and 1985 forest decline was observed, but with less intensity, in the Jura, in the northern Alps, and to the east of the Massif Central.

C. Creation of the DEFORPA Program

In the autumn of 1983, the Department des Recherches Forestières of the Institut National de la Recherche Agronomique (INRA), [8] the Direction

---

[8]Department of Forest Research of the National Institute for Agronomical Research.

des Forêts du Ministère de l'Agriculture,[9] and the Office National des Forêts[10] put into operation a forest survey network, only in the Vosges at the beginning, then expanded in 1984, 1985, and 1986 to the main French forestry regions. This was done with the help of the Centres Régionaux de la Propriété Forestière[11] in the Jura, the Alps, the Massif Central, the Pyrenees, the Lorraine plateau, the Ile de France, Normandy, and the North.

At the same time, a group of experts was established under the direction of the Ministry of Environment to undertake a research program into the causes of forest decline.

III. The Structure and Goals of the DEFORPA Program

A. Administrative Structure

DEFORPA (Dépérissement des Forêts et Pollution Atmosphérique[12] today operates under the jurisdiction of three ministeries: Environment, Research, and Agriculture. Since the autumn of 1985, it has been managed by three different bodies.

A *steering committee,* presided over by the head of the Research Service from the Ministry of Environment, with 13 members coming from the main public bodies whose role is to approve the research projects and to assure their financing.

A *scientific committee* of 15 members whose role is to define the large scientific directives, to give an opinion to the steering committee on the quality of the projects presented, and to evaluate the research results.

An *operational group* that initiates the research projects, puts the program into operation, and assures contact with the laboratories that carry out the work. It also consists of 15 members, 8 of them being responsible for a research group in a determined scientific discipline.

More than 40 laboratories, belonging to 20 scientific bodies, universities, and professionals, participate in this program: Centre National de la Recherche Scientifique,[13] Institut National de la Recherche Agronomique, Commissariat à l'Energie Atomique, Direction de l'Espace Rural et de la Forêt,[14] Office National des Forêts, Electricité de France,[15]

---

[9]Forest Direction of the Ministry of Agriclture.

[10]National Forest Service (the body which supervises the management of national and communal forests).

[11]The body representing private forest owners in the different regions.

[12]Forest decline Attributed to Atmospheric Pollution.

[13]National Scientific Research Organization.

[14]Administration for Rural Space and Forest.

[15]French Electricity Agency (equivalent to the CEGB).

Météorologie Nationale, Agence pour la Qualité de l'Air,[16] Institut National de Recherches Chimiques Appliquées, Association pour la Surveillance de la Pollution Atmosphérique en Alsace,[17] Centre de Recherches en Environnement du Département des Pyrénées Atlantiques,[18] Universités de Paris VII, Paris XII, Bordeaux, Clermont-Ferrand, Toulouse, Besançon, Strasbourg, Nancy, Lille, Orléans, Ecole des Mines[19] de Douai, Société Europoll.

B. Scientific Goals

The objective of the program is to conduct research into the possible causes of the forest decline and to find out whether it is due to atmospheric pollution or other anthropogenic or natural causes. The work is divided into four main parts.

1. Damage Observation

Ground damage survey using an annual examination of permanent sample plots at 1 km distances on the east-west axis and every 16 km from north to south. To this "blue network" was added a "red network," which followed the evolution of very characteristic stands.

Research into aerial and spatial teledetection.

Symptomatic studies and their relationship to ecological conditions.

The possible role of soil microflora : research into the possible role of root parasites in the causes of decline or in its worsening. There is also a study of the behavior and the activity of mycorrhizae.

2. Characterization of Atmospheric Pollution

Two air pollution monitoring stations are functioning in the Vosges. The main one is at the Donon Pass at an altitude of 750 m (SO_2, O_3, NO_x, rain quality by 1/10 mm samples). It was replaced in 1988 by a tower placed in a damaged stand on which the gaseous pollutants can be measured at three different levels. In addition to the measurements mentioned earlier, others are carried out on snow and aerosols in a sporadic manner. The second is near Aubure, 30 km to the south at an altitude of 1100 m at the top of an equipped watershed. The pollutants SO_2, O_3, and NO_x and rain are analyzed here.

Along with these measurements, the arrival trajectories of these pollutants in the Vosges, the Jura, and the Alps were determined, and a model

---

[16]Air Quality Agency.
[17]Association for Air Pollution Monitoring in Alsace.
[18]Environment Research Center of the Department of the Atlantic Pyrenees.
[19]Mining High School.

of the oxidation of SO_2 and NO_x, according to the meteorological conditions, is in progress.

3. The Direct Action of Atmospheric Pollutants

Research on the direct action of atmospheric pollutants was carried out using

a. Experiments on the Norway spruce using either open-top chambers (an exclusion experiment in the Donon Pass and a simulation experiment with SO_2, O_3, and $SO_2 + O_3$ in the Pyrénées Atlantiques), or climatized chambers, as well as an experiment on the effects of acidic fog.
b. Physiological work, in order to characterize the functioning of the spruce used in the experiments as compared to the healthy or declining spruce chosen in forest stands.

4. The Effects on Soil as an Intermediary in Acidic Pollution

This fourth part of the program consists of three distinct themes:

a. A study on the ionic aluminum content of the soil water on two different soils in the Vosges.
b. A study on the cation loss due to acidic deposition, dry or wet, on the throughfall and on the behavior of airborne sulphur and nitrogen. These studies are carried out in a watershed near Aubure on the Alsatian side of the Vosges.
c. Fertilization trials : Several fertilization trials were installed in the Vosges, either on adult fir (6 trials) or on spruce (2 trials) or on young plantations of discolored spruce. The fertilizers used were mostly calcium- or magnesium-based. The experimental design was the same on all adult stands. For the young spruce plantations, the best fertilizer supply was chosen after needle analysis.
   The whole program was put into place at the end of 1984.

IV. The First Results of the DEFORPA Program

Only the results of a part of the program will be described here because many of the experiments (open-top chambers and fertilization trials) have not been operating long enough to provide clear results.

A. The Actual State of Forest Decline

Forest damage is mainly characterized by foliar loss and needle or leaf discoloration.

Defoliation has been classified, as in Germany, into five different classes:

1. Healthy trees: foliar loss (or more exactly, lack of foliage) in relation to an ideal state, less than 9%.
2. Lightly defoliated trees: foliar loss between 10% and 24%.
3. Moderately defoliated trees: foliar loss between 25% and 59%.
4. Severely defoliated trees: foliar loss between 60% and 99%.
5. Dead trees.

Figure 12–2 shows the percentage of moderately or severely defoliated and dead trees for the main forest regions at the end of the summer of 1986. It is obvious that it is primarily the forests in eastern France that are seriously affected: mainly mountain forests, the Vosges, Jura, and Alps, but also some regions on the plains (Alsace plain and Lorraine plateau) where damage is more than negligible. In the Paris basin, Normandy, and the North, the forests are very slightly damaged. The

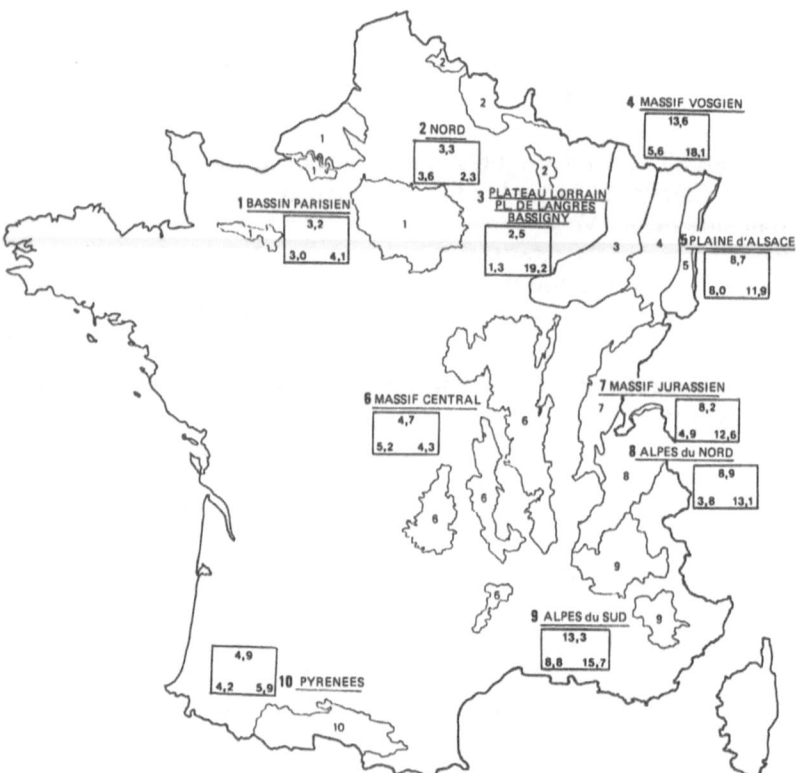

Figure 12–2. Loss of foliage in the main forest regions in autumn 1986. Upper number: all species; lower number: *left*—broad-leaved species, *right*—coniferous species (from Buffet, 1987).

Massif pyrénéen is only slightly affected by decline, except for 1000 ha near Luchon where the silver fir is discoloring increasingly. This damage has a tendency to spread toward the west.

Figure 12–2 also shows that in all regions the conifers are more damaged than broad-leaved trees, but whereas damage remained stable on conifers between 1985 and 1986, the damage on broad-leaved trees increased slightly. Since 1986 the defoliation level has remained about the same with slight fluctuations from year to year, except in the Pyrenees.

It is very important to note that these percentages, which concern foliar deficit only, include at the same time forest trees that have been well known for a long time to have a low vitality (low altitude fir, old stands, those overexposed to wind or snow) as well as trees that have lost some of their foliage in the last few years.

1. Vosges

Defoliation has affected all parts of the Vosges, but to different degrees. Generally, it is more widespread between altitudes of 650 m and 900 m, but there are quite a few exceptions, notably on sandstone in the Basses Vosges. Two regions are more affected than most: the crests in the north and south of the Donon Pass and in the south of the Bonhomme Pass. The tops and the middle parts of the slopes are more affected than the bottoms where, for the most part, the stands are in very good health. Since 1985 the condition of the fir has tended to improve; the situation for the spruce has remained stable. This defoliation is not generally accompanied by a high mortality rate except in some areas.

Discoloration, still rare in 1983, spread spectacularly from 1984 to 1985. In 1986, it rested stable overall. It affects all species and ages, but is more frequent in the spruce. Defoliation is observed in all ecological conditions but discoloration is strongly concentrated on the poorest geological substratas, very acidic granites, and vosgian sandstone. It is the same discoloration as already seen in the Black Forest. The youngest needles remain green while those of preceding years turn yellow. In broad-leaved trees, especially in the beech, the edges of the leaves and the spaces between the nerves become yellow while the part nearest to the nerve stays green.

2. Jura

It is in the Haut-Jura that damage among resinous trees, especially the fir, is the most important (13% of spruce and 27% of fir have a needle deficit of more than 25%). The damage is equally important for the fir on the "intermediate slopes," a region situated a little to the west of the Haut-Jura. There is important discoloration, also in the Haut-Jura, either scattered or concentrated in certain areas like the Faucille Pass and the Bugey for the fir. A decline has also been noticed in the oak on the "First Plateau," at relatively low altitude.

3. Alps

In the Alpes du Nord, three mountainous areas are particularly affected in the neighborhood of Grenoble: Chartreuse, Vercors, and Belledone. The first two regions are the most affected. A local SO_2 pollution is partly responsible for the damage in the Chartreuse area. Discoloration is scattered, especially for the fir in these three areas.

In the Alpes du Sud damage affects mostly Scotch pine in the Haute-Provence region.

4. Massif Central

Defoliation in the spruce in the Monts Dôme region is obvious, but the fir trees keep their abundant foliage. On the other hand, there is a discoloration in the fir in the Cantal and the Forez. Discoloration in the spruce is observed on the highest part of the Plateau de Millevaches to the west of the Massif Central.

5. The Pyrenees

The state of the forests in the Pyrenean chain, which was very satisfactory in 1986 except for a strong discoloration in the silver fir in the central Pyrenees near Bagnère de Luchon, has worsened between 1986 and 1988.

B. The Relationship Between Soil, Nutrition, and Decline

Needle analysis has been carried out in the Vosges since 1984 on approximately 40 fir and spruce adult stands situated on different soils and parent rocks and covering all the degrees of decline. It has been found that the magnesium level in the needles is clearly the weakest in the most declining stands (Figure 12–3b, Table 12–1). In trees with a deficit of more than 40%, the magnesium level of current year needles is in the order of 0.07% in the fir and 0.06% in the spruce. It goes down to 0.045% on the average for current year needles (still green) in discolored spruce trees. There is a similar relationship for calcium (Figure 12–3a) and for zinc (Table 12–1).

These results are exactly the same as German researchers have found in the Black Forest. Levels of nitrogen, phosphorus, and potassium, usually sufficient or even high, do not vary according to the intensity of the decline. Manganese is more abundant in needles from deficient trees than from healthy ones. Sulphur is always at a moderate level and does not change with the health of the tree. Table 12–1 gives the levels of the principal elements for the autumn of 1984 (Bonneau and Landmann, 1987; Landmann, Bonneau, and Adrian, 1987).

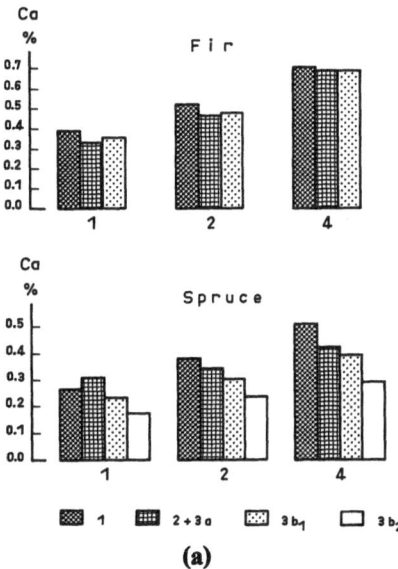

Figure 12-3a. Ca content in the needles of silver fir and Norway spruce of different crown transparencies in the Vosges in autumn 1984. 1: 0% to 9% of needle loss (healthy trees); 2 + 3a: 10% to 39% of needle loss (slightly damaged); 3b¹: 40% to 59% of needle loss (moderately damaged); 3b²: strongly yellowing trees; *Age of the needles*: (1) first year needles, (2) second year needles, (4) fourth year needles (from Landmann et al., 1987).

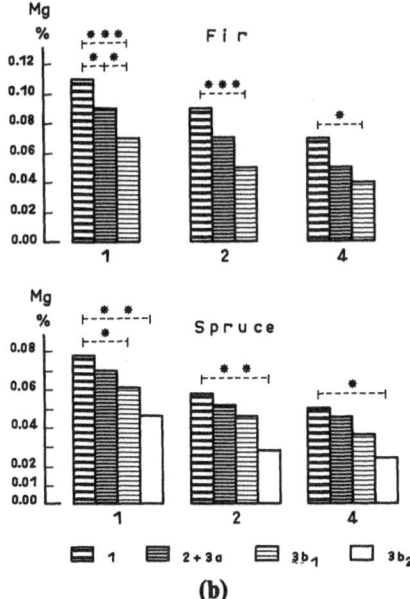

Figure 12-3b. Mg content in the needles of spruce in the Vosges in 1984 (see caption of Figure 12-3a).

Table 12-1. Nutrient contents of the first- and fourth-year needles of white fir (F) and Norway Spruce (S) in the Vosges in autumn 1984 (Landmann et al., 1987).

Needle loss, tree species, and age of needles		N a	P a	K a	Ca a	Mg a	Mn b	Zn b	S a	Cl a
0%–9%										
F	1 yr	1.35	0.18	0.72	0.39	0.11	750	31	0.11	0.10
	4 yr	1.40	0.12	0.49	0.70	0.07	1.200	20	0.13	0.09
S	1 yr	1.50	0.16	0.69	0.27	0.08	550	27	0.11	0.10
	4 yr	1.40	0.10	0.41	0.51	0.05	800	16	0.10	0.08
10%–34%										
F	1 yr	1.35	0.19	0.65	0.34	0.09	950	28	0.12	0.09
	4 yr	1.38	0.12	0.47	0.69	0.05	1.900	25	0.12	0.08
S	1 yr	1.38	0.16	0.65	0.31	0.07	980	26	0.09	0.14
	4 yr	1.25	0.11	0.42	0.43	0.04	1.300	16	0.10	–
35%–60%										
	1 yr	1.40	0.16	0.70	0.35	0.07	1.300	25	0.12	0.11
	4 yr	1.40	0.10	0.48	0.70	0.04	2.100	23	0.12	0.08
S	1 yr	1.42	0.16	0.70	0.20	0.05	1.050	21	0.10	0.10
	4 yr	1.20	0.10	0.43	0.40	0.03	1.250	14	0.11	0.11

[a] % of D.M.

[b] mg kg^{-1}

Soil analysis carried out in the same sites as the needle analysis shows that the affected stands, whose needles were lacking in magnesium, are situated on poor soils. They have a notably low Mg exchangeable/Al exchangeable ratio: 0.03 in Al horizon and 0.01 in the mineral horizons for unhealthy stands versus 0.1 and 0.03 respectively, for healthy stands.

Nutritional characteristics in young discolored spruce plantations are the same as for the adult trees.

In the Monts du Forez in the Massif Central, there is a similar situation, the most declining trees showing a marked shortage of magnesium.

In the Jura and on the basalt soils of the Massif Central the relationship between decline, nutrition, and soil characteristics seems to be much less clear. In the Jura, declining stands can be seen equally on soils that are only slightly acidic, rich in calcium (brown calcic soils, humo-calcic soils) as on brown leached, more acidic soils. On the other hand, it seems that the most affected stands are on the shallowest soils (Bruckert and Boun Suy Tan, 1987). The nutrition is very different from that in the Vosges. Magnesium and calcium levels in the needles are always satisfactory but nitrogen, phosphorus, and potassium levels are weaker. This tendency for weaker levels is often found in stands whose foliar loss reaches or is higher than 40%. Table 12-2 shows the needle analysis results in the Jura.

In the Luchon region on schists, the highly discolored firs are characterized by an acute deficiency in potassium, with a current year needle

Table 12–2. Nutrient contents of the first- and fourth-year needles of White fir and Norway spruce in the Jura in autumn 1985 (Bonneau and Landmann, 1987).

Needle loss and age of needles		N a	P a	K a	Mg a	Mn b
0%–9%						
	1 yr	1.33	0.12	0.53	0.10	496
	4 yr	1.27	0.09	0.44	0.08	686
15%–59%						
	1 yr	1.24	0.12	0.50	0.09	340
	4 yr	1.11	0.08	0.35	0.07	330

[a] % of D.M.

[b] mg kg^{-1}

level of 0.30% only, as against 0.53% in the healthy trees. In the fourth year needles, the potassium level is still very weak: 0.17% in discolored trees and 0.39% in the healthy trees (Cheret, 1987; Cheret, Dagnac, and Fromard, 1987).

Thus it appears that in a certain number of cases, notably on the sandstone and certain granites in the Vosges, the gneiss and the granites of the Massif Central, and the schists of the Luchon region in the Pyrenees, there is a direct correlation between the nutrition level in certain elements and the degree of decline; but the deficient element varies from region to region. This observation does not contradict the hypothesis that the decline could be due to a leaching of elements on the foliage or in the soil under the effect of acidic deposition. The food supply of the tree would be seriously compromised vis-à-vis the elements that are rare in the soil.

The aluminum level in gravitary or capillary soil water was studied on two sandstone soils in the Vosges. It is mostly the gravitary water that is strong in this element, but the aluminum level varied enormously depending on the time of year and is in direct correlation to the level in nitrates. During a rainy period that follows a long period of drought, like the autumn of 1985, one can have up to 3 mg and even 8 mg of ionic aluminum L^{-1} in the seepage water, with a very low level of calcium (ratio Ca/Al much less than 1). Therefore it cannot be ruled out that in certain cases the tree roots could be damaged by aluminum under certain circumstances (Gras, 1987).

C. Results on the Growth of the Fir in the Vosges Since the End of the Nineteenth Century

When the first work began on forest decline, certain results led to the supposition that the degradation in the health of the forests was the end of an old process affecting all forests, which was accompanied by an important growth reduction. In France the results of the National Forestry Inventory showed a drop in production of 40% for conifers in the Vosges in 1972 and therefore seemed to confirm this idea.

A systematic study on the ring width of 1200 firs (6 trees on 200 sample plots representing all the ecological conditions in the Vosgian Massif) has shown that in reality circumstances were different (Becker, 1987).

In this work it was necessary to use methods to remove the effect of the age of the tree because the more a tree ages, the narrower the ring it produces. Figure 12–4 shows the width of the rings, according to age, for a selection of fir of the same age (85 years in 1985).

Figure 12–4 also shows that the Vosgian firs have undergone many important crises throughout their history of 100 years that correspond to cycles of drought years: 1920 to 1925, 1943 to 1952, 1960 to 1965, 1973 to 1982. The last of these crises was most acute in 1976, and was followed immediately by a period of decline. Even though the decline continues (with an improvement, nevertheless, in the health of the fir) the growth rate of most stands has again reached a normal level and sometimes even higher. The drop in the growth rate shown by the forest inventory reflects a period of unfavorable climate.

The construction of the curve showing the width of the rings according to the date and corrected by age effect, shows that, for the samples used, growth is much better today than in the nineteenth century. It shows a maximum around 1935, which coincides with a maximum in the average

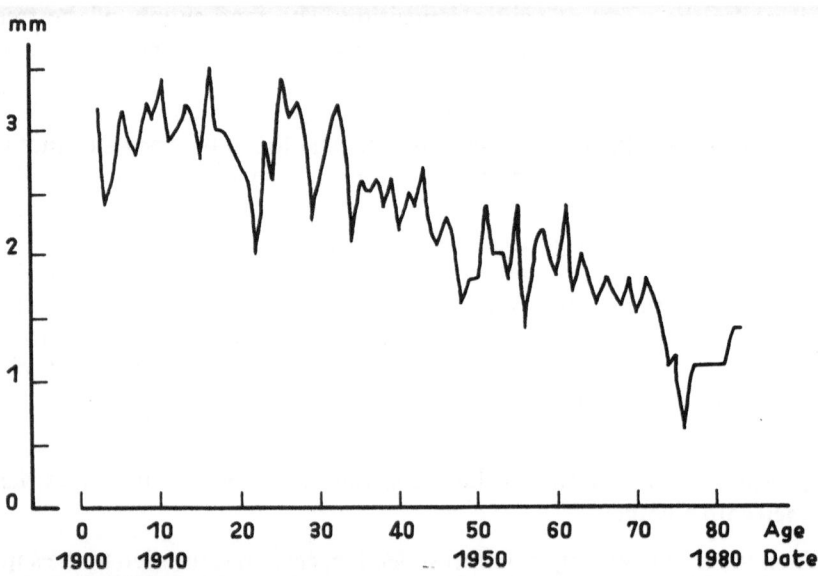

Figure 12–4. Annual ring width of silver firs in the Vosges that were 80 years old in 1980 (from Becker, 1987).

temperatures in the Northern Hemisphere. Then the growth curve diminishes slightly at the same time as the temperature, clearly climbing again in 1982 (Figure 12–5) (Becker, 1987).

There is, therefore, a nondefinitive drop in the vitality of the fir in the Vosges as compared to past decades for a sufficiently numerous sample selection. It is true, however, that if we consider two subpopulations from the original sample selection, one presenting a needle deficit of less than 40% and one presenting more than 40%, we find that the second subgroup has a much narrower ring width than the other. This divergence of curves in these two subpopulations dates back to the beginning of the 1950s (Figure 12–6).

Applying the same method to the width of the sapwood (proportional to the mass of foliage) shows that firs whose sapwood is less than the average have a smaller ring than those whose sapwood is larger than average. The difference started to become obvious between 1920 and 1925 (Becker 1987) (Figure 12–7).

This leads us to think that the most affected firs are those whose vitality started to diminish a few decades ago. The beginning of these crises coincides with the drought years, but could also, for the period 1945 to 1950, correspond to an increase in the atmospheric pollution brought about by increasing industrialization. This highlights a probable responsibility for the start of decline in the drought as a predisposing factor. The growth recovery (Figures 12–5, 12–6, and 12–7) allows us to think that the last dry period might have played the role, with a certain delay, of an inciting factor of decline. Very recent results comparing growth and the ecological conditions of healthy and declining sample plots (pairs consisting of one healthy sample and one unhealthy sample next to each other, and therefore submitted to the same pollution) show that unhealthy samples suffered greatly during the successive droughts of 1959 and 1976, and that they are in some cases found on soils with weaker water storage capacity than the healthy samples.

D. The Actual Level of Pollution in France in Rural Areas

1. The Whole of France

Knowledge of rain composition in rural zones in France is taken essentially from the results of measurements made at the EMEP stations as well as the BAPMON stations (Figure 12–1).

It is clear that the rain in the east of France is much more acidic than that in the west. The Rostrenen Station in Brittany recorded rains with an average pH (the value corresponding to a cumulative frequency of the monthly averages of 50%) of 5.1 (Figure 12–8). At Phalsbourg in Lorraine the average pH was 4.5. This is explained easily by the fact that the dominant winds coming from the Atlantic are loaded progressively with pollutants as they cross France.

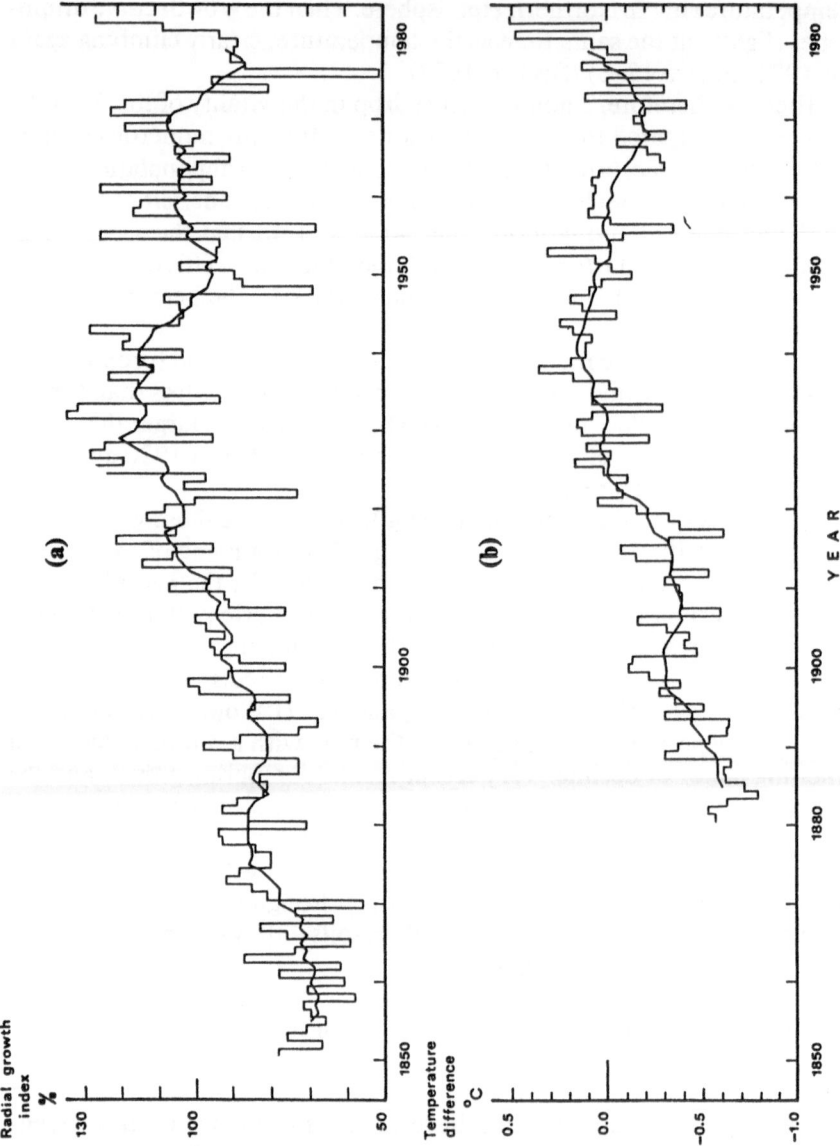

Figure 12–5. Corrected (without age influence) annual ring width of silver firs in the Vosges (a) from 1850 to 1980, and evolution of mean annual temperature of the Northern Hemisphere (b) from 1880 to 1980 (from Becker, 1987).

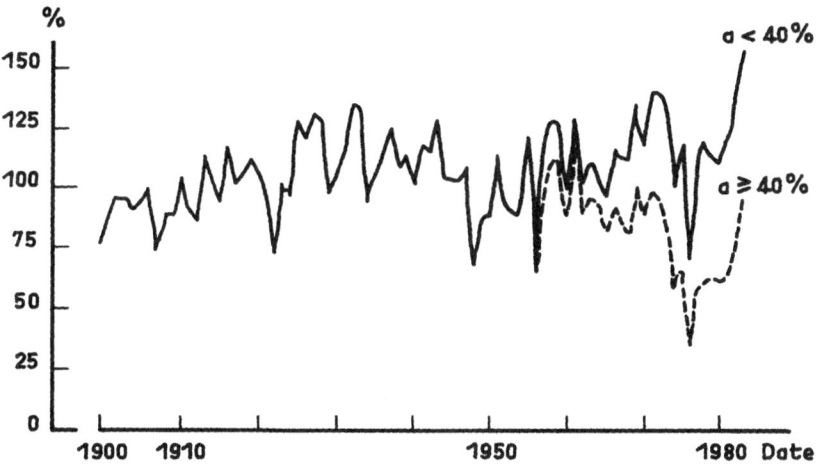

Figure 12–6. Corrected annual ring width of two subpopulations of silver fir in the Vosges with two different degrees of needle loss, a (<40% or ≥ 40%). Until 1950 the growth curves of these two subpopulations were not different (from Becker, 1987).

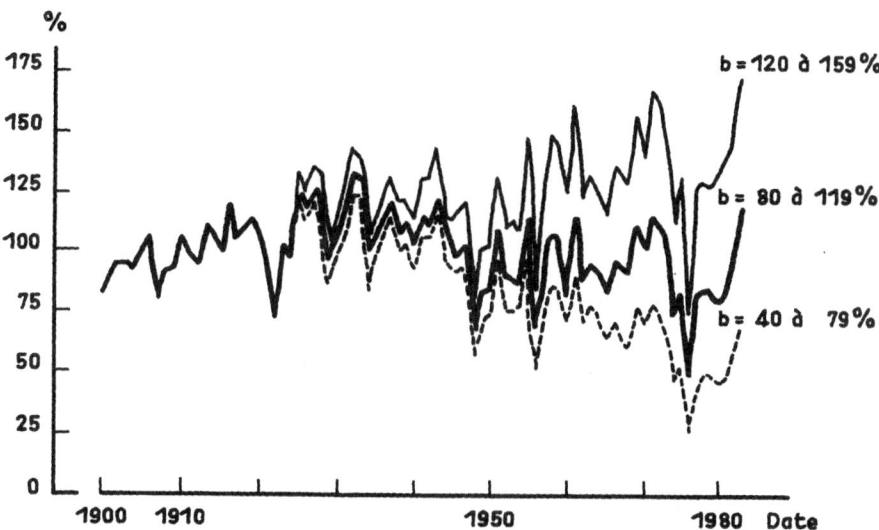

Figure 12–7. Corrected annual ring width of three subpopulations of silver fir in the Vosges with three different relative sapwood width b Lower curve: 40% to 79% of the mean sapwood width of the total population; middle curve: 80% to 119% of the mean sapwood width; upper curve: 120% to 159% of the mean sapwood width. Until 1922 the growth curves of the three subpopulations were almost identical (from Becker, 1987).

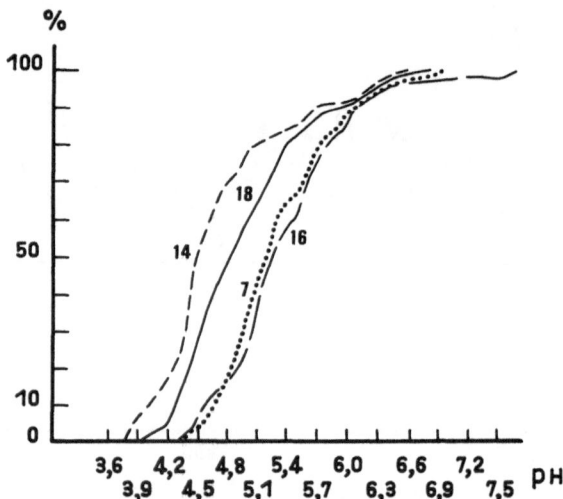

Figure 12–8. Cumulative frequency curves of rain pH at four monitoring stations: (7) Gourdon; (14) Phalsbourg; (16) Rostrenen; (18) Vert-le-Petit (from Hennequin et al., 1984).

In the different measuring stations the monthly pH range is important: to the order of two units for 95% of the measures (4.2 to 6.2 at Rostrenen, 3.8 to 6.2 for Phalsbourg).

From 1977 to 1983 the acidity of the rain barely increased. The concentration of the three major anions in rainwater, SO_4^{2-}, NO_3^-, and Cl^-, varies according to station. From west to east the sulphate and nitrate levels increased while the chloride level decreased. In addition to this, sulphates from 1977 to 1983 had a tendency to decrease while nitrates increased progressively (Hennequin et al., 1984) (Table 12–3). The annual wet deposition of sulphur was of the order of 0.8 g m^{-2} at Rostrenen and 1.2 g m^{-2} at Phalsbourg. That of NO_3^-–N is 0.2 g and 0.4 g m^{-2}, respectively, per year.

2. Massif Vosgien

The two measurement stations set up in 1985 and 1986 already provide a relatively precise knowledge of the pollution climate in the Massif Vosgien.

The situation for rain is similar to that already described in Phalsbourg. From November 1985 to September 1986 average pH measurements at Aubure for periods of 10 days varied from 4.1 to 6.4. At the beginning of a rain shower this pH can drop to 3.6. Table 12–4 shows the average composition of the rain and annual estimated supply for the most abundant anions and cations (Ambroise, Fritz, Probst, and Viville, 1987).

Table 12-3. SO_4^{2-}, NO_3^-, and Cl^- concentrations in the rainwater at two monitoring stations from 1977 to 1983 (Hennequin et al., 1984).

Years	Rostrenen[a]			Phalsbourg[b]		
	SO_4^{2-}	NO_3^-	Cl^-	SO_4^{2-}	NO_3^-	Cl
1977	46	8	118	62	23	20
1978	38	9	169	82	35	17
1979	44	17	121	74	28	13
1980	50	18	112	78	28	18
1981	44	18	112	70	36	20
1982	46	19	98	78	37	11
1983	46	11	126	68	34	17

$\mu eq\ l^{-1}$
[a] west of France (see Figure 12-1)
[b] east of France (see Figure 12-1)

Table 12-4. Ion concentrations in the rainwater of the Aubure catchment from November 1985 to August 1986 and estimated annual deposition by rain (Ambroise et al., 1987).

Ions	Concentration a	Deposition b	Deposition c
NH_4^+	23.6	4.6	3.6
Na^+	10.0	2.4	2.4
K^+	4.4	2.9	2.9
Mg^{2+}	4.6	0.7	0.7
Ca^{2+}	18.4	4.4	4.4
H^+	34.3	0.3	0.3
Cl^-	13.1	5.8	5.8
NO_3^-	26.2	13.5	3.1
SO_4^{2-}	48.1	22.3	7.4

[a] $\mu eq\ l^{-1}$
[b] $kg\ ha^{-1}$ (of ions)
[c] $kg\ ha^{-1}$ (of elements)

Sulphates are less abundant than in Phalsbourg. This can be partly explained by the difference in dates (1983 for Phalsbourg, 1985–1986 for Aubure), SO_2 emissions from 1982 to 1985 having dropped, for the whole of France, from 2.4 million to 1.8 million t yr^{-1}. Nitrate levels are slightly less, perhaps due to geographical position, although nitrogen oxide emissions are slightly increased (2.5 million t yr^{-1} estimated for 1987).

The in-depth study of the composition of rain at the Donon Pass (1/10 of mm per 1/10 of mm) has shown two types of evolution during rainfall. For rainfall of a stormy nature the pH is relatively high at the beginning of the storm (5.3) and the levels of NO_3^- and SO_4^{2-} are high (roughly 10 mg

L^{-1}). This is followed by a rapid drop in pH (3.7 after 20 min). The levels of nitrates and sulphates drop as quickly (3.5 mg L^{-1} after 10 min, 1 mg L^{-1} after 20 min). On the contrary, rainfall related to a perturbation from the northwest has a low and stable pH throughout the time of rainfall (4.4 to 3.7) as well as constant levels of SO_4^{2-} and NO_3^- (roughly 2 mg L^{-1}) (Derexel and Masnière, 1987).

A detailed study was also carried out on snowfall. For the same type of weather the snow is more acidic than the rain: the pHs recorded vary from 3.05 to 5.3. Also, when the wind comes from the north, northwest, east, or northeast, the snow is more acidic, richer in nitrates and sulphates, and the ratio of nitrates/sulphates is higher than if the wind comes from the south or the west (Table 12–5) (Colin, 1987).

The pH of streams in the Massif Vosgien is variable. For most of them it stays near neutral, which proves that the soils are capable of buffering the wet and dry acidic deposits. Some are very acidic and rich in aluminum; the trout population has greatly diminished in these streams.

3. Atmospheric Pollutants at the Donon Pass

The situation at the Donon Pass is characterized by an extremely variable SO_2 air level (Figure 12–9): practically nothing in summer, but daily maximums of 300 µg m^{-3} and hourly peaks of 375 µg in calm weather or with a slight wind from the east have been measured in winter. The average annual level is around 15 µg m^{-3}. Trajectographic studies have shown that during a drought year, such as 1976, where winds came more often from the northern and eastern sectors, the SO_2 emissions could be 20% more than those of a normal year (Strauss and Cuiller, 1987).

Table 12–5. pH and composition of snowfalls at the Donon Pass related to the wind direction in March 1985 (Colin, 1987).

		Wind coming from			
		N or NW	N or NE	W	S
pH	m	3.6	3.2	4.4	5.0
	e	3.45 4.51	3.05 3.36	3.81 5.21	4.51 5.26
NO_3^{-} [a]	m	12.3	15.7	2.5	0.9
	e	9.8 17.0	13.1 21.1	1.1 5.1	0.6 1.9
SO_4^{2-} [a]	m	6.3	9.9	2.4	0.8
	e	3.4 9.8	6.4 12.7	1.3 6.1	0.5 1.0
$\dfrac{NO_3^-}{SO_4^{2-}}$		1.9	1.6	1.0	1.1

[a] 10^{-5} mol L^{-1}
e extreme values
m mean values

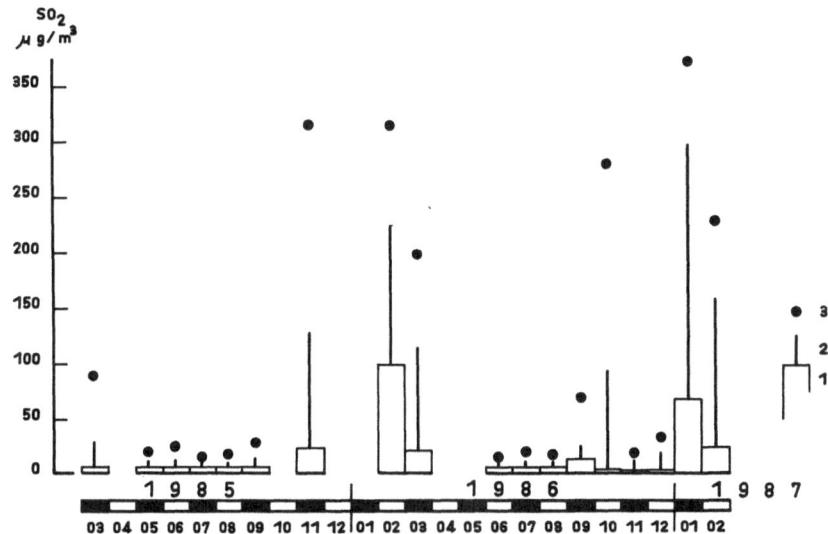

Figure 12–9. SO₂ air content at the Donon Pass monitoring station (Vosges). (1) monthly mean; (2) maximum daily mean of the month; (3) maximum hourly mean of the month (from Drach and Target, 1987).

These conditions can therefore be detrimental for the health of the trees in two different ways and thus the effects accumulate, as was shown in the physiological studies.

The ozone levels are more regular. The annual average is roughly 70 µg m⁻³ but the summer monthly average is close to 100 µg with daily maximums of 150 to 175 µg and hourly peaks of 200 µg (Figure 12–10).

The ozone level at the Donon Pass is likewise characterized by a relatively weak daily fluctuation. The maximum is to be found around the middle of the day, but the night levels still reach 60% to 70% of the daytime level (Drach and Target, 1987).

E. The First Experiments and Physiological Research into the Decline

To study the effect of acidic rain and atmospheric pollution, many experiments were undertaken, two of which took place in open-top chambers, one in a greenhouse, and one in a phytotronic chamber.

One of the open-top chambers experiments was installed at the Donon Pass next to the measuring station. It used three spruce clones from Istebna (Poland), Lake Constance (Germany), and Gerardmer (Vosges). Three plots of 15 plants were used (5 of each clone). One plot was raised in the open air, the second in a chamber supplied with surrounding air and receiving natural rainwater, and the third in a chamber where air was filtered on activated charcoal and which was watered with water of a pH

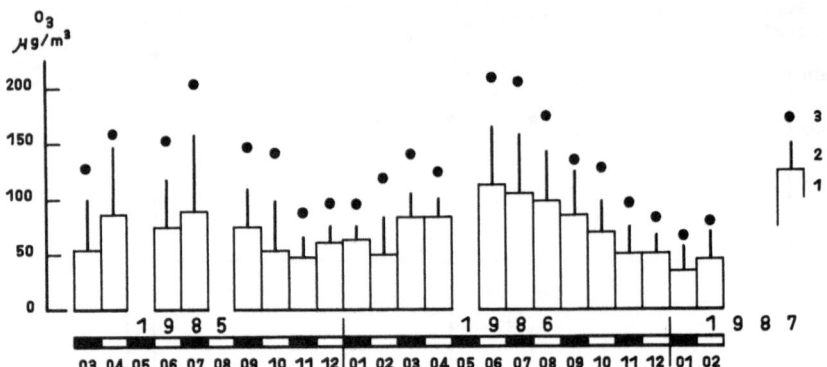

Figure 12–10. O₃ air content at the Donon Pass monitoring station 1, 2, 3: See Figure 12–9 (from Drach and Target, 1987).

of 5.5 at a quantity and rhythm equivalent to those trees receiving the rainwater (an automatic roof closed the chamber when it rained).

Another experiment in open-top chambers was carried out at Montardon near Pau (in the south of France). It used the same three clones of spruce as at the Donon Pass. Six groups of plants were compared: (1) in the open air, (2) in an open-top chamber with surrounding air, (3) in a chamber with filtered air, (4) in a chamber with filtered air + O_3, (5) in a chamber with filtered air + SO_2, and (6) in a chamber with filtered air + SO_2 and O_3. The originality of this experiment is the hourly reproduction, but with a week's delay, of the air levels of SO_2 and O_3 measured at the Donon Pass (Bonte, Cantuel, and Malka, 1987).

In a greenhouse in Nancy, an experiment took place on the same plant material, the effect combining water stress (plants kept at the field capacity with a potential of 3 bars, compared to some plants that were left to dry out as much as 20 bars and then rewatered) and acidic fog at a pH of 3.4 (two-thirds of H_2SO_4 and one-third of HNO_3) applied once a week from June to August.

In a phytotronic chamber at Gif-sur-Yvette, the same clones and young spruce seedlings were submitted to diverse conditions: SO_2 was applied at a concentration of 200 µg m⁻³ during the growing season in combination or not with water stress.

These experiments were designed to examine the treatment effects on the growth and the appearance of the plants. At the same time the young spruce are or will be the object of various physiological tests.

Simultaneously, diverse populations of young spruce between 10 and 15 years old in the region of the Donon Pass in the Vosges that showed either a normal appearance or symptoms of decline were studied. The same techniques were used in order to be able to compare their physiological characteristics with those of the young spruce raised under controlled conditions, as described earlier.

These experiments have only been in operation for one or two years, and the plants are not developed enough to undergo complete physiological tests. The same is true of the results from the field where the trees were mostly studied through only one or two seasons of vegetation. A certain number of results, however, can be presented.

Photosynthesis is generally depressed in the trees most suffering from decline in the field. This reduction in photosynthesis is difficult to measure exactly due to variations in light intensity, and it is necessary to use laboratory methods (cut shoot technique) in order to show this. Using this method the drop in photosynthesis is confirmed although it remains moderate (Gounot, 1987).

In the phytotronic chamber, the fumigation with SO_2, even to 200 μg m^{-3}, has no effect if the spruce is fed enough water. However the drying of the soil rapidly diminishes photosynthesis. The plants undergoing this drying without SO_2, if watered after a few days, quickly recover a nearly normal rate of photosynthesis (Figure 12–11a). Those that undergo a SO_2 fumigation at the same time as drying recover much less quickly, only reaching half the normal rate of photosynthesis 10 days after their rewatering. The largest reduction in photosynthesis in plants undergoing dehydration in the presence of SO_2 is due to a supplementary lack of water (Figure 12–11b), the relation of photosynthesis efficiency to the level of water in the plants remaining practically linear (Figure 12–11c) (Pierre et al., 1987).

In the experiment studying the effects of acidic fogs, transpiration is lower, no matter which hydrological regime was followed on the spruce trees tested (Aussenac, Clément, and Guehl, 1987).

In the phytotronic chamber, when dehydration and SO_2 pollution are combined, the young spruce dry and rehydrate with much more difficulty than those which receive only pure air. This sensitivity to dehydration is shown particularly in the roots (Pierre et al., 1987).

Enzyme activity in declining spruces in situ in second year needles is sometimes higher in relation to that of healthy spruce (phosphoenolpyruvate-carboxylase, isocitrate-dehydrogenase) and sometimes reduced (glutamate-dehydrogenase, malate-dehydrogenase).

In the phytotronic chamber the combination of dehydration and SO_2 pollution brings with it a much stronger diminution in enzyme capacity than pollution alone.

Soluble protein levels studied in situ are 30% lower for trees showing a needle deficit of 25% and some discoloration on the needles in relation to those of healthy spruce with completely green needles. For trees that are more affected (a needle loss of 50% and discolored old needles), this decrease in soluble proteins reaches 40% in the needles of the current year and 55% in the second year needles (Pierre et al., 1987).

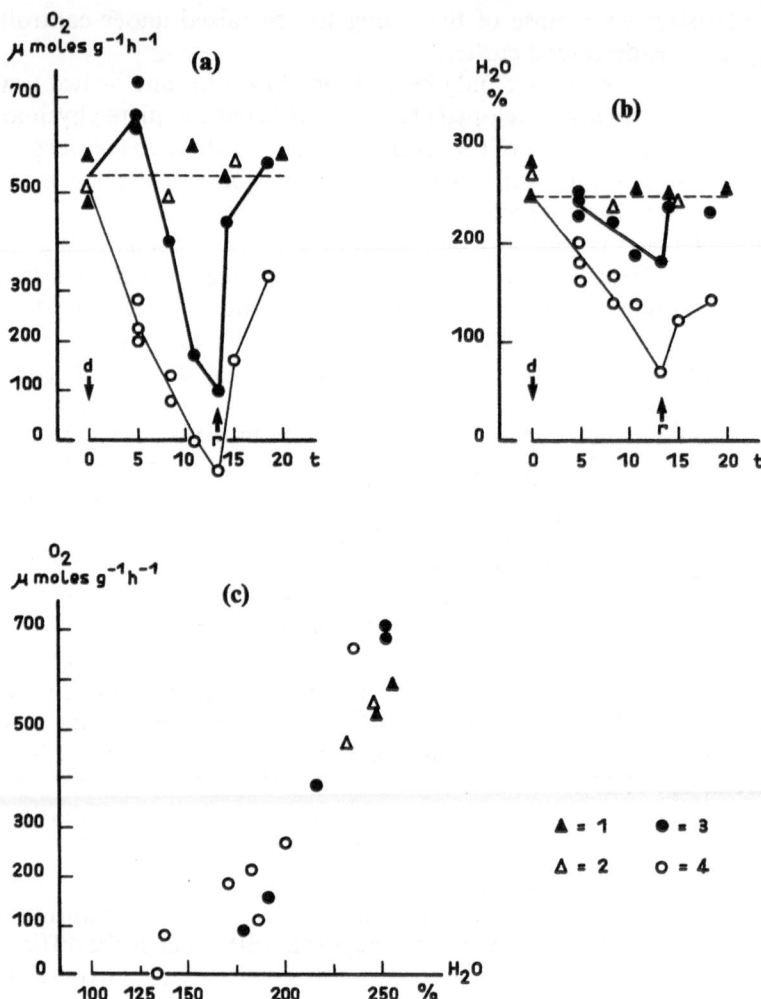

Figure 12–11. Effect of SO_2 fumigation (200 µg m^{-3}) and soil drying on the photosynthetic activity (a), water content (b) of Norway spruce seedlings; relationships between water content of the seedlings and photosynthetic activity (c). (1) control: no fumigation, no drying; (2) fumigation without drying; (3) drying without fumigation; (4) drying and fumigation; (d) start of drying; (r) rewatering; (t) time (in days). (From Pierre et al., 1987.)

The terpene needle level was also studied on the field. Although it is difficult to draw conclusions due to the great variability between trees, it seems that this level increases in damaged trees, especially at the end of the growing season (Saint-Guily, 1987).

In the field the needles from healthy trees contained 52 mg of polyphenols g⁻¹ of dry material, but those trees in class 2 (a needle deficit of 20%) contained 73 mg g⁻¹. In the trees from class 3 (35% to 50% needle deficit) there was 99 mg g⁻¹ in the needles that were still green, and 124 mg g⁻¹ in the discolored needles.

In the open-top chambers (combined action of SO_2 and O_3) it was established that spruce trees grown in the surrounding air, therefore more or less charged with SO_2 and O_3 outside the chambers, also had needles richer in polyphenols than those grown in the filtered air chambers (79 mg g⁻¹ versus 59). In the treatment with filtered air and ozone, the needles did not contain more polyphenol than those in the treatment with filtered air with no addition of pollutants. Ozone alone, at the levels measured in the summer at the Donon Pass, seemed to have no effect on this part of the metabolism of the spruce (Louguet, Malka, and Contour-Ansel, 1987).

The declining trees in the field were also characterized by an accumulation in free sugars (fructuose and glucose) in their needles (Table 12–6) (Villanueva, 1987).

The study of amino acids and polyamines was also interesting. The declining trees contained more histidine and tryptophane than the healthy trees (Table 12–6). In the healthy trees there is more arginine and less putrescine than in the unhealthy trees. In the young spruce grown in the presence of 200 µg of SO_2 m⁻³ in the phytotronic chambers, and that had not developed visible signs of illness, one found a tendency toward an increase in arginine when compared to those spruce grown in nonpolluted air. This increase in arginine and decrease in putrescine could therefore be interpreted as the result of a change in the physiological functioning, revealing a defense mechanism against pollution. In the forest, trees with a normal appearance could be those which are capable of triggering this sort of defense mechanism, whereas those trees in decline have been incapable (Villanueva, 1987).

All these physiological results need to be confirmed in the future because the trials in open-top chambers or in greenhouses must still be continued over a sufficiently long period of years.

Table 12–6. Free sugars, tryptophane, and histidine content in needles of healthy or declining spruce trees near the Donon Pass (Villanueva, 1987).

| | Fructose | Glucose | Tryptophane | Histidine |
	a	a	a	a
Healthy trees	10	8	47	17
Declining trees	27	22	85	27

a: µ mol g⁻¹ (D.M.)

V. Conclusions on the Role of Acidic Precipitation in Forest Decline

The process of forest decline is obvious, in France as in other European countries, and not yet completely understood. The possible factors responsible are numerous: the cycles of drought years, sometimes of long duration, which have helped to weaken the trees; the action of gaseous pollutants, SO_2 or ozone, as well as acidic deposits; and the lack of nutrients in the soil in certain regions.

The level of ozone in the air in the Vosges in summer is, for the period April to September, of the order of 75 to 100 μg m^{-3}, reaching, or slightly exceeding, the level of harmfulness indicated by Prinz in the second summary report which was published in Germany by the Federal Research Council on the decline of forests. However the corresponding levels, applied experimentally in the open-air chambers, do not seem to have caused damage in the young spruce plants.

The peak levels of SO_2 in winter, notably from November 1986 to February 1987, seems on the other hand to have caused, in the same experiment, visible symptoms of damage, SO_2 being capable of generating, after a dry deposit and oxidization, an important acidity on the very surface of the needles. Therefore attention should be directed toward acidic pollution.

This conclusion would seem to be confirmed by the large deficiency in nutrients that has been shown in the last few years by the discoloration of the fir, spruce, and beech trees on acidic soil in certain areas: a deficiency in magnesium and calcium in the Vosges, in the granite and gneiss regions of the Massif Central, and a deficiency in potassium on the schists in the Luchon region in the Pyrenees.

One can therefore formulate the hypothesis that the increase in acidic deposits, dry or wet, has led to a washing out of the alkaline and alkaline-earth cations from the foliage. This process has been more active in recent years. The cumulative effect of these deposits has also weakened the soils more and more, with an exchange of these same cations with H^+ and Al^{3+}. Old nutritional insufficiencies in regions where the soil was already desaturated are therefore accentuated into clear deficiencies. The speed at which these deficiencies are translated into accentuated discoloration still remains insufficiently explained.

Some experiments under controlled conditions on young plants show that there is a positive interaction between the SO_2 effect and drought. The study on the width of the annual rings of adult fir in the Vosges during the last century confirms that the drought cycles, be they old or recent, have contributed to the weakening of certain trees.

Thus one arrives at the hypothesis that the weakening of trees in drought periods will be, for the most part, responsible for a drop in needles a few years later. This phenomenon is not new and has existed several times in the past, but it is perhaps worsened these days by the high levels of SO_2 at certain periods of the year. The dry deposits of SO_2 or NO_x, as they bring about acidic pollution on the surface of the foliage, together with the wet deposition of sulphuric and nitric acids, may have brought about, in certain sites, an exaggeration in the old nutrient deficiencies. This leads to a discoloration of the foliage from the second year of vegetation.

The different aspects of forest decline can therefore be explained by a worsening of the effect of certain natural stresses on the forest ecosystems due to acidic pollution.

References

Ambroise, B., B. Fritz, A. Probst, D. Viville. 1987. In *Programme DEFORPA. Etat des recherches à la fin de l'année 1986,* Vol. 3, 215–231. Ministère de l'Environnement, Paris.

Aussenac, G., A. Clément, J.M. Guehl. 1987. In *Programme DEFORPA. Etat des recherches à la fin de l'année 1986,* Vol. 3, 41–58. Ministère de l'Environnement, Paris.

Becker, M., 1987. In *Programme DEFORPA. Etat des recherches à la fin de l'année 1986,* Vol. 1, 83–96. Ministère de l'Environnement, Paris.

Bonneau, M., and G. Landmann. 1987. In *Programme DEFORPA. Etat des recherches à la fin de l'année 1986,* Vol. 1, 187–204. Ministère de l'Environnement, Paris.

Bonte, J., J. Cantuel, P. Malka. 1987. In *Programme DEFORPA. Etat des recherches à la fin de l'année 1986,* Vol. 3, 13–39. Ministère de l'Environnement, Paris.

Brucker, S., and Boun Suy Tan 1987. In *Programme DEFORPA. Etat des recherches à la fin de l'année 1986,* Vol. 1, 133–147. Ministère de l'Environnement, Paris.

Cheret, V. 1987. La Sapinière du Luchonnais (Pyrénées Hautes—Garonnaises) *Etude phytoécologique, recherches sur le phénomène de dépérissement forestier.* Thèse de Doctorat. Université Paul Sabatier, Toulouse. 287 pp.

Cheret, V., J. Dagnac, F. Fromard. 1987. In *Revue Forestière Française* XXXIX: 12–24.

Colin, J.L. 1987. In *Programme DEFORPA. Etat des recherches à la fin de l'année 1986,* Vol. 2, 55–65. Ministère de l'Environnement, Paris.

Derexel, Ph., and P. Masnière. 1987. In *Programme DEFORPA. Etat des recherches à la fin de l'année 1986,* Vol. 2, 47–53. Ministère de l'Environnement, Paris.

Drach, A. and A. Target. 1987. In *Programme DEFORPA. Etat des recherches à la fin de l'année 1986,* Vol. 2, 5–15. Ministère de l'Environnement, Paris.

Gounot, M. 1987. In *Programme DEFORPA. Etat des recherches à la fin de l'année 1986,* Vol. 3, 87–103. Ministère de l'Environnement, Paris.

Gras, F. 1987. In *Programme DEFORPA. Etat des recherches à la fin de l'année 1986,* Vol. 3, 195–214. Ministère de l'Environnement, Paris.

Hennequin, C.L., G. Hervouet, C. Brun, N. Cenac, D. Cheymol, and M. Zephoris. 1984. In *Livre Blanc sur les Pluies Acides. Première Approche Scientifique du Problème en France,* 145–160, Secrétariat d'Etat à l'Environnement et à la Qualité de la Vie, Paris.

Landmann, G., M. Bonneau, and M. Adrian. 1987. *Revue Forestière Française* XXXIX:5–11.

Louguet, P., P. Malka, and D. Contour-Ansel. 1987. In *Programme DEFORPA. Etat des recherches à la fin de l'année 1986,* Vol. 3, 105–124. Ministère de l'Environnement, Paris.

Pierre, M., A. Sieffert, A. Savoure, O. Queiroz, G. Cornic, C. Hubac, and V. Macrez. 1987. In *Programme DEFORPA. Etat des recherches à la fin de l'année 1986,* Vol. 3, 59–86. Ministère de l'Environnement, Paris.

Saint-Guily, A. 1987. In *Programme DEFORPA. Etat des recherches à la fin de l'année 1986,* Vol. 3, 161–173. Ministère de l'Environnement, Paris.

Strauss, B. and G. Cuiller. 1987. In *Programme DEFORPA. Etat des recherches à la fin de l'année 1986,* Vol. 2, 119–132. Ministère de l'Environnement, Paris.

Villanueva, V.R. 1987. In *Programme DEFORPA. Etat des recherches à la fin de l'année 1986,* Vol. 3, 141–159. Ministère de l'Environnement, Paris.

Index